51.00

# SEED BIOLOGY

# PHYSIOLOGICAL ECOLOGY

*A Series of Monographs, Texts, and Treatises*

EDITED BY

## T. T. KOZLOWSKI

*University of Wisconsin*
*Madison, Wisconsin*

# SEED BIOLOGY

*Edited by*

## T. T. KOZLOWSKI

DEPARTMENT OF FORESTRY
UNIVERSITY OF WISCONSIN
MADISON, WISCONSIN

## VOLUME II

Germination Control, Metabolism, and Pathology

ACADEMIC PRESS    New York    San Francisco    London    1972

A Subsidiary of Harcourt Brace Jovanovich, Publishers

ACADEMIC PRESS, INC.
111 Fifth Avenue, New York, New York 10003

*United Kingdom Edition published by*
ACADEMIC PRESS, INC. (LONDON) LTD.
24/28 Oval Road, London NW1 7DD

LIBRARY OF CONGRESS CATALOG CARD NUMBER: 71-182641

PRINTED IN THE UNITED STATES OF AMERICA

# CONTENTS

4   **Physiological and Biochemical Deterioration of Seeds**

AREF A. ABDUL-BAKI AND JAMES D. ANDERSON

5   **Seed Pathology**

KENNETH F. BAKER

# LIST OF CONTRIBUTORS

Numbers in parentheses indicate the pages on which the authors' contributions begin

AREF A. ABDUL-BAKI (283), Seed Quality Investigations, Market Quality Research Division, Agricultural Research Service, United States Department of Agriculture, Plant Industry Station, Beltsville, Maryland

JAMES D. ANDERSON (283), Seed Quality Investigations, Market Quality Research Division, Agricultural Research Service, United States Department of Agriculture, Plant Industry Station, Beltsville, Maryland

KENNETH F. BAKER (317), Department of Plant Pathology, University of California, Berkeley, California

TE MAY CHING (103), Department of Farm Crops, Oregon State University, Corvallis, Oregon

DOV KOLLER (1), Department of Agricultural Botany, The Hebrew University of Jerusalem, Rehovot, Israel

T. A. VILLIERS (220), Department of Plant Biology, University of Natal, Durban, South Africa

# PREFACE

Man's existence and health are directly or indirectly dependent on seeds. This fact has for many years pointed out the urgent need for a comprehensive coverage of information on seed biology. The importance of this work became even greater during the recent years of rapid population increases throughout the world. It was with these thoughts in mind that this three-volume treatise was planned to bring together a large body of important new information on seed biology.

The subject matter is wide ranging. The opening chapter outlines man's dependency on seeds as sources of food and fiber, spices, beverages, edible and industrial oils, vitamins, and drugs. Harmful effects of seeds are also mentioned. Separate chapters follow on seed development, dissemination, germination (including metabolism, environmental control, internal control, dormancy, and seed and seedling vigor), protection from diseases and insects, collection, storage, longevity, deterioration, testing, and certification. These books were planned to be readable and interdisciplinary so as to serve the widest possible audience. They will be useful to various groups of research biologists and teachers, including agronomists, plant anatomists, biochemists, ecologists, entomologists, foresters, horticulturists, plant pathologists, and plant physiologists. The work has many practical overtones and will also be of value to seed producers and users.

These volumes are authoritative, well-documented, and international in scope. They represent the distillate of experience and knowledge of a group of authors of demonstrated competence from universities and government laboratories in England, India, Israel, South Africa, and the United States. I would like to express my deep personal appreciation to each of the authors for his contribution and patience during the production phases. The assistance of Mr. W. J. Davies and Mr. P. E. Marshall in index preparation is also acknowledged.

T. T. KOZLOWSKI

# CONTENTS OF OTHER VOLUMES

# 1

# ENVIRONMENTAL CONTROL

# OF SEED GERMINATION

*Dov Koller*

## I.   Environmental Control of Germination and Its Biological Significance

### A.   *The Dormant State and Its Biological Implication*

One of the most universal characteristics of practically all existing plant species is that at least once, and in many instances several times, during their life cycle they produce specialized cells, or multicellular bodies, that exhibit the phenomenon which has become known as dormancy. With few exceptions, these dormant structures are a developmental antithesis of the tissues from which they develop. The latter are at the peak of differentiation and at their lowest potential for resuming active growth. On the other hand, the former are at their highest potential for resuming active growth, to the extent of initiating an entirely new life cycle, or a new phase of the existing cycle. At the same time, they are essentially at their most undifferentiated state, although they may be encapsuled in structures and tissues of the highest degree of complexity and differentiation.

There is a dramatic quality to this dormancy, since formation of the dormant structures is commonly associated with a remarkable intensification of metabolic, synthetic, and morphogenetic activities, all of which come to an abrupt and almost simultaneous end. Onset of dormancy nevertheless appears to be much more than a mere cessation of the aforementioned activities. It apparently involves a sequence of events as orderly as that which is encountered in the formation of the dormant structures. It is tempting to visualize the operation of sequential gene activity as postulated by Heslop-Harrison (1963) for the process of differentiation of a floral apex. However, our understanding of the mechanism which controls the sudden cessation in all these activities is both fragmentary and elusive, and will be dealt with in other chapters of the present volume.

The remarkable fact about this onset of dormancy is that it takes place at a time when the external environment to which the plant is exposed is in no apparent way unfavorable for continued metabolic, synthetic, or morphogenetic activities. The imposition of dormancy is therefore most likely to be controlled endogenously. The ultimate control is probably located in the tissues of the mother plant. Nevertheless, there is ample evidence that in many if not most plants the formation of the dormant structures is initiated under a more or less specific combination of environmental variables. Thus, even when the morphological development of these structures and their physiological transition to the dormant state are under endogenous control, the primary timing mechanism which initiates these events depends on perception of and response to environmental signals, such as photoperiod or cold in the case of many flowering plants, availability of minerals or light in many algae, etc.

The drastic reduction in physiological activities which is integrated in the dormant state is commonly associated with development of external protective tissues and (at least in seed plants) with a drastic reduction in hydration of the cytoplasm. All these combine to make the dormant structure much more resistant to unfavorable environments than the plant which produced them. Dormancy thus becomes associated with resistance against, or tolerance of adverse environmental conditions. Dormancy must therefore have evolved primarily as a solution to the periodic, as well as nonperiodic, changes in the environment, which makes it difficult if not impossible for the plant to function properly, or even exist, during certain periods. The life span of most higher plants is limited by their very nature (monocarpic species), by limitations in their environmental resources, or by pathogenic agents. The investment in dormant cells or seeds which possess the highest potential for resuming growth and forming a fresh individual is in the nature of general insurance for survival of the species. Dormancy also appears to be a natural evolutionary consequence of most forms of plant dispersal, with the exception of fragmentation and the like, since the environments through which dispersal takes place are almost invariably hostile to active growth. Moreover, the time passed in transit through the hostile environment is relatively indeterminate.

### B.  Termination of the Dormant State

The state of dormancy is overtly terminated when active metabolism, synthesis, and finally growth are resumed. In seeds (and spores) the resumption of these activities is identified with the term germination. Such postgerminative activities can only take place in environments within which the parent plant could function properly. The requirement for photosynthetically active radiation and for external mineral supply may be deferred for a length of time which depends on the size of the organic and mineral reserves within the seed. This leaves the immediate postgerminative growth activity with virtually few and simple requirements from the environments. One of these is an adequate moisture supply, yet which would not interfere with the gaseous exchange which is essential for aerobic respiration and adequate supply of metabolic energy. Another such requirement is for "normal" temperature, i.e., within the range which is suitable for normal growth of the more mature seedling. Ostensibly, therefore, exposure of the seeds to environments consisting of adequate moisture, aeration and "normal" temperature, should suffice also for germination to take place. However, experience of foresters, horticulturists, and agronomists through the years has shown that germination under such conditions can turn out to be negligible or is at best a slow process and erratic to the point of unpredictability. Rapid, simultaneous germination in high percentages under such conditions is the exception rather than the rule. By far most exceptions occur among species which have been longest under cultivation, such as cereals which originated in the Old World (e.g., Tang and Chiang, 1955; Takahashi and Oka, 1957), for the obvious reason that they had been continuously selected for high, rapid, and simultaneous germination. The phenomenon of erratic, incomplete, or otherwise unsatisfactory germination presented considerable difficulties for agriculture and therefore resulted in the establishment of "seed testing" or "seed analysis," which was primarily aimed at certifying seed quality in terms of expectancy of maximal potential field performance. The vast amounts of experimental data which were gathered by the efforts of numerous seed analysts showed that in a great number of species germination can be hastened, its rates increased and its final percentages improved by a variety of means. The capacity for germination was in many instances lowest soon after harvest and afterward improved with time. As a result, the failure of apparently mature seeds to germinate when provided with the moisture and temperature which are normally adequate for growth was treated as a manifestation of dormancy, with the idea that it was most probably a direct continuation of the primary dormancy which the seeds enter in the normal course of their maturation, as described above. Out of these studies also grew the realization that the dormancy which is involved in preventing germination from taking place readily under apparently "normal" conditions may be an evolutionary safeguard against the uncertainty of the plant's environment. Catastrophies which may occur will involve only a fraction of the seeds which had been produced, and since the risk of annihilation is minimized, their effect on the population of the species as a whole will therefore be temporary.

## C. Environmental Control

The proceedings of various national seed testing associations, and of the International Association (*Comptes Rendus, Association Internationale d'Essais des Semences*) provide a wealth of detailed information on the different methods which were found most effective in obtaining the maximal germination response in the different plant species. Some of the treatments which were successful in inducing germination of such dormant seed were clearly agrotechnical, in the sense that they were entirely unavailable to the seed without man's technical assistance. Such was the case, for instance, with seeds which responded to pretreatment in concentrated sulfuric acid, or in boiling water, or to mechanical damage to their coats. Other treatments, on the other hand, were such that the seed could conceivably have been exposed to them under natural conditions. Such was the case, for instance, with seeds requiring moist pretreatment at near-freezing temperature (a practice which became known as stratification), exposure of the imbibed seed to light, to diurnally fluctuating temperature, etc. It thus became increasingly apparent that the dormancy which is expressed by inhibited germination under seemingly normal environmental conditions is released by exposure to a much more specific complex of environmental conditions. Once this release has been effected, further (postgerminative) growth of the seedlings is no longer similarly restricted and can take place normally within a much less specific environmental complex. This observation suggested that seed germination in a great many species has all the earmarks of an activation process, which is at least in great part under environmental control, in the same manner that in many species activation of dormant buds, or the transition from the vegetative to the reproductive state, are triggered by environmental signals and are therefore under environmental control. In these two developmental processes, the environmental control of activation has a relatively clear-cut role in enhancing the survival potential of the species. Where bud dormancy is released, or flowering is initiated by prior exposure to near-freezing temperatures, the new and actively growing shoots or flowers are made to avoid the damage by such temperatures to which young and actively growing tissues are commonly particularly susceptible. Where flowering is initiated by transition to a specific photoperiodic regime, its subsequent stages are thereby made to synchronize with other favorable environmental components of a specific season in that particular habitat such as temperature, radiation climate, moisture regime in the soil and atmosphere, and presence of the suitable pollinating vector. Clearly, in both cases described above the activation process involves perception of and response to certain components of the changing environment. These environmental components have become selected during evolution, not because they themselves are favorable for the physiological activity which they initiate, but because they are much more reliable and effective indicators of the advent of the favorable environment than any actual component of such environment. Thus, the advent of mild weather with clear sky may serve as a false indicator of spring if it is of short duration. The quality of mild weather as an indicator of spring improves as its duration increases, but the useful part of spring becomes correspondingly shorter. In comparison, a suitably long period of near-freezing temperature is a reliable indicator of the advent of spring, without wasting any of it. A similar argument can be applied in the case of photoperiodic control.

The remarkable thing about environmental control of bud dormancy and flower initiation is the small number of environmental signals which are employed, namely photoperiod and low temperature. This is in marked contrast to the large variety of environmental variables which have been identified as involved in the control of seed germination, to be dealt with below. There are several possible explanations to account for this difference between bud dormancy and flowering on the one hand and seed germination on the other. In the first place, dormant buds and shoot meristems which have achieved a state of "ripeness to flower" are situated on mature plants whose very existence as established individuals is proof that their environment can and does normally supply their requirements for sustained growth. In contrast, seeds contain entire plants in an embryonic state. In order for them to perform the function for which they have evolved, they should be capable of providing the optimal conditions for the very establishment of the embryonic plant first as a seedling and later in its more mature stages. Therefore, at some time before the seed has committed itself to the irreversible terminal processes of germination in a certain site, it must be capable of discriminating the potential of that site to provide the necessities which would increase the probability for successful completion of the life cycle. In the second place, the mass of the seed and of the plant which bears it differs by several orders of magnitude. Furthermore, the mass of the seed is characteristically compressed into the most economically compact space, whereas that of the mature plant is equally characteristically extended into the thinnest and most tenuous processes that are possible mechanically and functionally, so as to occupy as large a volume as possible of the available space. As a consequence, the mature plant is able to sample a much larger section of the environment than the seed. By integrating all this information, the mature plant can obtain a much more reliable estimate than the seed, of what the environment can provide in the way of the factors which are essential for its sustained existence. The mature plant is commonly described as living within its microenvironment. It is equally justified to state that the seed exists within a submicroenvironment. It is from this submicroenvironment that the seed must obtain as much information as possible about the actual microenvironment. It is most probably for the reasons stated above, and possibly also because of other reasons of which we are unaware, that environmental control plays such a vital role in the regulation of seed germination, and seed germination is so responsive to environmental manipulations.

## D.  *Dormancy and Survival*

In closing this section, the reader must not be left with the impression that the environmental control of germination is the sole means which serves the survival of the species. It may not be the most important one or even the most prevalent one. As a matter of fact, of the many cases in which certain environmental factors have been clearly shown to exert control over germination in varying degrees of specificity or strictness, relatively few have been interpreted in terms of the value to species survival. In even fewer cases have there been any serious attempts to study this possibility. A noteworthy exception is the work of Harper and McNaughton (1960) on *Papaver*. Rigorous proof is hard to come by. One major difficulty may be that when the physiology of germination is studied under laboratory conditions, there is probably very little similarity between this artificial environment and that which the seed encounters in nature. Few researchers actually appreciate the extent of this dissimilarity and this is quite understandable in view of the dearth of any precise information on the components of the submicroenvironment of the seed in nature. Another major difficulty is the fact that species which share the same habitat rarely have a common denominator in the environmental control of their germination (Pelton, 1956). However, this is not so surprising, since different species are able to share the same habitat only by occupying different microenvironmental niches in it. In the subsequent sections we shall try to speculate, where possible, on the potential role which the capacity of the seed to preceive and respond to certain environmental factors may have in providing it with information of survival value.

## II.   Environmental Indicators and Plant Perception

### A.   *Requirements and Indicators*

The autotrophic plant depends for its existence on relatively few components of the environment. These are water, energy, carbon dioxide, certain soil physical conditions, a balanced supply of minerals, and freedom from excessive competition.

### 1.   SUPPLY OF WATER

In most habitats, an adequate supply of water to the root system is at best seasonal. As aridity increases, potential evapotranspiration outstrips precipitation. As a result, plant life becomes increasingly dependent on the elements of chance and of nonuniformity. Both play an essential role in providing situations (microenvironments) where the balance between evapotranspiration and supply of moisture is temporarily and locally sufficiently favorable to plant growth. Rainfall is seldom uniformly distributed in time and space. There is usually also an element of chance in this nonuniform distribution. As a result, amounts which are much higher than the mean may be precipitated locally and unpredictably to a great extent. In addition, nonuniformity in surface topography and physical properties of the soil, sometimes even when minute, may lead to redistribution of the precipitated rainfall by accumulating runoff from catchment areas and storing adequate amounts as available soil moisture. Thus, where availability of soil moisture may become a factor which limits growth, the plant will benefit if its seeds are equipped with some mechanism which will enable them to sense and assess the availability of soil moisture and control their germination accordingly. As availability of soil moisture gets to be more of a limiting factor it will become increasingly important for the seeds to be capable of discriminating between presence of water and its availability in sufficient amounts to sustain further growth and establishment.

Soil moisture is depleted both by drainage and evaporation. The supply of soil moisture may therefore be ephemeral if soil physical conditions and the evaporative demand of the external environment are such that either or both of these processes become too rapid. This situation is not restricted to arid habitats and is actually much more widespread than is generally realized. By far the greatest proportion of the seed population in the soil is situated in the topmost surface layers, since most seeds do not penetrate deeply unless the soil is disturbed by the action of weather, animals, or man. These layers are the first to become moistened by rainfall, but at the same time their water content is the most ephemeral, since they lose the most water by evaporation, as well as by percolation. Thus, from the point of view of water relations there appears to be a distinct disadvantage to germination from very shallow depths. Yet there are other disadvantages to germination from great depth. One disadvantage arises because of limitations in the supply of stored food, which must sustain the seedling until it reaches the open air and becomes autotrophically established. Second, the shoot has to overcome mechanical impedance as it grows upward through the soil, and this impedance increases with decrease in soil moisture content (Arndt, 1965) (Fig. 1.1). After rain has ceased falling, the soil moisture content decreases (and mechanical impedance increases) with time and as the soil surface is approached. For both these reasons, it may be of distinct advantage for the seed to be capable of perceiving its distance from the soil surface.

## 2. SUPPLY OF EXOGENOUS ENERGY

The supply of energy is of vital importance to the plants from at least two aspects. One is concerned with the photosynthetically active parts of the solar spectrum. The other is the thermal energy components of the environment which determine the temperature of the different plant organs by radiation, sensible heat transfer, and convection.

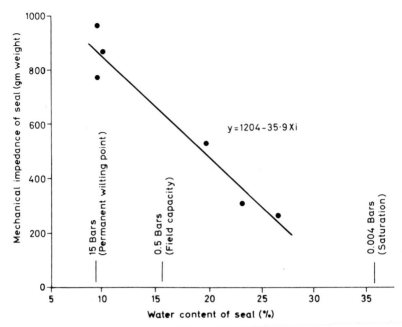

FIG. 1.1. The inverse relationship between mechanical impedance and moisture contents of the soil between field saturation and the permanent wilting point. From Arndt (1965).

a. VISIBLE RADIATION.    The availability of photosynthetically active radiation in different macroenvironments depends on their geographical latitude and amount of cloud cover. In different microenvironments it depends also on the amount of shading by optically opaque objects on the one hand, and by the selectively transmitting plant canopy, on the other. Within each environment, the availability of such radiation is subject to changes of a more or less seasonal nature, as cloudiness, the angle of the sun, the length of day, and the density of plant cover go through their annual cycles. Plants have become adapted to such specific radiation climates, on the one hand, by synchronizing their developmental life cycle to the cyclical changes and, on the other hand, by adjusting the capacity of their photosynthetic apparatus to the normally available light intensity. The light in the submicroenvironment of the seed is rarely uniform. Clearings occur even in densely vegetated areas, and open areas are not devoid of patches of shade. The latter vary in the spectral composition of available light, depending on the spectral transmittance of the shading object and on the spectral reflectance of neighboring objects. They may vary in density, depending on the optical density of the shading object and on the amount of reflected light. They may also vary in their daily duration and timing. For all these reasons it may be of obvious value for the survival of a species to have seeds which are equipped with some means which will enable them to sense and assess the light-environment, both quantitatively and qualitatively, and to control their germination accordingly. For the seed, the availability of photosynthetically active radiation has an additional immediate significance of vital proportions, because after their dispersal most seeds come to rest at various depths within the soil, where such radiation is practically absent. Their stored foods suffice for a limited amount of nonautotrophic growth of the seedling, and their photosynthetic apparatus must reach a suitable light-environment to make them autothropic before these stores are depleted. It thus becomes of crucial importance for germination to be controlled by some mechanism which enables the seed to perceive the distance which must be traveled by the tip of the epicotyl until it reaches light.

b. HEAT.   The temperature of different plant organs evidently affects the rates of their various metabolic activities. However, the reduced metabolic activity which results from temporary departures from the optimum rarely has long-lasting effects which significantly reduce the survival potential of the species. Long-term departures from the optimal temperature, on the other hand, may strongly reduce the survival potential of the plant. Therefore, plant survival is more strongly dependent on the temperature regime of the microclimate of its habitat than on the occasional temperature changes of the weather. Root temperature is determined by soil temperature, which for a given soil is mainly controlled by net exchange of radiation. Leaf temperature is dependent on net exchange of radiation as well as on temperature of the ambient air. However, the effect of air temperature depends on wind velocity and leaf size and is generally less important than that of radiation. Furthermore air temperature is controlled to a large extent by radiation. Therefore, the environmental component which has the strongest effect on plant temperature is solar radiation. The greatest variations in this component are seasonal. As a result, temperature usually becomes a limiting factor for plant survival only during certain seasons. For this reason it would be of benefit for the plant to have seeds equipped with a mechanism which controls their germination to take place only during certain seasons. The importance of such a mechanism for survival increases markedly where temperatures during the unfavorable season rise above or descend below the lethal limits. Under such conditions it becomes essential for germination to be deferred until the probability of encountering such dangerous temperatures becomes sufficiently low.

3.  SUPPLY OF CARBON DIOXIDE

Carbon dioxide is required only for photosynthesis, and under natural conditions its concentration never falls below the limits for survival, because of its rapid diffusion in air and high solubility in water. It is probably for this reason that to my knowledge there is no known mechanism in seeds which regulates their germination so as to allow the plant to take advantage of a more abundant supply of $CO_2$.

## 4.   SOIL PHYSICAL CONDITIONS

Soil physical conditions are of importance for survival because they determine soil–plant–water relationships, soil aeration, and the mechanical impedance to root and shoot growth. A given combination of soil type and water regime may provide optimal water relationships for one plant and hazardous ones for another. For instance a coarse-grained, porous soil may be an excellent substrate for deep-rooting species under arid conditions, because precipitated rainfall will rapidly percolate to deep soil layers and be protected from evaporative loss. The same substrate under the same conditions will be distinctly disadvantageous for most shallow-rooting species. Soil aeration may become a limiting factor for growth and the normal physiological function of the root system in certain types of soils and moisture regimes. Exceptions to this are plant species which have become adapted to soil anaerobiosis by anatomic, morphological, or other physiological modifications. Such plants use these habitats as ecological refuges from the competition of other species which are less tolerant of anaerobiosis. Mechanical impedance of the soil to growth of shoots has already been dealt with to some extent in relation to the progressive drying of the surface layers of the soil. Clearly, this type of limitation is not restricted to such dynamic situations. Impedance is undoubtedly encountered by both roots and shoots (Barley *et al.*, 1965; Arndt, 1965) and may create a problem for survival even when soil moisture remains virtually unchanged. For all the reasons listed above, and for some which may have been overlooked, it is of benefit for the plants to have seeds which are able to sense and evaluate the relevant physical properties of the soil and control germination accordingly.

## 5.   SUPPLY OF MINERALS

A balanced supply of minerals is obviously essential to the successful growth of any plant. Plant distribution is sometimes determined by the availability of a single essential element, or group of elements, or by presence of a toxic element. It is therefore quite surprising that to date there have been no reports of seeds with a capacity of regulating their germination so as to ensure a suitable mineral environment. Seeds are usually equipped with the essential mineral elements to make the seedling independent of external supply for a considerable amount of growth, but no provision is apparently made for increasing the probability of supplying the mineral requirements of subsequent growth. One exception to this is in the case of the nitrate ion. This ion has been frequently reported as exerting a marked stimulation on seed germination of a fairly large number of species. It is, however, doubtful that the sole information about the ionic composition of the environment which is of survival value for so many species is the presence of nitrate. On the other hand, certain species have become adapted to highly concentrated ionic environments, mostly saline, which are unsuitable for the majority of other species. Such adapted species use these niches as ecological refuges. It may be of distinct advantage for them to have seeds equipped with the means of sensing these specialized mineral environments and controlling their germination in relation to their availability.

## 6. AVOIDANCE OF COMPETITION

Physiological adaptation to ecological refuges which are inhospitable to other species is one means by which certain plants avoid competition. Such species utilize naturally available refuges. Other species apparently create their own, by modifying their immediate environment so as to exclude newcomers, by killing off plants at their seedling stage, or even earlier by inhibiting the germination of their seeds. The simplest environmental modification is that of filtering out the photosynthetically active radiation by creating a dense, uninterrupted canopy. More sophisticated modifications are created by excreting from living, senescent, or decomposing organs and tissues, chemicals which are either toxic to seeds or seedlings of other species or which merely inhibit germination of their seeds. This phenomenon is known as allelopathy (see review by Evenari, 1961a). It may be argued, however, that seeds which are inhibited but not killed by the allelopathic agents have developed germination-regulating mechanisms which enable them to survive allelopathic environments by evasion.

## B. Perception

### 1. CRITERIA

This brings us to the crux of the problem of environmentally regulated germination. A seed which germinates only (or optimally) in a more or less specific environmental complex out of the great variety which are normally available to it, is not *ipso facto* environmentally regulated. The acid test is whether or not (or to what extent) its viability is impaired during its exposure to the environmental complex in which it does not germinate. If viability is impaired, the environmental effects are of no survival value and cannot be regarded to operate as control mechanisms. If viability is not impaired while the seed is prevented from germinating because its germination requirements are not entirely fulfilled by the environment, one of two things can happen to it. In the simplest case the seed remains physiologically unchanged and will germinate as usual when its normal germination requirements are fulfilled. In many species the outcome is more complex. The original germination-regulating mechanism is apparently replaced by another, inasmuch as the seeds no longer respond to the environmental complex which would have previously caused optimal germination, but will germinate in response to an entirely different set of environmental conditions. The change in the nature of the germination-regulating mechanism, which is induced by exposure of the seed to one or more environmental factors that are unfavorable for its germination, has been termed "secondary dormancy."

## 2. SPECIFICITY AND THE FACTORS INVOLVED

The perception and characterization of the environment by the seed appears to depend on relatively few components, mainly temperature, light, and water. However, by restricting perception to specific levels and regimes of these components, and to specific combinations or interactions of these few factors, a high degree of characterization of the environment may be achieved. An analogy may be drawn with the genetic code, in which highly specific information may be expressed by using only a few nucleotides, but in different relative amounts and arranged in specific sequences.

In the following sections, various responses of germination to environmental factors and complexes which have been described in the literature will be considered. The discussion will be restricted to environmental factors within the range of values which seeds in general (though perhaps not the specific seed in question) are likely to encounter in nature. In the course of doing this, we shall try to identify the natural environment which a specific germination regulating mechanism can aid in characterizing. This does not mean that regimes, combinations, and sequences of these factors which are unlikely to be naturally encountered will be excluded, since the responses to them can sometimes clarify the nature of the germination-regulating mechanism. It will be worth bearing in mind throughout that physiology of germination is almost exclusively studied under conditions which are (at least in part) practically nonexistent in the natural environment of the greatest majority of seeds. Seeds germinate (or do not) in contact with a soil–water–atmosphere system which normally surrounds them on all sides, and not more or less immersed in water. In Petri dishes the water usually evaporates and has to be replenished, exposing the seeds to a highly unnatural cycle of water availability and aeration. The composition of the atmosphere, the water potential, its components and the changes in them with time in the natural submicroenvironment of the seed, are not duplicated in laboratory studies, with the exception of those few which are specifically designed to examine their effects. It will become apparent that strong interactions may occur which greatly modify the effectiveness of a given environmental factor in controlling germination and may result in misleading interpretations and conclusions.

### III.  Immediate Responses

From an agricultural viewpoint germination is a process which starts when the dry seed is planted in wet soil and ends when the seedling emerges above ground. However, since it was found that failure to emerge could be a result of seedling mortality or soil impedance, the concept of germination was narrowed down to exclude the problems which confront the seedling. As a result, the physiological concept of germination became that of a process which starts with supply of liquid water to the dry seed and ends where the growth of the seedling starts, most commonly by protrusion of the embryonic radicle through the seed coat.

The criterion of radicle protrusion as the end of germination has generally proved to be a very useful one, but a few exceptions necessitate caution in applying it indiscriminately. Thus there have been a few instances of apparently nonviable seeds in which the radicle (or cotyledons) protruded as a result of imbibition and rupture of the seed coat, without any subsequent growth taking place. Therefore, where such behavior is suspected, it is customary to classify seeds as germinated only after active growth has been established, for instance by observation of geotropic curvature.

The established physiological criterion for the start of the germination process has also become blurred. On the one hand, uptake of water by imbibition is a purely physical process, of which nonviable seeds are also capable. It is nevertheless a process which is essential for germination. Yet imbibed seeds can usually be dried to their original water content without any aftereffects. On the other hand, in many cases environmental conditions to which the seed is exposed before ever being rehydrated by imbibition do have aftereffects on its subsequent germination. As we shall presently see, some of these enviromental effects can be traced as far back as embryogenesis. On these grounds, it seems advisable for the purpose of the present discussion to avoid definitions and semantics by treating germination from a functional viewpoint and deal only with that germination which had reached successful completion in the physiological sense, but at the same time to examine environmental control of this germination as far back as possible in the ontogeny of the seed. The discussion in the present section will deal with environmental effects on germination in the reverse order in which they affect the physiology of the seeds. First, effects on seeds which have been presented with water (so-called immediate effects), then effects on the mature seed (another ambiguous term) between its dispersal and its subsequent germination, and finally environmental effects on the developing seed.

## A. Temperature Regime

### 1. CONSTANT TEMPERATURE

a. OPTIMA, RANGES, AND KINETICS.    The concept of the existence for every plant species of upper (maximal) and lower (minimal) temperature limits, outside of which germination of its seeds cannot take place, was first formulated only just over a century ago by J. Sachs (1860) who first studied the effects of constant temperatures on seed germination of several cultivated plants. The term "optimal" temperature was used to describe that intermediate temperature at which best germination was obtained. With increased insight into the properties of enzymes and the nature of enzymic reactions, it became self-evident that any physiological process in which an enzyme-mediated step was an essential component, can take place only between certain temperature limits. Its performance will improve as temperature is increased from the lower limit until a certain temperature is reached above which any further increase in temperature will result in reduced performance, reaching zero at the upper limit. It was also recognized that a "time factor" was involved in such processes. When the reaction was measured as a function of temperature over a short time interval, the optimal temperature was frequently considerably higher than when it was measured over a longer interval. Kinetic study revealed that at temperatures below the "long-term optimum" the reaction rate increased with temperature and remained stable with with time. As temperatures were increased above this long-term optimum, the stability with time was lost. The reaction rate was initially higher, but declined more rapidly and to a lower level. This decline was ascribed to a time-dependent thermal denaturation (progressive change in configuration) of the enzyme, which was more rapid the closer the temperature was to that of protein denaturation. In reviewing studies on temperature relationships of seed germination, Edwards (1932) cites several instances where the optimal temperature as well as the width of the temperature range are strongly dependent on duration of the incubation period. As incubation is prolonged and more germination takes place, the optimum shifts to lower temperatures, becomes less sharply defined, and the interval between the minimal and maximal temperature becomes greater. The response of *Pinus rigida* seeds is quite typical. After one day, maximal germination percentages were obtained at 47°C, and no germination at all was evident outside the range of 38°–55°C. With increasing time (and progress of germination), the maximum shifted to progressively lower temperatures, reaching 23°C after 14 days. At the same time, the range within which germination had occurred widened to 17°–57°C (Haasis, 1928). When such data are translated into kinetic terms it means that germination started earlier and was initially more rapid at higher temperatures, but the rates decreased to zero sooner than at lower temperatures. The similarity with the time factor in temperature dependence of physiological processes in which enzymic reactions are involved is clear and can be extended to include the temperature dependence of germination kinetics. The analogy between enzyme kinetics and those of germination is not as farfetched as may appear. Although the concept may be hard for a biochemist to accept, each seed contains all the substrates and the biochemical apparatus which are required to perform the variety of interrelated biochemical processes of germination, the complex end product of which is the seedling. The activity of many enzyme reactions is characterized by a lag phase, the length of which may be temperature dependent. Germination too is characterized by a similar lag phase, the length of which may be temperature dependent. Part of the processes which contribute to the presence of the lag phase in germination are strictly physical (for instance the initial steps of imbibition) and are not very dependent on temperature. Other preparatory processes are biochemical, and their own temperature dependence will endow the lag phase with temperature dependence. Similarly, if an enzymic process takes place in presence of a limited amount of substrate in the reaction vessel (as is the seed population in a germination vessel) the reaction rate (germination rate) will eventually decrease, as the substrate molecules (seeds) become scarcer. At the same time, the reaction rate will also depend on availability of active enzyme molecules, and this may decrease with time (or even cease entirely) by thermal inactivation. Accumulation of inhibitory by-products may reduce the reaction rate (germination) to zero at maximal temperatures which are much too low to cause thermal inactivation. Under such limiting conditions the substrate may not be fully consumed in the reaction (germination less than 100%).

The foregoing discussion makes it quite clear that much more information can be obtained about the temperature dependence of germination by kinetic analysis of germination than by merely recording germination percentages at arbitrarily selected times after the start of incubation. The kinetic data are obtainable from the time course of germination, as determined by frequent counts (and removal) of the newly germinated seedlings. (In many species great care must be taken to count and remove seedlings without interfering with the constancy of the temperature regime or inadvertently triggering the light-sensitive apparatus which controls seed germination.) A curve is then obtained which is S shaped, but not always perfectly sigmoidal, and is therefore difficult to express by a mathematical formula. If a mathematical formula cannot be applied, the curve can be characterized by at least three parameters which describe the length of the lag phase (to some arbitrary point), the maximal rate of germination (the tangent at the inflexion point) and the final percentage (no additional germination). A crude way to do this has been described for curves which are approximately sigmoidal (Koller, 1957). Additional parameters which characterize the process can possibly be obtained by further analysis, for instance, of the phase of decreasing rate of germination.

b. FIELD CONDITIONS.   Desirable as such information may be, there is very little of it available in the numerous publications dealing with dependence of germination on the level of the constant temperature to which the imbibed seed is exposed. In what follows, we shall deal with a few aspects that are fairly well established. One fact which is easily demonstrable is that seeds may differ greatly in the width, as well as in the actual limits of the range of constant temperature within which they germinate. The upper and lower temperature limits within which seed germination has thus far been found to occur are close to freezing on the one hand and approach the lethal limit, for higher plants, on the other. All conceivable variations of width of the temperature range and of its place on the temperature scale within these overall limits have been shown to be represented in the temperature-dependent germination of different plant species: narrow or wide ranges, spanning the low, intermediate or high temperatures between these limits. Lauer (1953) has studied nearly a hundred species of weed seeds from cultivated fields in Europe (mostly in Germany) and has classified them according to their temperature dependence. Extreme examples from near the lower temperature limit are also given by Sifton (1927), Aamodt (1935), and Schroeder and Barton (1939), and from near the upper limit by Haasis (1928), Kadman-Zahavi (1955), and Gutterman (1969). The question can now be asked, what significance can be ascribed to these marked differences in the temperature dependence of germination in seeds of the different species? An apparently clear-cut correlation, albeit circumstantial, has frequently been found between the temperature-range within which germination could take place under laboratory conditions and that which normally prevails in the natural habitat of the plant during the season in which it germinates. In certain cases, differences in temperature dependence of seed germination, which are related to climatic differences of the natural habitat, were observed within the same species. Western hemlock (*Tsuga canadensis*) has a wide latitudinal distribution in North America. Studies by Stearns and Olson (1958) showed that seed germination of this species was controlled by the photoperiod, the prevailing temperature, and the degree of pretreatment at near-freezing temperature (stratification). Photoperiodic control of germination and "stratification" will be discussed in subsequent sections of this chapter. With respect to temperature, it was shown that when the results with stratified seeds at all photoperiods were pooled, seeds from plants growing in northern latitudes (Maine, Quebec) germinated best at the lower temperatures within the range tested ($7°-12°C$) and were increasingly inhibited as temperatures became higher and approached $27°C$. Seeds collected in Connecticut responded in a similar manner, but had a wider optimum ($7°-17°C$) and were less inhibited at higher temperatures. Seeds collected in Indiana had a higher optimum ($12°-17°C$) and were inhibited by both low and high temperatures, while those from Tennessee had a still higher optimum ($17°-22°C$), but were much less inhibited by the low or high temperatures. Pioneer attempts to relate germination in nature with responses in controlled laboratory experiments were made by Went (1948, 1949), Went and Westergaard (1949), and Juhren *et al.* (1956), on desert annuals of the Joshua Tree National Monument and Death Valley in southern California. These studies established the striking difference between the floristic composition of the annual vegetation which arises after the light, prolonged rains which are characteristic of winter, and that which arises after the infrequent thunderstorms which occur in summer and result in heavy, short, and localized rainfall. A third variant of annual vegetation, also differing in composition, was characteristic of autumn rains. Samples of topsoil were collected, irrigated, and placed at different temperatures. At the highest temperature the germination which occurred was predominantly that of the summer annual group, at the lowest temperature germination of winter annuals predominated, while at an intermediate temperature germination of the autumn annuals (and perennials) predominated. Went (1957) interpreted these results as indicating the crucial role of temperature in determining the floristic composition of the desert annual vegetation through its control of germination. This interpretation was supported by the behavior of the group of autumn annuals, which was of particular significance since the germination in nature was dependent on elevation, rather than on season. At low elevation, it followed late autumn rains, while at increasingly higher elevations it occurred after progressively earlier rainfall. This indicated that temperature rather that photoperiod or some other seasonal variable was the environmental factor controlling germination, provided of course that rainfall was adequate. Work of similar nature, with similar general conclusions, was carried out by Lauer (1953) on weed vegetation of cultivated fields in Europe and, more recently, by Hammouda and Bakr (1969) on desert plants in Egypt.

Went (1957) also states that in nature the separation between groups was more extreme, and there was less overlapping than in the laboratory. This finding throws additional light upon the role of temperature in the environmental control of germination in nature. There was obviously some additional environmental factor which under natural conditions narrowed down considerably the temperature limits within which germination could take place. One can speculate about the possible nature of this environmental factor and assume that under desert conditions it could probably be the availability of soil moisture. In nature the water potential at any specific point in the soil decreases after rainfall and this decrease is characterized by a time course, which is dependent on several environmental factors. Germination evidently cannot take place when the soil water potential decreases below some critical point (as will be discussed in a subsequent section). Seeds which had not germinated by the time the soil water potential in their vicinity had dropped below the critical level, gradually became dehydrated, rarely with any significant loss in viability (Milthorpe, 1950; Negbi and Evenari, 1961), though possibly with some change in their germination requirements which we now know by the name "secondary dormancy." Seedlings which germinate before the soil water potential in their immediate vicinity drops below the critical value, will not be affected by this drying out, as long as the positively geotropic growth of their radicles can keep pace with the gravitational descent of the soil water profile, so that their absorbing surfaces retain contact with the zone of available soil moisture. Temperature may, and usually does have considerable effect on the kinetic aspects of germination, besides those on its final percentage, such as duration of the initial lag phase and the subsequent rate. There thus exists a race for time between the two processes. The eventual outcome of this race under a specific set of conditions of decreasing soil water potential will therefore depend on the manner in which the specific kinetic aspect(s) of germination is (or are) affected by temperature (Fig. 1.2). If, for instance, the duration of the lag phase is greater, or the rates of germination are lower, or both, at temperature $T_1$ than at temperature $T_2$, the final percentage in nature may be smaller at $T_1$ than at $T_2$, even though they may be identical under laboratory conditions, in absence of any change in water stress. On the other hand, if duration of the lag phase is longer but the rates of germination are more rapid at $T_1$ than at $T_2$, the final outcome under natural conditions will depend solely on the rate of decrease in soil water

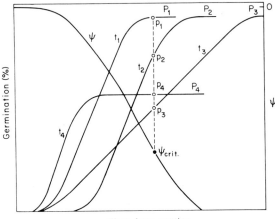

Time from wetting

Fig. 1.2. Combined effect of temperature-dependent time course of germination and of decreasing water potential (assumed to be independent of temperature) on germination percentages (schematic). At $\psi = 0$ (e.g., Petri dish), germination at temperatures $t_1$, $t_2$, $t_3$, and $t_4$ will be $P_1 = P_2 = P_3 > P_4$. In the field, $\psi$ will decrease to its critical value for germination, $\psi_{crit}$, causing germination to stop at $P_3 < P_4 < P_2 < P_1$, because of differences in kinetics at the various temperatures (it is assumed that these kinetics are not appreciably affected above $\psi_{crit}$). From Koller (1969).

potential; if drying is rapid, the effects on duration of the lag phase will predominate. Germination will reach higher percentages at the temperature which hastens its onset, even when the reverse may be true under laboratory conditions, i.e., in absence of any change in water potential. On the other hand, as drying becomes slower, the effects of temperature on rate of germination will become relatively greater and may even outweigh those on duration of the lag phase. Germination in nature may thus eventually reach higher percentages at the temperature which accelerates its rate, whereas under laboratory conditions the results may be quite different. For these reasons, the temperature relationships of germination which are observed under laboratory conditions are not simple, straightforward indicators of the degree to which germination is temperature-regulated under natural conditions. In fact, they can at times be quite misleading.

2. CIRCADIAN FLUCTUATIONS

a. PHENOMENOLOGY. It has been known for quite some time that seeds of many plant species will germinate better in a regime of diurnal alternations between a lower and a higher temperature than at constant temperatures. Numerous cases are cited in the seed-testing procedures recommended by the Association of Official Seed Analysts. The phenomenon was first reviewed extensively by G. T. Harrington (1923), who also carried out some rather extensive studies on the subject. Seeds of Bermuda grass (*Cynodon dactylon*), for instance, hardly germinate at all in darkness in constant temperature regimes between 20° and 35°C, but do so readily when any of these temperatures alternates diurnally with one that is several degrees lower, even though the latter itself does not cause any germination when applied in a constant regime. Within the range tested, germination improved with increase in the high temperature part of the cycle, and for any high temperature it improved with decrease in the low temperature part of the cycle. The response was apparently saturated at the 35°/20°C alternation. These results were confirmed and extended by Morinaga (1926b), who found the maximal response of *C. dactylon* at an alternation between an even higher temperature (38°C) and an even lower one (10°C). In his experiments, seeds of *Typha latifolia* behaved similarly in not germinating at constant temperatures and in responding well to diurnal alternations. Their behavior appeared to differ from that of *C. dactylon* by their relative apparent insensitivity to the high temperature part of the cycle (22°–38°C) when germinated under water (which is more favorable for them than wet blotters!), provided the low temperature part was sufficiently low (10°C). A different type of response to diurnally alternating temperatures was observed in seeds of Johnson grass (*Sorghum halepense*). Germination percentages increased almost linearly from less than 10 to more than 75% with increase in constant temperature from 20° to 40°C. However, germination at 30°, 35°, and 40°C was increased when these temperatures alternated diurnally with *suboptimal* temperatures which were 10°–15°C lower (G. T. Harrington, 1923). Germination of seeds of celery (*Apium graveolens*), on the other hand, increased from zero to over 40% with decrease in constant temperature from 27° to 10°C. These seeds also responded favorably to a regime of diurnally alternating temperatures, but their overall temperature dependence was diametrically opposite to that of Johnson grass seeds, with respect to the level of constant temperature, as well as in their requirement for the alternation to be between a low (5°–15°C) temperature and a *supraoptimal* one (> 15°C). The actual level of the supraoptimal temperature was not critical until it exceeded 27°–32°C (Morinaga, 1926b). An additional striking case is that of *Bidens radiatus,* whose seeds, when freshly harvested, will germinate profusely at constant temperatures, provided they were even slightly higher than 25°C, and not at all at temperatures lower than 25°C, unless they alternated diurnally (e.g., 5°/22°C and 10°/22°C) (Rollin, 1959).

b. Hypotheses. There have been several attempts to account for the rather specific response of seed germination to diurnally alternating temperatures. Treatments which change the physical properties of the coats in which the embryo is enclosed — properties such as permeability and tensile strength — or treatments which remove these coats altogether frequently result in loss of the specific requirement for diurnal temperature fluctuations. Germination of seeds thus treated can take place over a relatively wide range of constant temperatures (G. T. Harrington, 1923; Morinaga, 1926b). This has led to the belief that the site of response to temperature fluctuations resides in these coats rather than in the embryo. However, it was also suggested (Morinaga, 1926b) that the same results could be expected if the effects of the temperature alternations was to increase the growth processes of the embryo to the extent that it would rupture the coats, a feat it was incapable of with the weak force which it could develop in a regime of constant temperature. In their review, Toole *et al.* (1956) suggest that as a result of differences in the temperature dependence of the reaction rates, respiratory intermediates may accumulate during the high temperature part of the cycle, because it is unfavorable for germination. These intermediates thus become available to the germination process during the low temperature part of the cycle, which is not favorable for respiration. Analyzing his studies on the effects of temperature fluctuations on dark germination of light-sensitive lettuce seeds, Cohen (1958) came to the following conclusion on thermodynamic grounds. A complex macromolecular compound, such as an enzyme precursor, or a membrane which separates reactants exists in the seed and inhibits its germination in darkness. This hypothetical compound is thermo-labile and may be sufficiently modified by the temperature change to initiate germination. P. A. Thompson (1969) states that neither of the above hypotheses are compatible with the obligate germination responses of *Lycopus europaeus* to diurnal temperature alternations. First, no critical temperature for the response was found. A single 24-hour exposure to 15°C induced 81% germination at 35°C. A single exposure to 35°C caused 54% germination at 15°C if it was 7 hours long, and was entirely ineffective when it was 24 hours long. Shorter exposure periods were not tried in either case. When seeds incubating at 25°C were subjected to one or more 24-hour alternations, the minimal level of the high temperature part of the cycle was 20°C and the promotion was increased as this temperature was raised to 34°C. The minimal effective amplitude was 7°C. The minimal level of the low temperature part of the cycle depended on duration of daily exposure to it, 1° and 6°C when the duration was 16 and 8 hours per cycle, respectively (Fig. 1.3). Secondly, the effects of successive cycles were cumulative, the response being saturated with five cycles. The minimal duration of either part of the cycle was 6 hours. Increasing the duration of the low temperature part to 16–48 hours

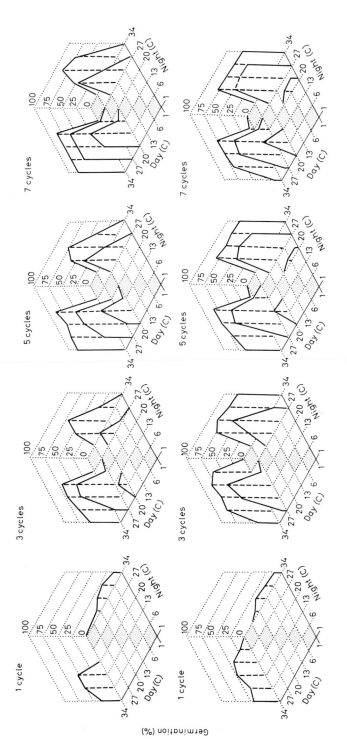

Fig. 1.3. Effect of various combinations of day/night temperatures on germination of *Lycopus europaeus*. Seeds were maintained at 25°C before and after being subjected to 1, 3, 5, or 7 completed cycles, comprising 16 hours at the higher temperature and 8 hours at the lower one (upper 4 diagrams), or vice versa (lower 4 diagrams). From P. A. Thompson (1969).

resulted in a maximal response, but longer exposures became ineffective. On the basis of these results, Thompson suggests that two successive changes of a complete cycle are linked metabolically and that a product resulting from the first temperature change plays a part in reactions following or during the second temperature change. Meneghini *et al.* (1968) present a tentative physicochemical model to explain the response of *Rumex obtusifolius* seeds to alternating temperature. According to their interpretation, the effect could result from the presence of an inhibitor as as intermediate in a closed system of two reversible chemical reactions. Their hypothesis is based on the fact that dark germination could be obtained at a constant 25°C by a single brief exposure to a higher temperature (32°C and above), as well as by a much longer single exposure to a lower temperature (15°C or lower). The more the alternation departed from 25°C, the more effective it became. Looked at differently, it is quite likely that germination in this species is capable of taking place over a relatively wide temperature range (4°–25°C), once it has been triggered by a single exposure to a supraoptimal temperature which is at least 10 degrees higher than the prevailing one. This exposure may but need not be quite brief, except at relatively high temperature. If this is indeed the case, it is similar to and more readily explainable by Cohen's model. In their review, Koller *et al.* (1962) suggest an additional possibility. This is based on the widespread existence among plants (and other organisms) of endogenous fluctuations in certain physiological activities, with a periodicity of approximately 24 hours. These have been termed circadian rhythms (or phenomena). It is generally believed that these rhythmic activities must be in phase with diurnal cycles (i.e., of 24-hour duration) in environmental variables, such as photoperiod or temperature, for the balanced functioning of the organism and, in particular, of processes involved in its growth and development. The response to diurnal temperature fluctuations has been termed diurnal thermoperiodicity and has been extensively studied and reviewed by Went (1957). Since the circadian rhythms have a periodicity of approximately 24 hours, it is quite likely that different processes in the same organism as well may have rhythms with a slightly different cycle length in the same way as the rhythmic activities of different individual organisms within the same population. Alternatively, the various rhythms may be out of phase with one another. It has been shown (Pittendrigh and Bruce, 1959) that in a constant environment, the individual rhythms will maintain their phase differences. Moreover, any slight differences in cycle length of different individuals will gradually amplify the differences in phase between them. On the other hand, a change in the light or temperature environment, unilateral or in the nature of a pulse, will result in a great deal of synchrony. If the change is made periodic, or the impulse is applied periodically, the synchrony will become greater, provided the cycle is of the proper length (i.e., close to 24 hours). The possibility suggested in the above-mentioned review rests on the assumed but unproven existence of circadian rhythmic changes in processes which take place in seeds. Where different processes in the seed are out of phase with each other to start with, yet are interdependent for their proper functioning in germination, exposure of the imbibed seed to constant temperature will increase their phase differences. As a result, the germination processes will function inefficiently or not at all. On the other hand, exposure to diurnal thermoperiod will synchronize the different interdependent processes and result in germination. There is hardly any evidence on the basis of which the merits of the different possibilities may be judged. One significant fact must be taken into consideration, namely that the effective duration of one part of the cycle can be quite short. Thus, for seed germination of *Cynodon dactylon* the response to the low temperature part of the cycle (at 15°C) starts when its duration exceeds 1 hour per daily cycle, (the remainder being at 38°C) and increases gradually until it becomes saturated at about 6 hours. The response to the high temperature part of the cycle (at 38°C) starts when its duration exceeds 2 hours per daily cycle, and continues to increase without reaching saturation even at 8 hours (Morinaga, 1926b). For seed germination of *Rumex obtusifolius* the response to the low-temperature pulse is saturated by 1–2 hours exposure, while the high-temperature pulse starts to become effective when its length is 4–8 hours and its effectiveness increases with duration beyond 32 hours (Meneghini *et al.*, 1968) (Fig. 1.4). Seeds of *Fraxinus mandshurica* var. *japonica* have a requirement for diurnal temperature fluctuations amongst other requirements for germination. Lengthening the high temperature part of the cycle at the expense of the low temperature part resulted in a shortening of the lag phase (hastening the onset), and a reduction in final percentages of germination (Asakawa, 1957b). The implications of such effects on the kinetics of germination under natural conditions have been discussed above in Section I. Seeds of *Oryzopsis miliacea* (from dispersal units which had been pretreated in sulfuric acid) germinate in darkness to a limited extent at constant temperature. Germination does not greatly exceed 30% at the optimal 20°C, and is practically zero at 30°C. Yet a single exposure of seeds incubating at 20°C to as little as 1 hour at 30°C (after all seeds which were capable of germinating at 20°C had already done so) increased percentages from 31 to 64, which means that this induced germination of about half the ungerminated seeds (D. Koller, unpublished results). This is about the maximal response which can be elicited from such seeds in a regime of

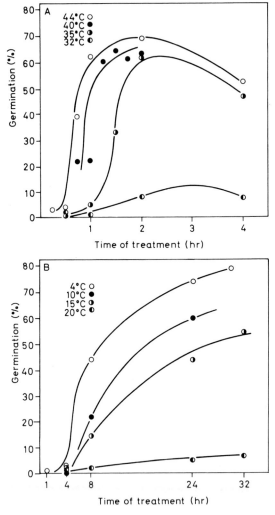

Fig. 1.4. Germination of *Rumex obtusifolius* at 25°C, as influenced by a single exposure, of different duration, to different levels of a higher (A), or a lower (B) temperature. From Meneghini *et al.* (1968).

diurnal alternation between 16 hours at 20°C and 8 hours at 30°C (Koller and Negbi, 1959). A similar response to a brief exposure to supraoptimal temperature was described in the Japanese tobacco varieties Kokubu and Bicchu. Neither oxygen deficiency, nor saturation with carbon dioxide during a 5-hour period, had any inhibitory effect on the response, unless the high-temperature pulse (15–30 minutes) was included in it (Ogawara and Ono, 1956).

c. BIOLOGICAL SIGNIFICANCE.   Response of germination specifically to a regime of alternating temperatures as compared to a lesser or even negligible germination in a regime of constant temperature is of very widespread occurrence among plants of diverse life forms, taxonomic groups, and ecological types. Such a response therefore merits evaluation in terms of any contribution it might make, as a biological control mechanism, to enhancing the survival potential of the species. Alternating temperatures in general can be regarded as providing much more information than constant temperatures. Whereas the latter can vary only by a single characteristic, namely their absolute level, the former can vary by two characteristics, namely the mean level around which the temperatures alternate and the amplitude of the fluctuation. A diurnal alternation in temperature is an ubiquitous feature of all environments which come under the influence of the diurnal cycle of the sun's radiant energy. Naturally occurring fluctuations differ from the artificial ones in the laboratory by their wave form. The former is a sine wave while the latter is an almost rectangular wave. This means that in nature there is also a time factor, which is a result of amplitude, to take into account. This time factor may be of greater biological significance than the actual limits of the amplitude. The latter may be reached only momentarily and thus have very insignificant effects (even when they reach well into the lethal range). On the other hand, results of laboratory studies, only some of which have been cited above, stress the critical importance of the duration of the diurnal exposure to the low or high temperature part of the (square-wave) cycle. Evidently, when the fluctuations are sine wave in form, the time during which the temperature passes through a given interval (within the amplitude) will be longer the closer that interval is to the limits of the amplitude. Therefore, a seed such as that of *Cynodon dactylon,* whose germination is conditional upon the extent to which its exposure to $38°\pm2°$ and $15°\pm2°C$ exceeds 2 hours and 1 hour, respectively, is likely to germinate better in a temperature regime fluctuating between 12° and 42°C than between 7° and 47°C.

These two characteristics of a diurnally fluctuating temperature regime, namely its mean value and its amplitude, are determined in any given spot on the earth's natural surface (with the exception of volcanically active ones) by the exchange of energy, whose primary source is the sun. Thus, all components of the environment which influence that energy exchange will affect the mean and amplitude of the temperature alternation. Among these components, some are extraterrestrial, such as the distance and angle of the sun. Some are characteristic of the macroenvironment, such as elevation above sea level, angle and direction of the slope, cloud cover, and atmospheric humidity. Others are characteristic of the microenvironment, such as plant cover, convection, conduction, and heat capacity of the medium. In this manner, the nature of the diurnal temperature alternation can provide specific information which may be of value in characterizing the season, the climate, and the microenvironment. Of these, the ability of the seed to characterize the season seems to be of little value to the plant from the point of view of survival, except where the most suitable season to start the life cycle is midwinter, or midsummer. Only in these seasons is the regime of diurnally alternating temperature unique, because they are the extremes. Regimes encountered in any other season, for instance in early spring, are likely to be encountered again, in this instance in late autumn. Evidently, only one of these two seasons is followed by a season which is suitable for seedling establishment, while the reverse is true for the other. It therefore appears that the survival value of the response of germination to diurnal temperature alternations as an indicator of season is limited to midsummer or midwinter, except when one of the two seasons with the same diurnal temperature regimes lacks one or more of the environmental prerequisites for germination, for instance, rainfall.

It seems likely that the response mechanism of germination to specific regimes of diurnally alternating temperature is of more significant value for the species in characterizing the environment, both macro and micro, than the season. More important yet is its probable role in characterizing the submicro environment. As already pointed out in an earlier section, seeds almost invariably come to rest after dispersal at various depths within the soil. Numerous studies, mainly with cultivated plants, have clearly demonstrated that for each species there exists a planting depth which is optimal for seedling emergence and survival. It is reasonable to expect that the same can be said for wild plants in nature. Indeed, the problem of optimal depth may even be more critical for the latter, because of the absence of man's helping hand in keeping the mechanical impedance of the superficial soil layers low and the supply of moisture regular by irrigation, cultivation, and mulching. For seeds of water plants, such as *Typha latifolia,* the difficulties which land plants encounter as a result of mechanical impedance do not exist. Instead, they face other restrictions, such as inadequate aeration, which limit the soil depth from which germination can take place. The diurnal alternations in the energy exchange between the soil surface and the environment result in diurnal alternations in soil-surface temperature. The latter are transmitted downwards through the soil, but because of the large heat capacity of the soil their phase is shifted and their amplitude is progressively damped with increasing depth. The quantitative aspects of this damping in relation to soil depth are determined by physical properties of the soil, its moisture content, and its homogeneity (Van Wijk and De Vries, 1963). Obviously, therefore, the requirements of a given seed for a specific diurnal thermoperiodicity in order to germinate can be met optimally at a specific depth in the particular soil of the habitat. It is tempting to assume that this is also the optimal depth for seedling emergence and establishment.

Response of germination to diurnal thermoperiodicity may play a special role in the biology and persistence of weeds in cultivated fields, their disappearance with fallowing, and their reappearance after fallowing. Cultivation is one means by which seeds get buried at much greater depth than in nature. Seeds of numerous weed species are obligate requirers of diurnal thermoperiodicity for germination, or germinate better under conditions of diurnal thermoperiodicity than at constant temperatures. When such seeds become buried at a depth which is out of reach of the required diurnal fluctuations they do not germinate but remain viable for extended periods of time. They will germinate when brought closer to the surface by future cultivation to within reach of the required diurnal thermoperiodicity. Galil (1965) describes a somewhat similar case in *Allium ampeloprasum*. The plant is a geophyte which infests cultivated fields and normally forms numerous bulbils at a depth of about 15 cm. These may remain entirely dormant for many years, but sprout readily when planted shallowly at a depth of 3 cm, as a result of the greater amplitude in the diurnal thermoperiodicity to which they become exposed. In this manner, when the parent plant gets uprooted by cultivation, its place is rapidly filled with numerous (vegetative) offspring. Studies by Warington (1936) showed that when soil samples from arable land were moistened, a much larger population of weed seeds such as those of *Papaver rhoeas* had germinated in a regime of diurnal temperature fluctuations when the amplitude of the fluctuations was large (greenhouse) than when they were small (in a cave or in an incubator). Furthermore, large-scale germination took place during the second year, when soil samples were transferred from the regime of smaller fluctuations to that of the greater ones. This indicates that viability of the weed seeds which had not germinated as a sole result of being deprived of the necessary temperature fluctuations had not been impaired thereby. On the other hand, studies by Asakawa (1957a,b) showed that seeds of *Fraxinus mandshurica* var *japonica,* which have an obligate requirement for diurnal thermoperiodicity (in addition to other environmental requirements) for germination, will enter a stage of secondary dormancy if that requirement is not satisfied. This particular difference may be related to the fact that the species studied by Warington were weeds while *Fraxinus* is not. It seems that for the former it would be of advantage to retain their original response mechanism to diurnal thermoperiodicity, so as to be able to germinate readily when brought closer to the soil surface by cultivation, whereas for *Fraxinus* the same response probably does not serve as an indicator of soil depth, but of some other environment which is favorable for its survival.

## 3. SEASONAL FLUCTUATIONS

a. PHENOMENOLOGY. As was briefly mentioned in Section I,A,2, many plant species have seeds which normally will not germinate unless they have previously been exposed to near-freezing temperatures for periods of time exceeding some threshold value. Study of this phenomenon was prompted by the fact that it was exhibited by many species of economic value for horticulture, such as deciduous fruit trees and ornamentals, or for afforestation. The so-called chilling requirement was met by nurserymen in practice by layering or stratification, the essential elements of which were exposure of the imbibed seed to the low temperature for the requisite period before planting. The phenomenon was studied intensively at the Boyce Thompson Institute for Plant Research in Yonkers, New York, among other laboratories. A great deal of the relevant data is reviewed by Barton and Crocker (1948) and by Crocker and Barton (1953). The physiological processes which take place during stratification are commonly referred to as afterripening and are treated as such. The experimental evidence, on the other hand, indicates that a distinction can be made between a type of stratification in which the processes which take place are involved in some sort of additional ripening, and another type, in which the processes which occur during exposure to cold are but partial processes of germination itself (Davis and Rose, 1912). It is for this reason that it is preferable and more logical to treat plants which belong in the second category as having seeds the germination of which is controlled by an alternation between a season of low (near-freezing) temperatures and one with warmer temperature. Hence the title of the present section. Plants which belong in the first category will be more appropriately dealt with in the section dealing with postmaturation environment.

The facts which indicate that in certain species it is seed germination itself which is controlled by a seasonal change or fluctuation in temperature are briefly the following. Ripening processes in seeds are associated with progressive dehydration. In contrast, only fully reimbibed seeds are capable of responding to cold treatment (Haut, 1932, 1938; Evenari *et al.*, 1948). Seeds which have become fully ripe are capable of retaining their germinability for extended periods in their nearly dehydrated state. In contrast, seeds which have been exposed to the required amount of cold, in the imbibed state will not germinate if dried to the original moisture content before being planted at the temperature in which they would have otherwise germinated (Haut, 1932, 1938). Seeds do not normally germinate of their own accord during ripening. On the other hand, fully imbibed seeds of quite a few species may actually germinate during exposure to the low temperature, provided it is prolonged sufficiently (Crocker, 1928; Joseph, 1929; Giersbach and Crocker, 1932; Evenari et al., 1948). There thus appears to exist in this type of seed a metabolic continuity between the processes which take place during germination and those which take place during the period of exposure to low temperature which precedes it. Such continuity appears to be absent between the processes of maturation and those which occur during the following period of exposure to low temperature.

b. HYPOTHESES. Detailed discussion of the probable nature of the physiological mechanism which is set in motion during exposure to cold is outside the scope of this chapter. Most, if not all of the metabolic changes which have been observed in such seeds during the exposure to cold, and were not observed in the same seeds during similar exposure to higher temperatures, are most probably the result of the actual response to cold, rather than its immediate cause. However, some of these cold-induced changes appear to be more significant than others. This is the case with endogenous growth-promoting and growth-inhibiting substances. Among the reasons why this is likely is the ability of excised embryos, from seeds which were denied exposure to cold, to grow slowly into so-called physiological dwarfs with stunted shoots and abnormal leaves (Flemion, 1959). Normal shoot growth takes place if the dwarfs are exposed to low temperature for a sufficiently long period (Flemion, 1933, 1934). The embryo itself shows signs of inhibited growth, inasmuch as only the cotyledon which is in contact with the moist substrate expands and becomes green in light. Exposure of the imbibed seed to low temperature increases the growth potential of the embryo, even in seeds whose excised embryos germinate freely into normal, undwarfed seedlings without being exposed to cold. Crocker et al. (1946) studied the capacity of walnut (*Juglans*) shells (among other nut-forming species) to withstand internal pressure of water which was being pumped into them. They found that this capacity was reduced with storage under moist conditions and that the rate of reduction was temperature-dependent. Intact walnuts germinate only after exposure to low temperature for a certain length of time, while the excised embryos show no such requirement for growth. Extrapolating from their results, these investigators calculated that embryos in seeds which had been incubated at 6° or 11°C could exert growth forces as high as 28 atm, while those in seeds incubated in temperatures higher than 17°C were unable to muster as little as the 18 atm required to split the shell. Working with *Fraxinus excelsior,* Villiers and Wareing (1965) found that an inhibitor in the embryo was responsible for its inability to germinate when unchilled. However, during chilling the inhibitor content remained unchanged, but a promotive substance appeared in increasing amounts and was capable of counteracting the effects of the inhibition. "Leaching" of unchilled seeds enabled germination to take place by diluting the endogenous (unleachable) inhibitor and resulted in growth of physiological dwarfs, so by itself presence of this inhibitor is not the cause of dwarfism. In other species which form physiological dwarfs from unchilled excised embryos, none of the known growth substances were capable of counteracting the dwarfing principle, with the exception of gibberellic acid, the effectiveness of which was neither total nor universal. Last, changes in growth substances may have been among the causes for the observed metabolic changes resulting from exposure to cold, by induction of enzyme synthesis, etc. These aspects are dealt with in more detail in other chapters.

The tantalizing problem which remains is the nature of the biochemical processes which have a temperature optimum so close to freezing. There are at least three striking phenomenological similarities between the effects of low temperatures on germination of stratifiable seeds and on flowering of vernalizable plants. The most obvious one is the surprisingly low temperatures at which the two processes can occur. The other is the relatively long time which is required for their completion. The third similarity is the gradual increase in the tolerance of higher temperatures with increasing duration of exposure to cold. In vernalization this is expressed by decreasing sensitivity to devernalization, while in stratification, if the initial exposure to low temperature at the required level is sufficiently long, the seeds will not only tolerate temperatures which are well above the level required for their stratification, but will actually germinate more rapidly upon transfer to these higher temperatures. This similarity is pointed out by Stokes (1965, p. 764), who also suggests that the mechanism, though not necessarily the participating reactions concerned, is most likely the same. She refers to Lang's (1965, pp. 1487–1488) explanation as a possible model. According to Lang's model, germination, like flowering, involves several competing processes, some of which are promotive, while others are inhibitory to the integrated process. All these processes are assumed to be "ordinary" biochemical processes, but they may differ somewhat in their temperature dependence. Taking the simple case of only one promotive and one inhibitory process, it suffices for the former's temperature-dependence curve to be shifted slightly more toward the lower end of the temperature scale to result in a net effect typical to vernalization and stratification. The integrated process will only be able to proceed where the inhibitory process does not overlap the promotive one, namely at low temperature. Naturally, the promotive process will be operating very far from its own optimal temperature and, therefore, at a rate which is much slower than the maximal. This may account for the long exposures which are required to achieve the full effect. In discussing the various possible mechanisms which may account for the optimum curve of temperature dependence of metabolic reactions in general and germination in particular, in Section III,A,1, one of the possibilities which was mentioned was inhibition by one of the reaction products. If a product of one of the partial processes is a substrate for another such partial process, but the rate constant of the latter is low, this would in turn prevent the former from taking place under any conditions which could lead to accumulation of the intermediate substrate. One way of avoiding such accumulation is by reducing the production rate of the intermediate, for instance at suboptimal temperature. This hypothetical mechanism is also compatible with the known phenomenology of germination of cold-requiring seeds or the induction of flowering in vernalizable plants.

c. BIOLOGICAL SIGNIFICANCE.    The requirement for an alternation between a season with near freezing temperature and one with a milder temperature and/or high elevation for seed germination is prevalent among temperate zone species and is common in many taxonomic groups whose origin is in such climatic regions, such as Coniferae and Rosaceae. The requirement appears to have a clear-cut value for survival of the species in such climates, protecting their seedlings from premature germination at the mild temperatures which prevail before the harsh winter, or occasional mild spells which might occur before winter is finally over. In this respect it might be significant that the response mechanism apparently measures time at temperatures somewhat above the freezing point, and that temperatures below freezing are ineffective, either when applied continuously or in alternation with the optimal low temperature (e.g., Davis and Rose, 1912).

d. ADDITIONAL PHENOMENOLOGY.    In nature, germination of seeds of the type so far dealt with would be controlled by a single seasonal alternation (cold → warm). Several cases are known where the germination requirements are satisfied only by a single seasonal alternation which incorporates an additional mild temperature season before the cold season. The additional season usually serves to accomplish one of two things in the biology of the seed, both of which are preparatory to the processes which take place during the subsequent cold period. One of these preparatory effects is modification of the structures which enclose the embryo, commonly reducing its mechanical resistance to expansion of the growing embryo. Various examples are cited by Barton and Crocker (1948). Crocker *et al.* (1946) found that higher temperatures were more effective than low ones in lowering the internal pressures required to overcome the mechanical resistance of the shell of walnuts (*Juglans*) and hickory (*Carya*) nuts in moist storage (Fig. 1.5). The response failed when

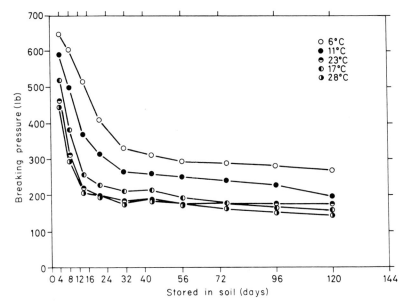

FIG. 1.5. Changes in breaking pressure (moving averages) of hickory nuts as a result of storage in moist soil at various temperatures. From Crocker *et al.* (1946).

the nuts were sterilized and treated in water, indicating that the effect was due to action of microorganisms. A similar conclusion was reached by Pfeiffer (1934) in the case of *Symphoriocarpos racemosus.* The other preparatory effect is found in species whose ripe seeds contain rudimentary, underdeveloped embryos, such as *Ilex opaca* (Ives, 1923), *Fraxinus nigra* (Steinabuer, 1937), and *Fraxinus mandshurica* var. *japonica* (Asakawa, 1957a). The embryo completes its growth and development and fills the cavity within the seed only during exposure in the imbibed state to mild temperatures (20°–30°C). When this process is completed the seeds become capable of responding to the two-season temperature alternation. The requirement for mild temperature for completion of growth and development of rudimentary embryos is not universal. In *Heracleum sphondylium* seeds, for instance, the optimal temperature for this process is 2°C. This requirement makes them respond to a two-season alternation, but in a manner which is quite different from seeds with a fully developed embryo, which have been dealt with so far. The processes which occur during this completion of embryonic development are mostly concerned with mobilization and transfer of nutrients to the embryo from the storage tissue in the seed. It has not yet been satisfactorily established whether the primary effect is the mobilization of the nutrients, which allows embryo growth as the nutrients become available, or whether it is the embryo itself which becomes active and thereby causes the mobilization of nutrients (Stokes, 1953).

It is not clear if there is any greater, or more specific survival value to control of germination by a response to a three-season alternation than to one by response to a two-season alternation. The same doubt exists regarding the possible role of "epicotyl dormancy" and "double protoplasmic dormancy" (Crocker and Barton, 1953) in the control of seed germination. Species exhibiting such phenomena will apparently produce fully developed seedlings in nature only after three and four consecutive seasons, respectively. However, as soon as growth has started, they lose the protection which the "uncommitted" state provides against the elements.

## B. Radiation Regime

Of all forms of radiant energy, visible radiation plays the leading part in the control of seed germination. The small size of seeds makes their temperature more directly dependent on the temperature of the environment than on the total radiation which they absorb. On the other hand, the natural environment of the seed is almost chronically deficient in the biologically active radiation of higher frequency. The effects of visible radiation (light) on germination of seeds have been the subject of very intensive studies, too numerous to review adequately within the framework of the present chapter. The subject has been reviewed rather extensively in recent years (Toole *et al.,* 1956; Evenari, 1956, 1965; Borthwick *et al.,* 1969). The present discussion will therefore focus on those aspects of light action which indicate the kind of role which it might play in the control of germination under natural conditions. Studies which were specifically oriented toward elucidation of the photochemistry of light action in germination and of the biochemical events which it sets in motion will be touched upon only where necessary and as sparingly as possible.

The original concept of the role of light in seed germination was that light could control germination by its mere presence or absence. Although this simple fact was discovered a relatively long time ago (about the same time that J. Sachs first published his results on effects of temperature on seed germination), it was not until fairly recently that attention was paid to reporting the presence or absence of light during germination experiments and even less attention was paid to attempts to control it. An uncomfortably large proportion of previous studies on germination has thus lost a great deal of its value. Additional data became questionable when it was discovered that light could affect germination not only by its presence or absence, but by three quantitative parameters, namely its intensity, spectral composition (or more precisely, spectral energy distribution), and duration. The tremendous impetus which these discoveries have given to photobiological research in seed germination has more than made up for past oversights, but great caution must be exercised in basing conclusions on the somewhat older studies.

The three characteristics of light which were found to have biological activity in the control of seed germination vary in nature within very wide limits. The total incident intensity (photon flux density) varies periodically within the course of a single day, with the maximum depending on geographical latitude and time of year. Of the total incident intensity at any given moment, the proportion which reaches the seed and its spectral energy distribution will depend greatly on attenuation by microenvironmental factors. There is no place on earth where sunlight (direct or indirect) is present continuously. Therefore, when dealing with germination within the framework of the natural environment, we have to take into account the effects of durations of continuous exposure to light which are shorter than 24 hours, but which are repeated periodically within 24-hour cycles. It is important to bear in mind that the three parameters which characterize the light environment are all quantitative. The more of them the seed can perceive and the greater the precision with which it can adjust its response to their level, the more precise will be its capacity to discriminate features of its environment which may be essential for survival of the species.

## 1. QUANTITATIVE ASPECTS

Of all the microenvironments in which the seed is liable to find itself under natural conditions, the one from which light is totally absent is the interior of the soil. Germination in total absence of light is therefore a possibility which most, if not all seeds may have to face at one time or another. There are quite a number of species which have never been reported to show the slightest response to light and will germinate in its total absence, provided of course that any other requirements which they might have for germination are met. On the other hand, more and more species which had previously been classified as insensitive to light, or "nonphotoblastic" (Evenari, 1956), were eventually found to respond to light under certain conditions. Almost without exception, the latent reaction to light which such seeds could be made to reveal was typical to the pigment system which is apparently universally involved in the control of germination of seeds which are normally capable of being stimulated by light (positively photoblastic). This is the reaction mediated by the phytochrome pigment system, which was discovered, thoroughly investigated, and to a great extent characterized at the Agricultural Research Service Laboratories of the United States Department of Agriculture in Beltsville, Maryland, by the efforts of H. A. Borthwick, S. B. Hendricks, M. W. Parker, E. H. Toole, V. K. Toole, and various collaborators. The properties of this system have recently been reviewed by Hillman (1967). Aspects which bear directly on germination are described in greater or lesser detail in some of the reviews listed in the beginning of Section III,B.

a. THE ROLE OF STRESS.   Quite a number of the conditions under which the latent requirement for light makes its presence felt are associated by a single common denominator, namely physiological stress. Thus in the classic material for the study of light sensitivity in seeds, the achenes of lettuce (*Lactuca sativa*) var. Grand Rapids, sensitivity to light is demonstrable at temperatures higher than 20°C, whereas at 20°C and lower, germination will take place as readily in darkness as in light (Evenari, 1952). Here, as in many similar cases, the requirement for light appears at supraoptimal temperatures, namely under a heat stress. The same seeds revealed their latent sensitivity to light even at temperatures where sensitivity is normally absent (14°C) in the presence of a germination inhibitor such as coumarin. At this temperature, coumarin concentrations as high as 80 mg/liter were incapable of causing significant inhibition in light but were totally inhibitory in darkness. On the other hand, at 26°C, the full promotive effect of light could be reversed by a coumarin concentration as low as 15 mg/liter (Evenari, 1965, Figs. 12 and 14). At 14°C the latent requirement for light expressed itself under a stress of chemical inhibition, whereas at 26°C, in presence of a heat stress, the chemical stress was effective at much lower levels. A further example was provided by A. Kahn (1960), who showed that when seeds of lettuce (Grand Rapids variety) were incubated in darkness at 21°C their high germination percentages (70–95) could be progressively reduced to zero by lowering the osmotic potential of the substrate to between −4 to −7 atm. This osmotic inhibition could be reversed by light, which is normally not required for germination at this temperature. It is noteworthy that (*a*) light was effective even when applied to the imbibed seed prior to lowering the osmotic potential, and (*b*) full germination was obtained in darkness if the osmotic potential was raised to 0 atm sufficiently early (before secondary dormancy had set in). The latent requirement for light was here expressed under condition of osmotic stress. It has long been known that requirement for light for germination could be almost invariably greatly reduced or even entirely eliminated by removal of coats surrounding the embryo. Germination of Grand Rapids lettuce seed at 26°C in darkness was increased from 15% to 94% by excision of the embryo, and the inhibition by coumarin was reduced (Evenari and Neumann, 1952). Thus the requirement for light at the supraoptimal temperature of 26°C did not appear to be absolute but seemed a result of some unidentified stress imposed by the coats.

b. CATALYTIC ROLE OF PHYTOCHROME. It remained for Scheibe and Lang (1965) to show that the promotive action of light was exerted on the embryo itself. Using half seeds (containing the embryonic axis and a small basal portion of the cotyledons) of lettuce, at 20°C they demonstrated that (a) rate of appearance of geotropic curvature was higher when phytochrome was activated (by red light) than when it was inactivated (by far-red light), (b) the light-sensitive Grand Rapids lettuce and the light-insensitive Great Lakes lettuce both responded to light in the same manner, and (c) in the presence of osmotic stress the difference in effect between red and far-red light was accentuated, the latter reducing the rate below that of the controls in darkness. It thus became clear that the promotive action of light on seed germination operated by increasing the growth potential of the embryo. In presence of stress the growth forces exerted by the embryo were insufficient to overcome the tensile strength of the coats unless they were amplified by activation of phytochrome.

This raises the possibility that the capacity of the embryo to grow is a function of its content of active phytochrome, and that any seed capable of germinating without the help of light can do so because it already contains a sufficient amount of prefabricated active phytochrome. This possibility is supported by results showing that light sensitivity can be induced in light-insensitive varieties of lettuce, as well as in species such as tomato, where light sensitivity is relatively rare, even in absence of stress of any kind, by dark incubation at supraoptimal temperature (in lettuce: Evenari and Neumann, 1953a; Borthwick et al., 1954) as well as by repeated or prolonged exposure to far-red light, both of which presumably reduce the content of active phytochrome (Macinelli and Borthwick, 1964; Mancinelli et al., 1966). However, the preexisting active phytochrome differs in at least two respects from recently photo-activated phytochrome. First, it is not susceptible to reversal by a single relatively brief irradiation with far-red light, even at high intensity. Second, the action spectrum for inactivation is characterized by a maximum at 716–

718 nm, not at 730 nm (Rollin and Maignan, 1966; Rollin, 1968; cf. Hendricks *et al.*, 1968) (Fig. 1.6). In spite of these indications, it is as yet too early to say just how general is the requirement for active phytochrome for the growth of embryos during germination, and whether or not it is essential for such growth or just stimulatory.

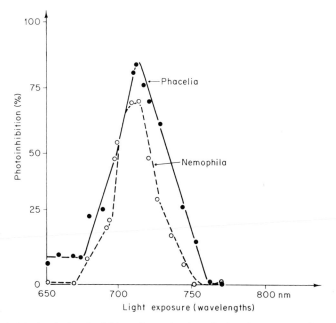

FIG. 1.6. Maximal photoinhibition of germination of *Phacelia tanacetifolia* and *Nemophila insignis* as a function of wavelength. From Rollin (1968).

c. BIOLOGICAL SIGNIFICANCE OF LIGHT REQUIREMENT.    A large proportion of the seed population in the soil is probably located at depths to which no light penetrates naturally. For light-sensitive seeds in the soil population which are located at such depths, the absence of light means that the limits of the environmental complexes in which their germination will be possible will be considerably narrowed around the optimum. The "noise" of the less favorable environments will be filtered off and the optimal ones will be more sharply defined. From the viewpoint of perception it is quite clear that absence of light increases the sensitivity of light-requiring seeds to the environment. Furthermore, awareness of conditions of stress, such as low soil–water potential, presence of toxic or inhibitory substances, will be amplified.

Other species with light-sensitive seeds may be obligate requirers of light for germination, and may be incapable of germinating in its absence under any naturally occurring conditions. Such seeds may be presented with an opportunity to germinate either before they reach soil depths to which light does not penetrate, or after they resurface by accident. The probability of resurfacing is greater in soils with a natural tendency for instability, such as shifting sand, as well as in soil which is artificially disturbed by cultivation. In both, the probability is also great for seeds to become buried at depths from which they could no longer emerge, were they to germinate. Sensitivity of germination of light may therefore play a significant role in the survival of plants in such specialized habitats. Thus, for instance, *Artemisia monosperma* is particularly well adapted to the conditions of constantly shifting sands in its natural habitat, and its seeds are absolutely dependent upon light for germination under all conditions (Koller *et al.,* 1964). The significant role which sensitivity of germination to light plays in the persistence of weeds is suggested by the prevalence of the more obligate type of light sensitivity in the germination of weed seeds of cultivated lands. Studies by Wesson and Wareing (1969a) have confirmed this suggestion by showing that germination of weed seeds from soil which had been used for pasture for several years was markedly increased when it had been disturbed by cultivation, but only if light had not been totally excluded from the soil surface. Somewhat similar results were obtained by Sauer and Struik (1964) with germination of pioneer species from recently disturbed soil of a deciduous forest, a conifer plantation, a tall-grass prairie and a muck field (usually planted to maize). Germination after brief illumination was three to five times as high as without it. Indeed, these results raise some doubts as to the validity of Warington's (1936) conclusions on the importance of diurnally alternating temperatures for germination and persistence of weeds (Section III,A,2,c), since these were based on experiments where light was much more intense in conjunction with the alternating temperatures (greenhouse) than with the more constant ones.

The depth at which the seed is buried is also of obvious importance for seeds in infrequently disturbed soils and this becomes more critical the smaller the seed and its food reserves. Generally speaking, stimulation of germination by light is more prevalent among small-seeded species and varieties, whereas most large-seeded ones are as capable of germinating in the absence of light as with its assistance. The reason may be that the smaller the seed the greater is the ratio of surface to volume. As a result, the balance between the growth forces which the embryo can exert and the tensile resistance of the coats which these forces have to overcome is shifted in favor of the latter.

d. INTEGRATION OF ENERGY AND TIME. Promotion of germination by light has almost invariably been traced to the activation of phytochrome.

It is one of the characteristics of phytochrome to be saturated by relatively low energies of light, provided it is of the appropriate spectral composition. This accounts for the fact that it usually takes very little light — short exposures of relatively low intensity — to stimulate germination. Seeds of *Lythrum salicaria* (Lehmann, 1918), certain varieties of tobacco (Kincaid, 1935), and *Artemisia monosperma* (Koller *et al.,* 1964) are examples of extreme sensitivity. This has raised some doubt as to the significance of such sensitivity for survival under natural conditions, where much higher light energies are ostensibly available. Two facts have to be borne in mind in this respect. First, natural light contains both the promotive (red) and inhibitory (far-red) spectral regions which are involved in activation of phytochrome and its inactivation. Consequently, in nature the amount of phytochrome which is activated is dependent on their net effect. Second, transmittance of soil to light is very low. As a result, even in full sunlight very long exposures may be required for saturation at relatively shallow depths in the soil. Indeed it is conceivable that beyond a certain depth the minimal effective exposure may become longer than the prevailing length of a single day. In such cases, more than one diurnal cycle of light and dark will be required, the number depending on the prevailing photoperiod. This creates an apparent photoperiodic response to long days.

A different quantitative aspect of light requirement is provided by seeds whose germination is dependent not so much on total energy as on the actual duration of incident light. Increasing the intensity may result in a corresponding shortening of the duration of exposure required for full promotion, but above a certain intensity this reciprocity no longer holds true and exposure time can no longer be reduced without loss in promotion. In laboratory experiments, germination of such seeds is progressively improved as duration of exposure to light (of a given intensity) is increased, until the response becomes saturated and further increases in exposure, up to continuous irradiation, become ineffective. When such seeds are exposed to light of insufficient duration to cause full promotion, its promotive action will be considerably increased if the total exposure time is split up into several portions which are separated by dark intervals and is thereby spread out to cover a longer period with the same total energy. Germination of seeds of *Kalanchoe blossfeldiana* (Bünsow and von Bredow, 1958). *Epilobium cephalostigma* (Isikawa, 1962), and *Lycopus europaeus* (P. A. Thompson, 1969) is typical of such a response. Conversely, the full promotion which is obtained by continuously irradiating the seeds can be equally achieved with a small fraction of the energy, by a series of short irradiations, applied at appropriate intervals. Such a response to "intermittent irradiation" has been studied in considerable detail by Isikawa and Yokohama (1962) and

Isikawa (1962) in seeds of *Epilobium* and *Hypericum,* and by Koller *et al.* (1964) in seeds of *Artemisia monosperma.* It is clear that since such seeds do not require light to be present continuously, each irradiation initiates a series of "dark reactions," which are apparently unaffected by the presence or absence of light. These dark reactions are inhibited by low temperature (Isikawa, 1962). The crucial question is why the photoresponsive pigment system, which is most probably phytochrome, has to be excited repeatedly. Borthwick and Cathey (1962) have observed a similar phenomenon in the photoperiodic control of flowering of *Chrysantemum.* They conclude from the results of their studies that reactivation of phytochrome is required, because after each irradiation, active phytochrome is partially inactivated (thermally, or through exerting its physiological action). Isikawa and Yokohama (1962) offer a different hypothesis to account for their results. They suggest that the amount of phytochrome which is capable of being photoactivated is limited because it exists in dynamic equilibrium with its precursor. Additional amounts can be released only when the equilibrium is disturbed by photoactivation of phytochrome and its subsequent utilization in germination. Koller *et al.* (1964), who studied a similar phenomenon over a wide range of temperatures, also suggest a model which is based on an equilbrium reaction between a substance and its precursor. Their model is more generalized than that of Isikawa and Yokohama, as the substance in question need not necessarily be phytochrome, but can just as well be a substrate with which active phytochrome interacts. The equilibrium concentration of the substance is a function of temperature, as is also the rate constant of the reaction by which it is reached. The former is higher and the latter is smaller at the optimal temperature than at a supraoptimal one. According to this model, availability of the substrate for phytochrome action may be the limiting factor and phytochrome nonlimiting.

From a functional viewpoint, seeds whose germination is controlled by an obligate requirement for an exposure to light which exceeds some critical duration may germinate to the extent that this requirement is met within the framework of existing daylength. They differ from seeds belonging in the previous category by the fact that if they require an exposure to light longer than the prevailing day length, they will germinate only after being exposed to the required number of photoperiods, irrespective of whether the light which reaches them is more or less intense.

2. PERIODIC ASPECTS

In Section III,B,1 above, two types of response were described which under natural conditions, particularly those of the submicroclimate, could cause a requirement for several photoperiodic cycles for full promotion.

In one response type, the energy is integrated and the effect could become photoperiodic only if the intensity of incident light becomes sufficiently low. This may actually be the case described by Isikawa (1962) for seeds of *Hypericum japonicum* f. Yabei. Although light intensity was not varied, the overall response to light appeared to be proportional to the total amount of energy, even when divided up into several exposures. When the intervening dark periods became excessively long, the response was progressively reduced. In seeds with such a response type, the energy for full promotion may have to accumulate over a period which is longer than that of the prevailing length of a single day. In the other response type, time is integrated and the period might exceed the prevailing length of a single day even when intensity of radiation is not limiting. This would also be expressed as a response of germination to photoperiod, which differs from the previous one, by not being conditional upon the intensity of light. An additional characteristic of the "time-integrating" seeds, results from the fact that the duration of the optimal dark interval between two successive irradiations depends on the rate constant of the "dark" reactions. Consequently, if the dark period is longer than the optimum, the germination response will be accelerated by insertion of an additional exposure to light during the over-long "dark-period," preferably towards its middle (e.g., Vaartaja, 1956). The analysis by Koller *et al.* (1964) indicates the manner in which both the optimal duration of the dark intervals and the number of saturating irradiations which is required could be affected by temperature. Similar responses are evidently not to be expected in the energy-integrating seeds.

a. LONG-DAY SEEDS.   The properties of the time-integrating type of seeds has led to their classification as "long-day seeds" (Isikawa, 1954). This concept gained support from the promotive action of a brief irradiation in the middle of over-long dark periods, because of its basic similarity to that of "light-break" treatments on flowering of long-day plants in short days (dark periods longer than the critical). This was, for instance, the case with seeds of *Begonia evansiana,* whose germination is promoted by photoperiods of 12 hours or longer, up to continuous irradiation. Photoperiods which are too short to be promotive (8 hours), become promotive when the dark period is interrupted with a brief irradiation (Nagao *et al.,* 1959). A similar response was observed in seeds of *Betula* (Black and Wareing, 1955; Vaartaja, 1956), in which the degree of inhibition (lack of promotion) was correlated with length of the dark period, but not closely with the amount of light in the light period. However, the response differed qualitatively from the one involved in the photoperiodic control of flowering. One striking aspect is its temperature dependence. The critical requirement for long days was expressed at low temperatures. At intermediate temperatures the requirement could be satisfied also in short days, even after a single, relatively short exposure to (red) light. At high temperatures, no light was needed for nearly maximum germination. Another significant difference is the apparent absence of inhibitory features to the dark period.

b. "SHORT-DAY" SEEDS. The analogy between photoperiodic control of germination and of flowering was extended by the finding that seed germination in certain species is promoted by short photoperiods, but is progressively inhibited as the duration of the daily exposure to light is lengthened (Isikawa, 1954, 1962; Stearns and Olson, 1958; Black and Wareing, 1960). Studies by Kadman-Zahavi (1960) with *Amaranthus retroflexus* showed that germination was strongly promoted by a short irradiation and that this promotion was mediated by phytochrome. Prolonged exposure to far-red light produced two inhibitory effects. It inhibited germination of the dark-germinating fraction of the seed population. It also reduced the responsiveness of the light-requiring fraction of the seed population to the promotive action of a subsequent short irradiation with red light (Fig. 1.7), unless the latter was separated from it by a

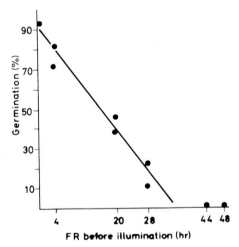

FIG. 1.7. Germination of *Amaranthus retroflexus* seeds. Inhibition by prolonged irradiation with far-red light (FR) of the promotive action of white light ($3 \times 10^3$ ft-c·sec). Seedlings which had germinated during initial 48-hour incubation in darkness were removed at time of transfer to far-red light. From Kadman-Zahavi (1960).

dark interval. Presence of red light with the far-red (as in white light) overcame both these inhibitions, but after considerable delay. On the basis of these results, Kadman-Zahavi suggests that most of the photoperiodic responses of "short-day seeds" may be due to the following sequence of events: inhibition of dark germination by prolonged far-red, stimulation of germination by red light activation of phytochrome at the beginning of each cycle and its immediate reversal by short far-red followed by progressive reduction of sensitivity to red light and progressive recovery of this sensitivity during the intervening dark period. These suggestions are supported by the analysis of light action in the germination of *Atriplex dimorphostegia* (Koller, 1971), which exhibits typical short-day photoperiodic resonses (see Fig. 1.15 in Section III,F), as well as the same type of inhibition by prolonged irradiation which were shown to exist in *Amaranthus*. Studies by Hendricks *et al.* (1968) on the opposing actions of light in seed germination of *Poa pratensis* and *Amaranthus arenicola* showed that action spectra maxima for suppression of germination were at 720 nm. The maxima were unchanged in position or magnitude in presence of radiation in the region of 600–670 nm which by itself was adequate to maintain phytochrome predominantly in the active (far-red absorbing) form.

   c. INHIBITION BY PROLONGED IRRADIATION. The mechanism by which prolonged irradiation exerts its inhibitory action is still far from clear. In Section III,B,1 above, the likelihood was discussed that the dark-germinating fraction of the seed population contain a form of active phytochrome. In view of this, it is probable that the inhibitions exerted by prolonged irradiation on the dark-germinating fraction, on the sensitivity to a subsequent promotive irradiation, and on the expression of the promotive action of a preceding promotive irradiation, are all mediated by the same system. This is supported by data of Chen and Thimann (1966), which show that inhibition of germination in *Phacelia tanacetifolia* by light, as well as by supraoptimal temperature, could be overcome by removal of the tip of the endosperm, but was reinstated by reducing the osmotic potential of the substrate. The similarity with A. Kahn's results with light-requiring lettuce seeds (1960) is quite striking. Results reported by McDonough (1967) showed that an apparent latent response to the inhibitory effects of continuous irradiation could become overt under the influence of a low (osmotic) water potential. Seeds of cucumber (*Cucumis*), radish (*Raphanus*) and sunflower (*Helianthus*), which normally germinate equally well in darkness as in continuous irradiation, were strongly inhibited by light in mannitol concentrations that had hardly any effect on germination in darkness. When the tip of the coat over the radicle was removed, the inhibition was partially removed. McDonough does not specify any precautions which were taken to prevent evaporation and the resultant increase in substrate concentration, which may have been much greater in light than in darkness. The promotive effects of partial decoating may have resulted from piercing of the semipermeable coat, as described below in Section III,C,2.

Spectral analyses of the effects of prolonged irradiation in inhibiting seed germination, as well as other photomorphogenic processes, failed to provide unequivocal proof of whether or not they were mediated by phytochrome itself, or some other pigment system (Hartmann, 1966; Wagner and Mohr, 1966; Rollin and Maignan, 1966; Hendricks *et al.*, 1968). The possibility cannot be ruled out that as a result of overlapping of the absorption spectra of the two end forms of phytochrome, prolonged irradiation will inevitably lead to cycling between them (Briggs and Fork, 1969), so that eventually all available phytochrome may be diverted into an inactive side product via one of the long-lived intermediates.

The photoperiodic response of short-day seeds is sometimes as strongly temperature dependent as that of long-day seeds. For instance, germination of *Hyoscyamus desertorum* is equally promoted by continuous and by intermittent irradiation at 25°–35°C. Both are equally ineffective at 10°C, but between these extremes only the latter is promotive (Roth-Bejerano *et al.*, 1971) (Fig. 1.8).

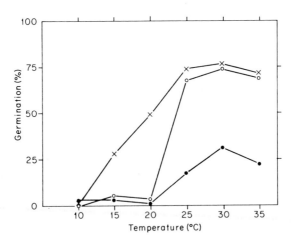

FIG. 1.8. Effects of temperature on germination of *Hyoscyamus desertorum* seeds in darkness (●), continuous irradiation (O), and in intermittent irradiation of 10 minutes light at 24-hour intervals (×). From Roth-Bejerano *et al.* (1971).

From what is known thus far about the photoperiodic control of seed germination, it appears that any similarity with photoperiodic control of flowering (and other morphogenetic processes) is probably only superficial. Phytochrome is involved in both processes, but its active form promotes flowering in long-day plants and inhibits it in short-day plants, while in germination its presence is equally essential for both long-day seeds and short-day seeds (Isikawa, 1962; Koller, 1971). There also appears to be very little in common between the response mechanism in short-day seeds and long-day seeds, except for the requirement for photoactivation of phytochrome, repeatedly under certain conditions.

d. Biological Significance of Photoperiodic Control. It is time to discuss the probable role which photoperiodic control of germination may have in survival of the species. From a functional viewpoint, a requirement for daylengths to be longer than a certain minimum may operate as in the control of flowering of long-day plants, by setting the seasonal time limits within which germination is possible. However, the temperature dependence of the photoperiodic requirements of long-day seeds such as those of *Betula pubescens* will apparently allow them to germinate at practically any time of the year, provided the temperature is suitable (i.e., optimal). Only when temperature is suboptimal does the requirement for long days approach limiting values, with respect to what is provided by nature. The virtual loss of photoperiodic control at a certain temperature reduces its value for survival as a means for perception of the season. So perhaps its value lies elsewhere. One possibility is based on the fact that the commonest submicroenvironment of the seed is within the soil. The natural daylength becomes exceedingly attenuated with depth, as optical path length in the soil and reflection from its surface change with position of the sun through the day. As a result, there will be a certain depth from which the seed will germinate only when the temperature is optimal. At shallower depths germination will be less strictly dependent on optimality of the temperature, but then the seedling will have a shorter distance to travel before it reaches the soil surface, and the need for optimal temperature may be correspondingly smaller. A similar reasoning is applicable in the case of short-day seeds, such as those of *Tsuga canadensis* (Stearns and Olson, 1958). Their germination will become progressively more dependent on season with decreasing depth in the soil. This may be of advantage in habitats where soil moisture conditions close to the surface are likely to be less favorable in certain seasons than in others. Where the requirement for short days is lost at high temperature (as in *Hyosayamus desertorum*) the most profuse germination in its natural habitat is likely to take place following rainfall in late spring, or in early autumn, which is infrequent but may be quite intense.

e. Negatively Photoperiodic Seeds. So far, three categories of photoperiod-sensitive seeds have been distinguished on the basis of their responses to conditions which they may encounter in nature. To these must be added a fourth which, for lack of a better term may be designated as negatively photoperiodic. In this category belong seeds which germinate optimally in total absence of light, but which are inhibited by continuous irradiation. The physiological background of this inhibition was described and discussed in the present section. The data of Mancinelli and Borthwick (1964) and of Mancinelli *et al.* (1966) suggest that such seeds are capable of germinating in absence of light because they contain preformed active phytochrome, in a form which is irreversible by a short irradiation with far-red light. The inhibition is apparently due to the susceptibility of this form of phytochrome to prolonged irradiation with far-red or white light. Seeds which exhibit this response will also be inhibited by light which is applied intermittently, if the dark periods are sufficiently short. Such a response has been described in *Nemophila insignis* by Black and Wareing (1960) and in several varieties of *Raphanus* by Isikawa (1962) (Fig. 1.9). Physiologically, such seeds are the

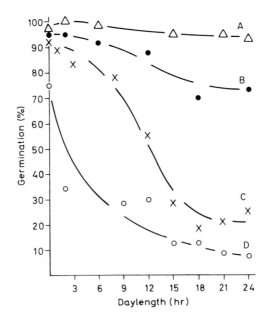

Fig. 1.9. The photoperiodic responses of various strains of *Raphanus* seeds exposed to 200 lux. A: Shôgoin, B: Nerima, C: Risō, D: Miyashige. From Isikawa (1962).

diametrically opposite of the long-day seeds without being short-day seeds, since they do not require any light. From a functional viewpoint, these seeds will not germinate in seasons, or at soil depths, which may expose them to subminimal night lengths. It might be argued that the survival value of this response as a seasonal indicator is restricted to cases where the optimal season for germination is midwinter. Otherwise, germination may take place over too wide a season (autumn or spring), during which the chances for survival change diametrically. It therefore seems more likely that this mechanism is concerned with preventing germination at too shallow depths in the soil, without a similar restriction being applied to deeper layers. To benefit from such a mechanism, seeds have to be amply supplied with stored nutrients, and live in habitats where shallow conditions are hazardous for survival. Two species which show this response fulfil these requirements, *Calligonum comosum* (Koller, 1956) and *Citrullus colocynthis* (Koller *et al.,* 1963), both having large seeds and inhabiting coarse sandy soils in the desert. *Citrullus* exhibited the negative photoperiodic response as predicted (Datta and Chakravarty, 1962; D. Koller, unpublished results). *Calligonum* has not been tested.

On the basis of these considerations it appears that the main function of photoperiodic control in survival is its potential to act as a soil depth gauge. It is significant that no correlation was apparently found between the photoperiodic response of the seed and the geographical latitude in which it grows (Stearns and Olson, 1958).

## 3.  QUALITATIVE ASPECTS

Thus far, the activity of light with respect to germination seems to be restricted to spectral regions in which phytochrome is known to absorb to greater or lesser extents. However, phytochrome exists in two relatively long-lived, interconvertible forms, only one of which is biologically active. Each has its own specific absorption spectrum, but the two spectra overlap almost over the entire range of visible light. As a result of this overlap, the biological activity of even the purest light source depends on the so-called photostationary state to which phytochrome is brought by the balance of the absorption by the two forms. This is what determines the absolute amount of active phytochrome which is formed. The same situation exists when impure light sources, or ones with broad spectra are used, with the difference that the photostationary states have to be integrated over the entire spectrum of the incident radiation. In general, however, it seems reasonable to assume that if the spectrum of the incident light is not too specific, its biological activity in terms of phytochrome activation will be proportional to the ratio of its energy content in the spectral region where the inactive form has peak absorption (red) to that where the active form has peak absorption (far-red). This is only approximately true, since the absolute level of incident energy in every region of the spectrum has to be taken into account. For instance, if the light source is much richer in energy in the shorter than in the longer

wavelengths, the relative energy in the main absorption peaks of phytochrome will have much lesser influence on the state of phytochrome than
that in the secondary absorption peaks, which are in the shorter wavelengths. This is not the case with sunlight, and the aforementioned assumption seems acceptable as a first approximation.

There have been a few noteworthy attempts to relate seed germination
to naturally occurring qualitative changes in the spectral energy distribution of sunlight, which result from specific absorption. One habitat is
the interior of the soil. Wells (1959), studying the possible role of sensitivity to red light for seed germination in ecological terms, assumed that
the transmittance of soil would be mainly determined by diffraction. The
latter is inversely proportional to the fourth power of the wavelength.
Therefore the light which is transmitted through the soil is relatively
richer in the longer wavelengths. Measurements agreed well with the pre-
diction (Fig. 1.10). The percentage transmission ratio 655 nm/450 nm
through a 5 mm layer of sand was 2.5 and increased to 6.0 when the thickness was doubled. The corresponding ratios for 735 nm/655 nm changed

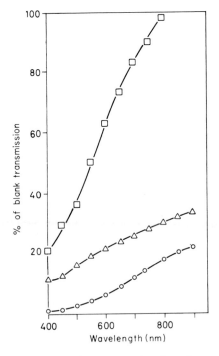

FIG. 1.10. Transmission spectra of quartz sand and a clay suspension. Top curve:
quartz sand (5 mm); middle curve: quartz sand (10 mm); bottom curve: clay suspension
(5 mm). From Wells (1959).

from 1.2 to 1.6. Wells suggests that these results indicate the adaptive value of sensitivity of germination to the promotive action of red light, because this light penetrates deeper. This adaptation allows the seed to germinate at nearly the maximal depth of penetration by visible light, where risk of early desiccation is diminished. The increase in the 735 nm/655 nm ratio with depth is too gradual to reverse the inhibition before the total energy of light becomes too low to be effective. Wells does not report any experiments which might indicate the depth to which physiologically active light actually penetrates. Koller *et al.* (1964) studied this problem with light-requiring seeds of *Artemisia monosperma* and found that germination was promoted by light filtered through a 2-mm thick layer of dry sand, when the source (mixed incandescent and fluorescent) supplied about 1.5% of full sunlight at filter level. Evidently effective levels will be encountered at considerably greater depths when the source is sunlight and the sand is not dry. The other habitat which was studied was below a plant canopy, where specific absorption of red light by chlorophyll results in a spectrum which is relatively enriched in far-red light, as compared to red. Cumming (1963) investigated the effects of various ratios of red to far-red spectral energies (R/FR) in the incident light on the photocontrol of germination in *Chenopodium* spp. Under noon conditions in early summer in Ottawa the ratios were 1.31 in the open, 0.70 in the shade of a soybean plant and 0.14 and 0.12, respectively, under a hardwood canopy and 20 inches within a dense stand of red clover. Germination of five species of *Chenopodium* was then tested at the optimal 30° C and suboptimal 15° C, in darkness, in continuous irradiation, as well as in daily exposures lasting 8 or 16 hours, the light having an R/FR ratio of 1.31 (identical with that in the open), or 0.14 (similar to that under the plant canopy). Promotive effects of light were consistently higher or equal in light with the higher R/FR ratio. Similar studies were carried out by Taylorson and Borthwick (1969), using light actually filtered through fresh leaves of tobacco, corn, and soybean. The transmitted light was relatively more deficient in the red than in the far-red spectral region. It was capable of reversing the promotive effects of an initial irradiation with red light on germination of several weed species, but had no inhibitory effect without prior promotion. Effects of prolonged irradiation with light transmitted through leaves were not studied. Cumming's (1963) data showed that germination of *Chenopodium* was reduced with increasing length of the daily exposure to artificial light, in both high and low R/FR ratios. When the exposure was continuous, germination was invariably lower than the dark controls even when the ratio was high. Even when the daily exposure was 16 hours, germination of seeds irradiated with the low R/FR ratio was in most cases lower than that of the dark controls. As we have seen (Section III,B,2), prolonged irradiation is a potent inhibitor of both newly formed and preformed phytochrome, especially when the light is rich in far-red. From an ecological viewpoint, prolonged irradiation is also the more significant radiation regime, since this is what the seeds are commonly presented with in nature. In any event, the results seem to justify the conclusion that the spectral composition of the light filtering through the plant canopy is altered to a degree which is sufficient to have different effects on the phytochrome system of the seeds than when it is unfiltered, and thereby affects their germination response.

## C. *Water Relations*

Water is essential for the rehydration of the seed as an initial step in its germination. The absolute amounts which are required are minute, not exceeding two or three times the dry weight of the seed. However, the subsequent growth of the seedling requires not only large amounts of water but also a sustained supply, and these requirements increase as transpiration becomes appreciable. As a result, control of germination by the water relations of the seed is of survival value only if the mechanism provides additional information on the environment, other than that there is sufficient water for adequate rehydration of the seed itself. Some aspects of the information which can be provided will be dealt with in Sections III,C,1; III,C,2; and III,C,3.

### 1. KINETIC ASPECTS

In the course of ecological studies in the deserts of southern California, Went (1949) was impressed by the fact that no germination (emergence) occurred when rainfall was less than 15 mm, except in micro-sites, such as roadsides and washes, where runoff water augmented the supply. This was suprising, since the soil horizon where most of the seeds are situated was equally wet by much smaller amounts of rain. Studying this problem under laboratory conditions, he found that germination from soil samples which were collected from the desert sites where the observations had been made was very poor although they were adequately supplied with moisture by subirrigation. Germination from similarly treated soil samples was profuse when they were first subjected to 20 mm of simulated rainfall. Extending these studies, Soriano (1953) subjected such soil samples (each containing approximately the same seed population) to various intensities of simulated rainfall (0–75 mm) during 24 hours, after which they were all subirrigated. Germination was maximal after a 7–15 mm rainfall (more than twice the number of seedlings were obtained than without rainfall), and fell off slightly after heavier rainfall. When the same total amount of rainfall (15 mm) was applied over different time intervals, the number of seedlings per container increased from 27 to 40 to 64 as the duration of rainfall was increased from 1 to 10 to 54 hours. The mechanism which underlies such specific responses to rainfall is undoubtedly related to the widespread occurrence of water-soluble substances in seeds, and more particularly in various external tissues of the dispersal units, which are inhibitors of germination of their own embryos. When such seeds come in contact with water, these inhibitors dissolve and start to diffuse away from the seed along the concentration gradient. However, under isothermal conditions, or under conditions which prevent convection (such as exist in the soil), diffusion through water is an extremely slow process. Its rate is increased if the concentration gradient is kept high by mass flow of water across the seed. This most commonly happens when water flows through the soil by gravity, that is after rainfall. Rate of diffusion will depend upon the flux of water through the soil. The total amount of the inhibitor which is leached away will therefore depend on both the rate and duration of flow and consequently on the intensity and duration of rainfall, as well as on hydraulic conductivity of the soil. As leaching continues, a critical concentration of the inhibitor will be reached, below which germination can take place. In this way the inhibitor content of the seed is used to meter the amount of water which a particular rainfall has made available for growth of the seedling in that particular soil, before allowing the seed to germinate. The metering will also take into account water which is added to or subtracted from the specific submicroenvironment by surface runoff. In habitats such as deserts, where rainfall is limiting plant growth and is also erratic, a correlation between the inhibitor content of the seed and the minimal requirement of soil moisture for completion of the life cycle (in annuals), or for establishment (in perennials) of the plant into which it germinates, is of obvious survival value. Water-soluble germination inhibitors are common in dispersal units of desert plants. A significant correlation has been found between the aridity of the habitat and the apparent inhibitor content of the seeds in two ecotypes of *Oryzopsis miliacea*. The ones from the 600 mm rainfall region require a greater amount of leaching in order to germinate than do those from the 100 mm rainfall region (Koller and Negbi, 1959).

## 2. DYNAMIC ASPECTS

a. FACTORS INVOLVED IN UPTAKE.   The forces with which water is held in the soil, or the amount of work which has to be done by the seed in order to extract them, are of obvious immediate importance to germination, since they determine whether or not the seed can become sufficiently rehydrated for germination to take place. The capacity of the seed to do so must as obviously be correlated with the ability of its seedling to survive such soil-moisture relations. However, since water has to move from the soil into the seed, the process of rehydration involves kinetic, as well as dynamic aspects. The following discussion is based, amongst others, on the studies by Sedgley (1963), Manohar and Heydecker (1964a,b), Collis-George and Williams (1968), and Hadas (1970).

The driving force for movement of water by mass flow and diffusion from pore spaces in the interstices of the soil into the tissues of the seed, is the difference in water potential at the two ends of the pathway. The flux of water is controlled by diffusivity of the medium along this pathway. All these parameters change with time. Part of these changes is a direct outcome of the process of water uptake itself, part is an indirect result of this process and part is entirely autonomous.

The direct changes involve an increase in water potential of the seed as it becomes rehydrated (seed moisture content increases), with a concomitant decrease in water potential of the soil from which this water is withdrawn (soil moisture content decreases). This decrease is most acute in the immediate vicinity of the seed and becomes progressively less with distance from the seed-soil interface. Diffusivity in a porous medium is a function of the water potential. Therefore, diffusivity of the seed will increase and that of the surrounding soil will decrease as a result of the changes in potential (and moisture content) in the two parts of the system.

The indirect changes involve processes of hydrolysis within the seed by which high molecular weight storage material becomes depolymerized and osmotically active, with a resulting decrease in the otherwise increasing water potential. At the same time, the diameter of the seed increases, with a resulting increase in the otherwise decreasing diffusivity of the soil in the immediate vicinity of the seed. An additional indirect change takes place during the initial stages of imbibition, as the temperature within the seed rises considerably and this increases its water potential as well as its diffusivity. These effects are transient at best, because of the rapid transfer of sensible heat, as a result of the small size of the seed and the relatively high heat capacity of soil with a high moisture content. On the other hand, the onset of metabolism in the seed releases heat of respiration in increasing amounts and this may set up a more sustained temperature gradient between the seed and the bulk soil.

The autonomous changes are all confined to the soil phase, the water potential of which changes independently of the presence of the seed, as a result of drainage and evaporation. In soils which are rich in clay, reduction in soil moisture content will also result in shrinkage of the soil matrix as the colloidal particles are pulled closer together, with a resulting increase in the isotropic mechanical forces which act on the seed and offer increasingly greater opposition to its increase in size. This compaction will usually occur after most of the swelling (and the increase in seed diameter) had taken place. Consequently, the germination process which is most likely to be affected is the growth of the radicle and hypocotyl, with which germination is terminated.

b. MATRIC POTENTIAL. Changes in soil moisture content affect mainly two components of the soil water potential, namely the osmotic (or solute) potential and the matric potential. Although there may still be some controversy on this point, the bulk of the evidence is in favor of the view that in nonsaline soils the changes which are of most significance to water uptake by the seed, and therefore to its germination, are those which take place in the matric component of soil water potential. The main reason for this is that change in matric potential involves a change in diffusivity as well, while a change in solute potential *per se* does not. A relatively small decrease in soil matric potential, which by itself may cause only a slight increase in the potential differences, may result in a strong reduction in flux. The actual magnitudes involved are character-istics of the soil in general, but particularly at the seed/soil interface. It is there that the potential gradient is maximal. More important still, because of the tremendous difference in size and shape between the seed and the soil particles which are in contact with it, there is an abrupt change in pore size distribution at this interface, the balance shifting in favor of pores with a larger effective diameter. These are the pores from which least energy is required to withdraw water and they will be the first to drain as soil moisture is depleted. The external surface of the seed is in part in contact with solid soil particles. Of the remainder, a considerable proportion is in contact with the larger diameter pores, the very ones which are the first to drain. Therefore, relatively small reductions in soil-moisture content are liable to result in disproportionately large reductions in contact area between the seed and soil water to which movement of water is restricted. The surface topography of the seed coat (e.g., spiny, reticulate, or smooth), and the size of the seed relative to that of the soil particles, will both have significant effects on the water relations of the seed (Harper and Benton, 1966). The situation is made more complex by the almost universal lack of uniformity in the structure and properties of the coat of an individual seed (which is characteristic for the species), particularly with respect to conductivity of water (Hyde, 1954; Bonner, 1968) and selective permeability. It may thus become of importance just where on the surface of the seed the water is in contact. For instance, in grains of *Panicum turgidum* germination percentages depend on whether the flat or the convex side is in contact with the moist substrate, and the difference is eliminated in presence of a wetting agent (Koller and Roth, 1963). In certain cases, such as peas, the seed coat exhibits selective permeability to solutes, except in the region of the micropyle (Manohar and Heydecker, 1964a,b). In such seeds the soil solution will exert its osmotic effects on germination only if it is not in direct contact with the micropyle.

It is an established fact that seeds must take up moisture until they attain a certain level of hydration before being able to germinate. Under limiting conditions, the internal distribution of moisture within the various tissues of the seed may also become quite critical. From what had been said above it is easy to appreciate, though quite difficult to calculate, the complex control of germination by the environmental factors which determine the dynamics of water relations of the seed. These factors include the physical properties of the soil which determine the retention of water and its movement, as well as soil compaction, the soluble matter in the soil, the climatic factors which determine the rate of supply and loss of moisture from the soil, and finally the nature of the contact between the soil particles and the surface of the seed (or dispersal unit).

c. SOLUTE POTENTIAL. The solute (osmotic) component of soil water potential can play a major role in the dynamic water relationships of seeds under saline conditions. In salt-adapted species, it may possibly act in the regulation of germination. Several features of germination responses to salinity which are specific to halophytes will be dealt with below.

There apparently exists a reasonably good correlation between the capacity of certain halophytes to germinate in a saline medium and the salinity of their natural habitat. However, this correlation failed between the various species within a single genus (*Tamarix*). The tolerance of the various *Tamarix* species was higher than of any of the other halophyte species which were tested, although some of the former inhabit much less saline habitats (Waisel, 1958). The difference may be related to the fact that *Tamarix* is noted for copious excretion of salts. There is also the possibility that the seed or fruit of *Tamarix* is freely permeable to the salts, whereas those of other halphytes are not (cf. Manohar and Heydecker, 1964a,b). Seed germination of certain succulent halophytes (*Suaeda depressa*, *Spergularia marina*) is apparently stimulated by the presence of 0.5 $M$ NaCl in the substrate. Two other such halophytes did not show a similar response. Seeds which remained inhibited for 30 days in high salinity germinated quite well upon transfer to water (Ungar, 1962). This response is clearly concerned with a low osmotic potential and thus differs from the one to reduced matric potential which has already been considered. A noteworthy response to salinity is exhibited by halophytic vegetation of marine salt marshes and sea shores. Germination of seeds of *Plantago maritima* at the optimal constant temperature (25°C) in light was about 35% in rainwater, and was progressively inhibited in increasing concentrations of seawater, up to complete inhibition in a dilution of 1:4. At a diurnal alternation of 5°/25°C, germination percentages were doubled and complete inhibition was obtained only in a 1:1.3 dilution. However, imbibition in full strengh seawater for 10 days or more, at 5° or 25°C, induced between 80 and 90% germination after transfer to rainwater at a constant temperature of 25°C in light, and complete inhibition was obtained only in a 1:1.3 dilution of seawater. Dark germination in rainwater was about 5 and 45% at 5° and 25°C, respectively (Binet, 1964). Additional studies on the effects of seawater on germination were carried out by Boorman (1968). Three response types could be distinguished. Seeds of *Sinapis alba* and *Aster tripolium* were capable of germinating very slowly in seawater, and continued to germinate slowly and nearly fully after transfer to distilled water. Seeds of *Rumex crispus* and *Limonium binervosum* were completely inhibited in seawater but germinated very rapidly and fully after transfer to distilled water. The response of the third type was the most striking. Germination of *Limonium vulgare* (and to a lesser extent also of *L. belledifolium*) was strongly promoted by prior incubation in seawater. The promotion was increased progressively with increase in the concentration of seawater, up to that obtained by evaporating to two-thirds of the original volume. Germination was progressively hastened with increase in the duration of incubation in seawater longer than 2–4 days. Germination percentages of *L. vulgare* and *L. humile* were progressively higher as duration of incubation was increased up to 60 days, but both species germinated slowly during the incubation. Boorman (1968) suggests that the promotive effect was a result of some kind of osmotic shock. The effects of gradually diluting the concentration of seawater were not studied. Finally, certain xero-halophytes accumulate considerable amounts of salt in their dispersal units, and those may participate in the chemical rain gauges, as described in Section III,C,1 (Koller, 1957).

d. PROMOTIVE ACTION OF STRESS. Most seeds have had to adapt their germination to a regime of water deficiency. Others have adapted to surplus of water and the concomitant low aeration (Morinaga, 1926a). Yet others have developed mechanisms which restrain them from germinating when water is too freely available. The latter are distinguishable by their preference for germinating under a mild moisture stress, one which is specifically obtained by lowering the matric component of water potential of the substrate with which they are in contact (see review by Koller *et al.,*

1962). The case of *Hirschfeldia incana* was studied by Negbi *et al.* (1966). In these seeds the inhibition arises by development of a mucilage layer from the epidermis of the testa. The development of this mucilage is progressively inhibited by slight reduction in the matric potential, and germination is improved as a result. This sensitivity to water is greatly reduced by increasing the temperature as well as by treatment with exogenous gibberellin, both of which hasten the processes leading to germination and thus allow it to win the race against the build-up of the inhibitory mucilage. The opposite was true for the growth-retardant CCC [(2-chloroethyl) trimethyl-ammonium chloride] and the germination inhibitor coumarin. Formation of a mucilage layer around the true seed of spinach (*Spinacia oleracea*), was responsible for inhibition of germination in excess moisture (Heydecker and Orphanos, 1968).

Witztum *et al.* (1969) have studied the requirement for a slightly reduced water potential for germination of the desert plant *Blepharis persica*. The seed is covered with mucilaginous hairs which have a specific function in establishment of the seedling (Gutterman *et al.,* 1967). As a result of a water-sensitive mechanism, the seeds are dispersed only during rainfall and thus alight on the soil surface when the latter is wet. The tightly coiled spiral thickenings in the hairs uncoil when wet, and their tips adhere tenaciously to the soil surface. Differences in length of the extended hairs raise the cotyledonary end of the seed above the soil surface, and thus poise the radicle at an angle of 30–45 degrees and touching the soil surface. Removal of these hairs prevented the radicle from penetrating the soil, and thus clearly indicated their role in seedling establishment. This is an essential feature for a seed that by virtue of its dispersal mechanism must germinate on the soil surface, and does so very rapidly. The mucilaginous nature of the hairs acts in much the same way as that of the seed coat in *Hirschfeldia incana,* inhibiting germination when water is present in excess by reducing the diffusive flux of oxygen across the thick layer of water. In this fashion, germination is prevented under conditions of excessive surface moisture, which would make the attachment of the seed to the soil surface precarious.

## 3. PERIODIC EFFECTS

Although there have been a few scattered reports of seeds whose germination was improved by allowing them to go through a cycle of wetting and drying, the phenomenon was never studied in detail. In the majority of cases only a single drying cycle was involved and its promotive effect may very well have resulted from removal of a water-soluble inhibitor. On the other hand, a study by Griswold (1936) showed that seed germination in several species of pasture plants was improved by alternate soaking and drying. A more clear-cut case is provided by seeds of *Citrullus colocynthis*. Intact seeds take up water freely, yet are apparently incapable of germinating above 5–10% under a variety of environmental complexes, even after pretreatments which greatly modify the structure and properties of the seed coats. The decoated seeds germinate readily and rapidly, given darkness and high temperature (Koller *et al.,* 1963). The only conditions under which germination of intact seeds took place at the optimal conditions was when they were immediately preceded by several 24-hour cycles of soaking and drying, with soaking carried out at a somewhat lower temperature than drying (Shur, 1965). An explanation does not readily come to mind. It is likely that the alternation somehow sets up strains in the coat which eventually weaken it along the suture connecting the two halves of the seed coat. Another possibility is that in this manner the cyclic changes in hydration in the embryo and seed coat, respectively, can become out of phase. As a result, the embryo is able to exert imbibition pressure against a dry and therefore shrunken seed coat. *Citrullus colocynthis* inhabits coarse sandy soils in the deserts of Israel, Sinai, and Egypt. Diurnal cycles of condensation of water vapor from deeper layers of the soil may occur in the subsurface layers of the soil as a result of the inversion of the temperature gradient. The specific germination response to diurnal cycles in hydration may therefore be an adaptation, not an experimental artifact.

### D.  *Chemical Environment*

The following discussion will deal specifically with control of germination by the exogenous chemical environment of the mature seed. The endogenous chemical environment which exists inside the seed may naturally be of crucial importance for germination, but its discussion is outside the scope of the present chapter.

Germination and the growth which it initiates involve a variety of biochemical reactions. It is therefore not surprising that various chemicals were found to affect the final outcome of germination. Such knowledge is of obvious importance in any attempts at elucidating the nature and properties of the partial processes. From the viewpoint of environmental control of germination, on the other hand, the interest must focus not only on naturally occurring substances, but also on those which the seed is likely to encounter in its natural environment. Indeed, we must be even more selective and confine ourselves to those substances which appear to have some regulatory action. This automatically excludes toxic substances, leaving only those whose effects do not reduce viability of the seeds and pass after removal of the seed from their influence. Such chemicals may be encountered by the seed in both liquid phase and gas phase with which it is in contact.

## 1. LIQUID PHASE

The dissolved substances which may affect seed germination fall into three general classes according to their origin: (*a*) inorganic mineral substances in the soil solution, (*b*) organic substances which originate in the immediate vicinity of the seeds (i.e., in and around the floral organs) and (*c*) organic substances that originate in plant organs which are far removed from the immediate vicinity of the seed. Control of germination by organic substances of plant origin has been extensively studied at the Department of Botany of the Hebrew University in Jerusalem, Israel. These studies have been reviewed by Evenari (1961b, pp. 599–606).

a. INORGANIC SUBSTANCES.    Most ions which can be expected to be in the soil solution do not seem to have any specific regulatory activity on seed germination. There are two notable exceptions. One is the nitrate ion, several salts of which have a widespread stimulatory effect on germination, particularly in species with light-requiring seeds. The mechanism is still quite obscure and no serious attempts to elucidate it have so far been reported. Similarly, the possible role which the nitrate ion may play in modifying the regulation of germination of such seeds under natural conditions has also not been studied. The other ion which has been reported to have a regulatory effect on germination is calcium. In studying the sensitivity of germination of *Blepharis persica* seeds to excess water (see Section III,C,2,d) Witztum *et al.* (1969) observed that the inhibitory effects of the mucilage is prevented and its water-holding capacity is greatly reduced by presence of divalent cations, such as $Ca^{2+}$ even in great dilutions ($5 \times 10^{-3} M$ $CaCl_2$) and this is reversed by monovalent ions, such as $Na^+$. This suggests that $Ca^{2+}$ acted by double bonding of carboxyl groups of the uronic acids which make up the mucilage of the primary wall, and that this was antagonized by $Na^+$. It is tempting to assume that this mechanism enables the seed to estimate not only the presence of excessive water, but also the concentration of sodium, which is the prevalent cation that contributes to salinity in its natural habitat.

b. ORGANIC SUBSTANCES WHICH ORIGINATE IN THE VICINITY OF THE SEED.    Substances which accumulate and in all probability are also synthesized in the organs surrounding the seed and which can, moreover, be shown to be capable of effectively controlling its germination even in minute concentrations are of obvious interest from the point of view of their natural function. There have been numerous reports of variously prepared extracts from such tissues in many species which inhibit germination. A thorough review of the subject was made by Evenari (1949). Thus far, no extract from a given species seems to exert a selective inhibition on its own seeds. On the contrary, in laboratory tests the action of such extracts appears to be remarkably nonspecific, although there may exist species (and varietal) differences in the sensitivity toward them. This may result from true differences in the reactivity of the embryos to them, but may equally be due to species differences in permeability of one or more of the coats to them, as well as to species differences in the mechanical restraint of these coats to expansion growth of the embryo.

The problem of the natural role of the substances in such extracts in seed germination is not a simple one. The crucial difficulty is that, in order for a substance to play a role in the control of germination of the seeds in their vicinity, the two parts of the system must have a natural opportunity of interacting. The mere fact that the substances exist in the vicinity of the seed in presumably the state in which they are extracted, is no proof that the two ever have access to each other outside the laboratory. This criterion has thus far only been applied to the water-soluble inhibitors in dry dispersal units, which have free access to the seed as soon as they are dissolved. Their probable role in plants which are faced with an uncertain supply of soil moisture has been discussed in Section III,C,1. However, equally active substances exist in dry dispersal units of plants which apparently do not seem to be faced with a similar problem in survival. A question therefore arises as to their function in such seeds. The answer may lie in the finding that germination-inhibiting substances are very common in the pulp and juice of many fleshy fruits and are even probably chemically related to one another and to inhibitors in dry fruits (Köves and Varga, 1959). Their accumulation occurs up to full ripening (Varga, 1957a). On the other hand, a reduction in their concentration in over-ripe fruits may be associated with presence of viviparous seed in such fruits (Varga, 1957b). These facts support the suggestion that the inhibitors in fleshy fruits may play a role in preventing premature germination of the seeds in the liquid environment in which they are enclosed. The inhibitors are possibly metabolically inactivated as part of the over-ripening process of the fruit, but it is more reasonable to assume that the seeds are physically released from their presence as the fruit rots and disintegrates, or is eaten by animals. Dry fruits also invariably pass through a phase at which they were more or less succulent, at which stage their seeds may have faced a similar problem. If this is indeed the case, then the presence of inhibitors in the dry fruits is no more than a carry-over from the succulent stage. In plants of arid habitats, natural selection may have maintained and quantified this property. In plants of more humid habitats, the inhibitors may be rapidly leached away without serving as rain gauges, they may be inactivated (Köves and Varga, 1959), possibly by microorganisms, or they may be discarded when the seeds leave the fruit. Evenari (1949) suggests that in certain plants with dry fruits, such as *Sinapis alba,* the presence of inhibitors may increase the probability of survival by delaying germination of seeds which remain enclosed in an indehiscent part of the fruit while the rest of the seeds are released by dehiscence of the remaining part of the fruit.

However, the crucial question is whether or not the inhibitors whose presence can be demonstrated in juice or macerated pulp ever come into direct contact with the seeds in fruits which remain succulent, or during the succulent stage in fruits which eventually become dry. As far as can be ascertained, rigorous proof is lacking to that effect, and the possibility exists that germination of seeds in permanently or temporarily succulent fruits may be otherwise inhibited, for instance by lack of adequate gas exchange. The problem of access becomes even more difficult to resolve in the case of inhibitors which are only demonstrable after complex extraction techniques, as is the case with many "naturally occurring" growth-inhibiting and growth-promoting substances.

c. Organic Substances Originating in Plant Organs which are Far-Removed from the Vicinity of the Seed.    It is a well-established fact that a great variety of organic substances are added to the soil solution by leakage or excretion from living underground plant organs, by leaching from leaves, by decomposition of such organs when they age and die, or by the activity of microorganisms which participate in this disintegration, or live on its products. Some of these substances are phytotoxic, either in general, or with distinct selectivity toward their own seedlings or those of certain other species (Al-Naib and Rice, 1971; Lodhi and Rice, 1971). Others are only indirectly lethal, rather than toxic, reducing the capacity for competition of other neighboring individuals by inhibiting their growth. Some of these substances apparently play a distinct and clear-cut role in plant sociology (see Evenari, 1961a, for review on allelopathy), but whether they are able to do so by control of germination (Went *et al.,* 1952) is not as certain (e.g., Funke, 1943; Lee and Monsi, 1963). In addition to the fact that few of these substances were specifically tested for germination-inhibitory activity, there is always a strong possibility that many of them become rapidly inactivated in soil by adsorption or microorganisms (King, 1952; Evenari, 1961b, pp. 599–606). On the other hand, it is an equally established fact that certain species are able to thrive, and therefore germinate in habitats where allelopathic action (or even so-called soil fatigue) is clearly established.

In closing this section, brief mention must be made of seeds whose germination is stimulated by organic substances which are produced by other plants. Although there are scattered reports of such seeds, the best-documented cases are those of parasitic angiosperms. The germination of representative species belonging to the genera *Striga* and *Orobanche* has been intensively studied by R. Brown and associates in England. The subject has been reviewed by Brown (1965), and the following summarizes the main findings and conclusions on the environmental control of germination in such seeds. The decisive environmental factor required for germination of parasitic seeds seems to be the presence of organic substances exuded from living roots. Although appearance of the adult parasites is conditional upon the presence of their specific hosts in their vicinity, the germination of their seeds is not host-specific and takes place in the presence of exudates from a variety of species which do not serve as hosts at all. Although the stimulating substance(s) has thus far not been definitely identified, it has been characterized to some extent. It is adsorbed on charcoal, more soluble in nonpolar solvent than in water, and is active in great dilution. Responsiveness of the seeds to the promoter increases gradually during their incubation. At its peak the response is saturated by a brief exposure to the promoter. There is some circumstantial evidence which suggests that the seeds themselves produce the promoter, but in subthreshold amounts. The promoter is apparently exuded exclusively from the zone of maximal extension growth of the root, although it is present in other zones as well. It is apparently non-specific, as far as either the donor plant, or receptor seed are concerned. Thus the presence of practically any actively growing root in the vicinity of the inhibited seed will cause it to germinate, depending on the degree of responsiveness it has reached. Infestation of the specific donor is a function of the seedling. This rather wasteful process is undoubtedly compensated by the enormous seed-bearing capacity of this type of parasite.

## 2.   GAS PHASE

The gas phase in the environment of the seed contains the usual constituants of the bulk atmosphere, though frequently the proportions are greatly modified by respiratory activity of the soil organisms. As a result of virtual absence of convection and advection, there is no turbulence and gaseous exchange with the ambient air takes place largely by the slow process of diffusion and partly by mass flow set in motion by temperature gradients inside the soil. The composition of the soil atmosphere may therefore show marked variations, depending, among other factors, on soil properties such as porosity, depth, and seasonal changes in the moisture content, the organic matter and the activity of soil organisms. In addition, the gas phase may contain a variety of volatile substances which are produced by different plant parts, sometimes in relatively high concentrations, especially in certain seasons. In fact, the seed itself may produce such volatile inhibitors of its own germination. This is apparently the case in seeds of *Vaccaria pyramidata,* which are incapable of germinating on moist filter paper but germinate readily when in contact with or in the presence of soil. The volatile inhibitor is presumably adsorbed on positively charged soil particles (Borris, 1936). Volatile substances probably play a decisive role in certain well-documented cases of allelopathy, and may do so by inhibiting germination. The difficulties of establishing this last fact are probably even greater than those enumerated above for the organic inhibitors in the soil solution. The present discussion will therefore be restricted to the metabolically active components of the atmosphere, namely oxygen and carbon dioxide.

Comparative studies by Bibbey (1948) showed that sensitivity to atmospheric composition may be of survival value. Seeds whose viability deteriorates relatively rapidly when prevented from germination by deep burial in soil (*Triticum vulgare, Brassica juncea*) germinated readily in subnormal oxygen and above-normal $CO_2$, whereas seeds whose viability is retained for extended periods under the same conditions (*Thlapsi arvense, Brassica arvensis*) germinated progressively less as oxygen concentration was lowered or $CO_2$ concentration was increased from their normal value in the open. On the other hand, seeds of hydrophytes, such as *Nelumbo nucifera* (Ohga, 1926) and *Peltandra virginica* (Edwards, 1933) are able to germinate well in total absence of oxygen. At least in *Nelumbo* there apparently is sufficient oxygen within the seed for germination. On the other hand, seeds of another hydrophyte, *Typha latifolia,* are actually inhibited by oxygen concentrations equal to, or higher than that in free air, and are stimulated to germinate by lower concentrations, down to one-tenth of atmospheric. This particular response disappears when the seeds are decoated (Morinaga, 1926c). A somewhat similar situation was observed in seeds of *Zizania aquatica.* Germination could only take place after prolonged submersion in water. When the water was not aerated, the response was considerably hastened and sensitivity of germination to supraoptimal temperatures was reduced (Simpson, 1966).

There are numerous reports of seeds which could be induced to germinate by increasing the partial pressure of oxygen over and above that of normal air, sometimes up to that of pure oxygen. The most notable example (Thornton, 1935) is the upper "seed" (fruit) within the dispersal unit of *Xanthium pennsylvanicum,* which is much more dormant than the lower seed. The response of both seeds to oxygen was temperature dependent. The minimum oxygen concentration at which the lower seed germinated was 6% at 21°C and 4% at 30°C, while the corresponding values for the upper seed were 60 and 30%. Evidently, such high partial pressures of oxygen are never encountered in nature, and the oxygen effect may be considered an artifact. This is supported by the fact that upper seeds which had been induced to germinate by high oxygen germinated abnormally, with elongation of cotyledons and plumules rather than of the radicle. However, the upper seed germinated in normal air at temperatures of 33°C or higher. Germination could also be induced at 25° even with 10% oxygen, provided $CO_2$ concentration was sufficiently high (10–40%), and what may be even more important, seedling growth was now normal. Subsequent studies by Wareing and Foda (1957) showed that the oxygen effect in the upper seed was due to the presence of a growth inhibitor which was prevented from leaching out of the seed by the semi-permeable seed coat and was inactivated by the oxygen. The results with $CO_2$ are not explained. Nor are any data presented to explain the fact that the lower, nondormant seed contains equal concentrations of the same inhibitor(s), yet does not show a requirement for oxygen. The implication of growth inhibitors in the germination of these seeds is supported by the studies of Esashi and Leopold (1968) on the physical forces involved (Fig. 1.11). The force required to pierce the testa of the upper

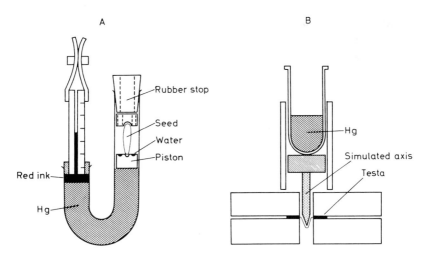

Fig. 1.11. Diagram of the apparatus used for determination of the thrust of germinating *Xanthium* seeds (A), and for the determination of the force necessary to rupture the testa (B). From Esashi and Leopold (1968).

seed was somewhat smaller than that for the lower seed (56 and 67 gr, respectively). However, the embryo of the lower seed was capable of generating more than twice the thrust of the one in the upper seed (84 and 41 gr, respectively) during the second phase of water uptake, when active growth of the embryo starts. As a result, the growth forces generated by the lower embryo are sufficient to pierce its testa, whereas those generated by the lower embryo are not.

High partial pressure of oxygen is also capable of forcing germination which is inhibited by unfavorable environmental conditions, for instance in *Phacelia tanacetifolia* incubated in continuous irradiation, but only when presented to the seeds from the start of irradiation. A germination inhibitor is apparently produced during continuous exposure to light and its activity is reduced in the presence of oxygen (Rollin, 1958). Again, the data fail to explain why the inhibition was inactive on grains of *Phacelia* itself in darkness. The conclusion that the inhibition by light is mediated by producing a block to oxygen penetration through the coats is not adequately supported by facts.

Additional manifestations of a relationship between light sensitivity and the effects of $CO_2$ on germination are evident in the responses of different varieties of lettuce seed (Thornton, 1936). The requirement for light at supraoptimal temperatures (26°–30°C), as well as the complete inhibition (in both light and darkness) at 35°C, could be overcome by high concentrations of $CO_2$, provided the oxygen level was maintained at the normal 20%. The minimal effective $CO_2$ concentration became higher as the temperature became more supraoptimal, reaching 40–80% at 35°C. The induction of a requirement for light by incubation at supraoptimal temperature is usually explained by assuming thermal inactivation of active phytochrome. On this basis, it would appear that $CO_2$ exerts a protective effect on active phytochrome in such seeds, possibly by inactivating the enzyme system which is responsible for the thermal inactivation.

The stimulation of germination by $CO_2$ which has been dealt with thus far involves concentrations which are too high to be encountered by seeds in nature, unless under exceptional conditions. Therefore, although study of the response may shed light on various aspects of the endogenous control of seed germination, it will bear little if any relationship to the environmental control of germination by the levels of $CO_2$ which the seed may encounter in its submicroenvironment. However, in recent years remarkable promotive effects have been recorded with $CO_2$ concentrations which are quite common in many naturally occurring situations. Germination of partially dormant seed populations of *Avena fatua* is strongly inhibited by light. This inhibition was entirely absent in dehusked grains in air containing about 3% carbon dioxide, but was retained in its absence in all concentrations of oxygen (Hart and Berrie, 1966). Dormant seeds of *Trifolium subterraneum* can be induced to germinate by $CO_2$ concentrations between 0.5 and 5.0% (Ballard, 1958), which was probably the reason why germination was higher in soil than in the laboratory (Morley, 1958). It is most significant that sufficient $CO_2$ could be generated by respiration of an aggregate of several seeds to induce their germination (Ballard, 1958). Under field conditions, mechanical impedance of the soil is frequently overcome much more effectively by an aggregate of germinating seeds than by single seedlings. Recent studies have shown that imbibed seeds of subterranean clover produce ethylene as well as carbon dioxide, and that their germination is promoted by both these substances, probably in an independent manner (Esashi and Leopold, 1969). The authors suggest that the self-regulatory activity of gases which are produced by the seeds may have an ecological advantage, by favoring germination within the soil, where their concentration can increase as a result of the high diffusive resistance. The results with *Trifolium subterraneum* were subsequently extended to other small-seeded legumes belonging to the genera *Medicago, Trifolium,* and *Trigonella.* Germination in the presence of $CO_2$ was equal to or better than after cold pretreatment and the former was effective even when the latter was not (Grant Lipp and Ballard, 1959). In subsequent studies it was found that seeds of *Trifolium subterraneum* which respond to $CO_2$ show a similar response when oxygen concentration was reduced below 5% and down to zero, during the initial phases of germination. The response was roughly proportional to duration of exposure, and integrated several separate exposures (Fig. 1.12). On the basis of the results it was suggested that the stimulation was mediated by enhanced production of an intermediate of glycolysis (Ballard and Grant Lipp, 1969). Inhibitors

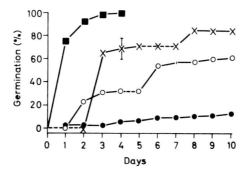

FIG. 1.12. Stimulation of germination in subterranean clover, CPI 19465, following exposure to nitrogen. Seeds were preincubated for 7 days. (●), Air (control); (■), 2.5% $CO_2$; (O), $N_2$ for two periods of one day each; (×), $N_2$ for two periods of two days each. Dotted lines indicate periods of nitrogen treatment. From Ballard and Grant Lipp (1969).

and uncouplers of oxidative phosphorylation (0.2 m$M$ DNP, 0.8 m$M$ sodium azide, 1.0 m$M$ sodium arsenate, 4.0 m$M$ sodium salicylate) also stimulated germination of these seeds. The stimulation by uncouplers was less than by $CO_2$ (Ballard and Grant Lipp, 1967). However, oxygen deficiency stimulated only during the initial stages of germination and higher oxygen concentrations were required during the terminal stages. It is therefore likely that oxidative phosphorylation is inhibitory only during the initial phase of germination and is required for the terminal phase. This may account for the lower responsiveness to the uncouplers, which were present during all phases of germination.

## E. Time

Time as such cannot be regarded as an environmental factor, yet it is an essential component of the action of the environmental factors. For instance, time is obviously involved in all periodic phenomena of light, temperature, and water relations in germination. Time is also involved in other aspects of the immediate environmental control of germination in a rather more complex fashion. First, a variety of partial processes take place in the imbibed seed. In a great many of them, possibly all, the rates are affected by one or more environmental factors within the biologically occurring range. Some of these processes are progressive and lead toward eventual germination. Others are retrogressive inasmuch as they inter-fere with the progress of germination, either by diverting intermediates, or by forming physical or metabolic blocks. This is apparently the situa-tion in light-sensitive lettuce seeds, in which the responsiveness of the phytochrome system at first increases with duration of incubation but then declines (Borthwick et al., 1954). An analogous situation has been described in Hirschfeldia incana (Negbi et al., 1966), where the rate of build-up of the inhibitory layer of mucilage was dependent on the matric potential, whereas the rate of the promotive germination processes was dependent on temperature. The same environmental factor may affect the rates of various partial processes in different ways. It may create a con-flict of interests by accelerating one progressive process while delaying another, or it may do so by simultaneously accelerating a progressive and a retrogressive process. Secondly, the environment is never quite stable, and as a matter of fact its various components may change at different rates, during the time when germination is still in progress, and is still under their control. Some of these changes may be of a favorable nature for the overall process of germination, others may not. The final outcome of germination will therefore depend on time relationships of the different kinds of internal processes, as well as of the various environmental factors.

In Section III,A,1,a the possible effects of temperature on the kinetic aspects of germination have been discussed in terms of the duration of the lag phase and the rate of germination and its final percentages. The implications of this temperature dependence on germination under conditions of increasing moisture stress have been described. A similar case can be made for deterioration of other environmental conditions, such as increasing mechanical impedance of the soil, increasing salinity, and decreasing aeration. The kinetics of germination are not necessarily determined solely by temperature dependence of the metabolic processes. The level of active phytochrome may determine any or all of the kinetic parameters of germination. The duration of the lag phase may depend on the time until the critical level of active phytochrome, or of its substrate, becomes available from preformed storage or by neoformation. The critical level is itself dependent on the mechanical resistance which the growing embryo has to overcome, and this resistance may change with time during incubation. The rate of germination may be correlated with the degree to which active phytochrome exceeds the critical level, in the same general manner that the rate of an enzymic reaction depends on availability of enzyme molecules. Final germination percentages, on the other hand, may depend on the rate of a subsidiary reaction, by which the level of active phytochrome is decreased.

The implication of time in the environmental control of germination is also expressed in certain phenomena of physiological heterogeneity in the seed population produced by a single individual. For instance, the dwarf annual *Gymnarrhena micrantha* produces a few large achenes in subterranean inflorescences and numerous small ones in aerial inflorescences just above the soil surface. The kinetics of germination of the two types of seed differ markedly in the following respects (Koller and Roth, 1964) (Fig. 1.13). The lag phase of germination is considerably shorter in the subterranean seeds at all temperatures. Rates of germination are nearly indentical in light, with a distinct optimum at 15°C. In darkness, however, the seeds differ. Rates of germination of the aerial seeds were maximal at 15°C, whereas those of the subterranean seeds increased almost linearly as temperature was lowered to 5°C. Final percentages were nearly the same in both types of seed at the lower temperatures, with no optimum in either light or darkness. As temperatures were increased, the aerial seeds exhibited a much higher sensitivity to inhibition by high temperature. These differences in kinetics result in different specificities of germination requirements in nature for the two types of seed borne on the same plant. These and other differences suggest that the aerial seeds were specifically adapted to the role of increasing dispersal of the species, while the subterranean ones were adapted to en-

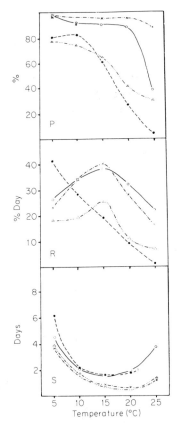

FIG. 1.13. Kinetics of germination of aerial (O, ●) and subterranean (×, Δ) fruits of *Gymnarrhena micrantha* in continuous irradiation (O, ×) and in darkness (●, Δ), as influenced by temperature. (S, start; R, rate; P, final percentages.) From Koller and Roth (1964).

suring survival of the species under the arid conditions of its natural habitat. A different example implicating time in the physiological heterogeneity of seeds borne by a single individual is provided by *Atriplex dimorphostegia* (Koller, 1969, 1971). This plant produces an array of morphologically differing dispersal units, varying between two extremes, flat and humped. The seeds exhibit typical short-day photoperiodic responses, but the two extreme types differ markedly in the length of the critical photoperiod (Fig. 1.14) which affects final germination percentages as well as rates, but not the duration of the lag phase. This difference is probably due to differences in the kinetics of the inhibitory prolonged irradiation processes. Presumably the intermediate types exhibit intermediate responses. Assuming that the photoperiodic response serves as a measure of soil depth, it is clear that the heterogeneity will lead to deliberate diversification in optimal soil depth, which presumably will increase the probability of survival.

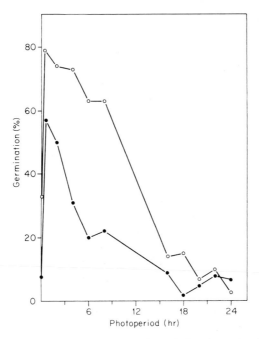

FIG. 1.14. Germination of flat (O) and humped (●) fruits of *Atriplex dimorphostegia* at 15°C, as influenced by photoperiod. From Koller (1971).

### F. Interactions and Conditioning

#### 1. INTERACTIONS

Interactions between environmental factors in the control of germination may arise either directly or indirectly. Direct interactions are those in which the same partial process is controlled by more than one component of the environment. Indirect ones are those in which different partial processes are affected by different components of the environment. For instance, temperature and dynamic water relations act on different partial processes in the germination of *Hirschfeldia incana*, yet a favorable temperature will permit germination in unfavorable water relations, and vice versa (Negbi *et al.,* 1966). Similarly, the germination of *Trifolium subterraneum* seeds at supraoptimal temperatures is low as a result of the formation of a water-leachable inhibitor within the embryo and not because the temperature as such is too high (Taylor and Rossiter, 1967). The requirement for light in germination of lettuce seeds is another good example. Light can overcome the inhibitory effects of a low water potential or of a chemical inhibitor such as coumarin, even at the optimal temperature, by increasing the growth potential of the embryo. These are most probably indirect interactions. On the other hand, a supraoptimal temperature can reduce the capacity of the seed to germinate in darkness by thermal inactivation of preexisting active phytochrome. The interaction between temperature and light is most probably direct, since light reestablishes active phytochrome at (or above) the required level. However, failure of the seeds to germinate even in light at higher temperatures most probably involves an indirect interaction, since it is likely that a metabolic reaction which is unrelated to light becomes limiting. Interactions are frequently exhibited between the effects of light and those of diurnally alternating temperature. These appear to be indirect, since within a single genus, such as *Rumex* (Steinbauer and Grigsby, 1960), can be found species which require light at constant temperatures, but not in alternating ones *(R. obtusifolius),* some which require only alternating temperature *(R. verticillatus* and *R. orbiculatus),* some which require light but show only a weak response to alternating temperature (*R. acetosella*). Similar results were obtained by Toole *et al.* (1955) with different species. In *Lepidium viriginicum* and *Verbascum thapsus* germination could not take place in the absence of light, but in its presence germination was promoted by alternating temperatures, whereas in two varieties of *Nicotiana tabacum* the absolute requirement for light existed only at constant temperatures, alternating ones inducing full germination also in darkness. Furthermore, a unilateral transfer of the seeds of *Lepidium virginicum* from 15° to 25°C was sufficient for full promotion, in conjunction with a brief irradiation with red light, but the transfer was equally effective when made as much as 20 hours before or after the irradiation. Seeds of *Lycopus europaeus* exhibit an obligate requirement for both light and diurnal temperature alternations. The response to light is of the long-day type. However, there was no precise requirement for any particular timing of the light/dark cycle with respect to the temperature cycle (P. A. Thompson, 1969). Marked interaction between light and temperature exists in the germination of photoperiod-sensitive seeds (Black and Wareing, 1955, 1960; Vaartaja, 1956; Stearns and Olson, 1958). The promotive photoreaction has a temperature coefficient close to unity and is therefore insensitive to temperature, but its immediate product is thermally unstable, and the reactions in which it participates are likely to be temperature dependent. For this reason many long-day seeds require longer (or more frequent) exposures to light (in an intermittent regime) when the temperature is supraoptimal. The same applies to short-day seeds, with the added possibility that the inhibitory action of prolonged irradiation may or may not be itself temperature dependent. There are indications that this is the case in *Hyoscyamus desertorum* (Roth-Bejerano *et al.,* 1971), but more definite experiments are required, possibly with seeds which exhibit negative photoperiodic responses, such as *Nemophila insignis* (Black and Wareing, 1960) (Fig. 1.15).

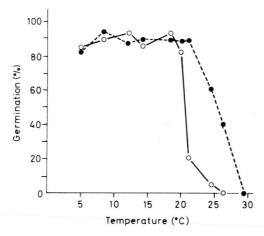

FIG. 1.15. Effect of temperature on germination of *Nemophila insignis* in light (O) and in darkness (●). From Black and Wareing (1960).

## 2. CONDITIONING

Germination is essentially a sequential process, although there may exist several alternate pathways in different sections of the reaction sequence. The sequential nature of the overall process sometimes expresses itself in a quantitative or qualitative change in the environmental control. This happens when a crucial reaction has specific environmental requirements, but depends also on the outcome (or product) or a preceding reaction which has its own specific environmental requirements. This raises no problem if the environmental requirements do not conflict. If they do, germination will take place only when the environment changes according to the requirements of the sequence. The response to one environmental complex is thus conditioned by a prior response to a different environmental complex. A relatively straightforward example is that of seeds with embryos which require a seasonal temperature alternation, and coats which are impermeable to water. The embryos are unable to respond unless they are rehydrated first. This imposes a prior requirement for an environment which will cause increased permeability of the coats. In nature this is achieved by activity of microorganisms, which takes place optimally at a much higher temperature than that required for satisfying the requirement for cold. The full response is therefore elicited by a three-season alternation: a cold one sandwiched between two with milder temperatures, as was already described in Section III,A,3,d. In seeds of *Zizania aquatica* a prolonged (6-month) submersion in water is a prerequisite for subsequent germination in air (Simpson, 1966). In the natural habitat (lakes in northern latitudes) such a prerequisite will effectively prevent germination in autumn, delaying it till the rigors of winter have passed. In short-day seeds, light-requiring processes take place when a sufficient amount of radiant energy has been absorbed by phytochrome in order to activate it. These processes initiate a series of dark reactions which no longer require light. Moreover, these dark reactions happen to be inhibited by the presence in these seeds of a light-sensitive system which is probably also mediated by phytochrome but is reactive only in the continuous presence of far-red containing light (Borthwick *et al.*, 1969) or by its repeated application at intervals (Mancinelli and Borthwick, 1964; Mancinelli *et al.*, 1966). The requirement for darkness is conditioned by the prior exposure to red-containing light, or by presence of preexisting phytochrome in the active, far-red absorbing form. Seeds of *Lepidium virginicum* will not germinate in total absence of light. At the optimal temperature of 20°C their responsiveness to the promotive action of red-light irradiation could be increased by a 2-hour exposure to the supraoptimal temperature of 35°C. However, when the high-temperature treatment followed the irradiation, the requirements for energy of red light were eight times as great as when the treatment preceded the irradiation (Toole *et al.*, 1955). Obviously germination requires the combined contribution of both active phytochrome and the temperature fluctuation. However, when the latter follows the former it most probably also causes partial thermal inactivation of active phytochrome and thus creates a conflicting side effect. The two responses do not appear to be related sequentially but become this way by the conflicting side reaction. A similar situation was found in seeds of *Atriplex dimorphostegia* (Koller, 1971). Here too a supraoptimal temperature (25°C) increases the responsiveness of phytochrome to light, since the germination percentages which result from a short irradiation increase progressively with duration of initial incubation at 25°C. However, at the same time there is a progressive delay in the onset (lag phase) and decrease of germination. Apparently, the supraoptimal temperature has deleterious side effects which conflict with its promotive conditioning of the phytochrome system. This possibility is borne out by the finding that incubation at the optimal temperature (10° or 15°C) in presence of a water potential which is sufficiently low to inhibit germination completely, also causes a promotive conditioning of the phytochrome system, but without any of the deleterious side effects caused by conditioning at the supraoptimal temperature. As a matter of fact, when the two conditioning treatments are applied together the rates of germination are greater than when either is applied separately. This indicates that the promotive conditioning by high temperature would actually be much more effective than the one at low water potential, were it not for the deleterious side effects. In combination with the low water potential these side effects are eliminated. The increase in effectiveness of the phytochrome system by the combined conditioning treatments is such that it permits germination to take place even at the supraoptimal temperature itself. The possible reason for this is that as a result of the increased efficiency of the phytochrome system, the products of the photoreaction are used in germination more rapidly than they are thermally inactivated. A more complex sequence was observed in germination of the light-requiring seeds of *Eragrostis ferruginea* (Isikawa *et al.*, 1961) (Fig. 1.16).

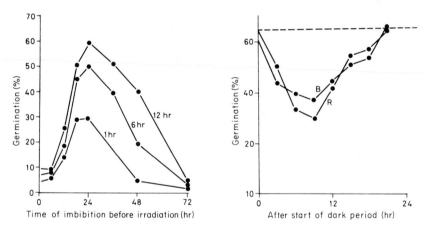

Fɪɢ. 1.16. Effects of light on germination of *Eragrostis ferruginea*. Left: promotive action of a single irradiation, lasting 1, 6, or 12 hours, as influenced by time of application. Right: inhibitory action of a single 10-minute irradiation with red (R), or blue (B) light, as influenced by time of application during initial (promotive) 24-hour dark period. From Isikawa *et al.* (1961).

Their sensitivity to light is conditioned by a preceding 24-hour incubation in absence of light. During the first half of the dark period a brief interruption with red or blue light is inhibitory (peak sensitivity at about the ninth hour) and the inhibition is reversible by far-red light. During the second half of the dark period a brief irradiation with far-red light is inhibitory, but its effects are not reversible by red light (possibly because the exposure is too long). The effects were not substantial (reductions from around 55% to around 35%), but appeared to be consistent.

From a functional viewpoint, interactions and conditioning responses seem likely to increase the probability of survival, because they reduce the probability of haphazard germination and thus increase the specificity and accuracy with which the environment can be sampled and assessed by the seed before it finally commits itself irrevocably to germination.

## IV. Postmaturation Conditioning

Because of their small size and concentrated supply of stored food, seeds disappear from view soon after dispersal, either by becoming buried in the soil, or by activity of harvesting animals. Seeds are therefore usually harvested by man very close to or frequently long before the time of dispersal because it is much easier to collect more seed that way. As a result of this practice, studies on germination are frequently carried out with seeds which have not matured sufficiently and which are derived from fruits which have not yet become separated from the mother plant by an abscission zone. In addition, the harvested seeds are usually stored under conditions which are far removed from any environment which they are likely to encounter in nature, as far as temperature regime, moisture, light, atmosphere, etc. are concerned. So long as germination research deals primarily with practical aspects of cultivated species or of weed seeds which contaminate harvested seeds of agricultural plants, there is nothing wrong with either of these procedures, since this is how such seeds are harvested and stored in normal agricultural practice. However, fundamental research into the physiology of germination in general, and particularly into its control by the environment, must not fail to take into account that seeds have become adapted to pass a certain length of time between their full maturity and the season which is most suitable for their germination, in a certain environmental complex, which also changes with time. It is with these thoughts in mind that one must approach and appreciate the numerous studies on "freshly harvested seeds" and on the effect of "storage conditions" on such seeds. This is not meant to detract from the scientific value of such studies, as they have provided invaluable information on the processes of the true "after-ripening," which are in all probability the terminal processes of seed ripening, though possibly they are somewhat modified by the premature harvesting and unnatural environment of storage.

The very fact that seeds do change their physiological behavior during storage and that these changes are frequently affected by the conditions which prevail during storage has led to gradual abandonment of the concept that the dry mature seed is indifferent to the environment because it is highly resistant to unfavorable environments. From this it is but a small step to start wondering just to what extent the environmental factors which are naturally encountered by the mature seed affect its subsequent germination response, and also whether such effects may be adaptive from the point of view of survival.

The type of dormancy which is exhibited by seeds when they are freshly harvested and which is probably at least in part due to premature harvesting, will be dealt with first in brief. In her review on seed dormancy, Barton (1965) deals with, among other forms, that which changes by after-ripening in dry storage. Such seeds are classified as dormant although they are capable of germinating in quite high percentages. This misleading classification originated in the early history of seed testing, when germination was tested and failed to take place under standard conditions which appeared to be normal. It very soon turned out that the standard conditions were actually not the most suitable ones and that the seeds were quite capable of germinating when provided with the appropriate and often highly specific environmental conditions. The term after-ripening was appropriately used to account for the fact that in this particular class of seeds the original state of dormancy, such as it is, was gradually relaxed during dry storage. Barton lists the periods in dry storage during which dormancy disappeared in forty-one plant species. These periods varied from two weeks (*Lepidium virginicum*) to several years (*Sporobolus cryptandrus, Cyperus rotundus*). The actual values are not quite meaningful in the majority of cases, since neither the initial moisture content of the seed nor the environment during storage were specified or controlled. The degree of maturity at harvest also was rarely specified. The changes which occur in the seed during the period of dry storage usually consist of a general relaxation in the degree of strictness of the environmental requirements for germination. The temperature range for germination, which might have been very narrow and confined to abnormally low or abnormally high temperatures, usually widens considerably to include higher or lower temperatures, or both. Specific requirements for diurnally alternating temperatures, for light, or for specific light regimes, are gradually weakened quantitatively, or disappear totally (e.g., Rollin, 1959; Steinbauer and Grigsby, 1960). We have already seen that in many cases, seeds whose germination exhibits specific environmental requirements may no longer require them, or the seeds may show a marked reduction in these requirements when the testa (or some other coat) is removed, punctured, or weakened. It is therefore likely that at least some of the changes which take place during dry storage are a result of changes in the coats which reduce their tensile strength or possibly their impermeability to water or respiratory gases. In other cases it is likely that changes take place internally to the coats, in the embryo itself or in its immediate chemical environment which result in an increased potential. This is probably the case in seeds in which after-ripening in dry storage can be effectively replaced by a relatively short incubation at near-freezing temperatures. An increase in the potential of the embryo to exert growth forces against the mechanical restraints of the coats may result from changes in the balance of endogenous growth regulators (Black and Naylor, 1959), and incubation at low temperature may cause such changes as in cold-requiring seeds (Villiers and Wareing, 1965). There is a tendency to attribute to the requirement of freshly harvested seeds for strict environmental conditions a protective role against precocious or accidental germination early in the season of maturation, when moisture and temperature are still favorable but become progressively and relatively rapidly less so.

There is increasingly more evidence that the changes in germination responses which the seeds undergo during their exposure to natural environments in the field after dispersal from the mother plant may be much more drastic than those which they undergo during artifical storage in the in the laboratory. An example is provided from studies on environmentally induced changes in seed coat impermeability to water in annual legumes from the Negev desert in Israel (D. Koller, unpublished results). Mature air-dried pods were exposed to outdoor conditions of their natural habitat, under a 10 mm layer of dry loess soil, protected only against rainfall, for periods of 5 and 17 months (corresponding to the time till the onset of the rainy season in the first and second year after maturation), and were soaked in water to test for seed coat impermeability. Pods stored in the laboratory were similarly tested at the same time. In the two-seeded pod of *Hymenocarpus circinnatus,* the percentage of permeable seeds in the distal position rose from the initial 8 to 13 to 18 in laboratory storage and to 69 and 95 in outdoor exposure. In the three-seeded pods of *Onobrychis crista-galli* var. *subinermis* the percentage of permeable seeds in the distal position rose from the initial zero to 16 and 21 in laboratory storage and to 83 and 99 in outdoor exposure. In this study it was also found that a gradient of susceptibility to weathering existed among the seeds of a single pod along its axis. In *Onobrychis* the increase in permeability occurred only in the most distal of the three seeds. In *Hymenocarpus,* on the other hand, the proximal seed started to increase in permeability only during the second year, and to a lesser extent than the distal one (43 *vs.* 95% after 17 months). An opposite gradient was observed in *Medicago tuberculata* where percentages of permeable seeds after 17 months' outdoor exposure were 100 in the proximal position and 95, 40, 7, 2, and 0 in the succeeding positions. Furthermore, intraspecific differences in susceptibility to weathering were observed in *Onobrychis crista-galli.* In *O. crista-galli* var. subinermis, which inhabits the more humid regions, 83% of the distal seeds became permeable after 5 months outdoors, whereas in *O. crista-galli* ssp. *eigii* only 3% were permeable after 5 months and 79% after 17 months. The survival value of this inherent heterogeneity within the seed population from a single plant is evident, leading to dispersal in time of the attainment of germinability. The intraspecific differences in susceptibility to weathering may also have a bearing on survival, since the time to attainment of full permeability of the seed population from a single growing season would be longer in the more arid region.

Studies by Lubke and Cavers (1969) on germination ecology of *Saponaria officinalis* from riverside gravel banks showed that seeds which retained their dormancy in laboratory storage lost it entirely after exposure to overwintering conditions when submerged in their natural habitat for 2 to 6 months. The effects of the various components of this environment were tested separately. The low winter temperature as well as the abrasive action of gravel were without effect, but the submersion itself was highly effective. In studies on germination of seeds of common weeds of the dry phase of low-lying, inundated land, Mall (1954) found that seeds of *Heliotropium supinum* and *Mollugo hirta* were incapable of germinating unless previously subjected to burial for 8 weeks in mud from the bottom of a pond. When the period of burial was prolonged up to 19 weeks, seeds of *Heliotropium* started to exhibit a requirement for a diurnal thermoperiodicity, germinating better at an alternation between 9° and 20°–31°C than in either higher or lower regime. Under the same conditions, seeds of *Mollugo* exhibited an increasing tolerance of high temperature (35°C). In both species the promotive effects of burial were almost entirely reversed by an intervening desiccation after 12 weeks, but not after 19 weeks. In none of the studies on the effects of burial was the mechanism studied in any detail.

In studies on effects of burial in wet soil on weed seeds, Wesson and Wareing (1969a,b) found that even species which originally showed little, if any, response to light (*Spergula arvensis, Stellaria media*), after burial failed to germinate in darkness and additional germination took place only upon illumination. This by itself is hardly surprising, since it could be argued that probably all of the seeds which could germinate in darkness had already done so during the period of burial. In controlled experiments, germination of *Spergula* seeds which had been buried in soil was lower than what could be expected from the mere absence of light. When the ungerminated seeds were dug up after 14 days, more germinated in light than in darkness. Again the experimental design would discriminate against the dark-germinating fraction, from which the seeds that germinated in soil were presumably deducted. However, in *Stellaria* hardly any germination took place during burial, and after it none at all took place in darkness, whereas in light the percentages were nearly identical to those in light prior to burial. Clearly, in this species burial suppressed entirely the germination capacity of the dark-germinating fraction, leaving untouched that of the light-requiring fraction. The former most probably contained preexisting phytochrome and were therefore more susceptible to enter a state of secondary dormancy, or even deteriorate, when prevented from germinating by burial. The authors offer data which show that the reduced germination of *Spergula* in soil is due to lack of aeration. They interpret their data as indicating that the effect of aeration is the removal of a gaseous inhibitor which is produced by the seed itself and is not carbon dioxide, but the evidence is not conclusive. In the following sections we shall deal with specific environmental factors as well as with some dispersal mechanisms which affect the seed after its maturation and thus modify its subsequent germination.

### A. Effects of Dispersal Mechanisms

Dispersal of seeds is dealt with in detail in Chapter 4, Vol. I and will therefore be treated here only with respect to its possible effects on subsequent germination.

Most small seeds are eventually buried in the soil, but their depth of burial is largely haphazard, depending on texture and stability of the upper strata of the soil. The critical role which depth of burial plays in seedling establishment has already been stressed in this chapter. So vital is this parameter that certain plants have evolved elaborate mechanisms enabling the seeds to be buried at a uniform depth, where the optimal submicroenvironment for seedling establishment probably exists. The wide distribution of some of these species may be at least partly due to this fact. The most familiar mechanism is exemplified by representatives of the genera *Erodium* and *Avena*. Briefly, the dispersal units are pointed at one end and equipped with a long beak at the other. The pointed tip is equipped with hairs which point backward (away from the tip) and act as an anchor. The proximal portion of the beak is anatomically structured to twist into a tight coil when dry, and untwist when wet. The distal portion of the beak remains straight. When the dispersal unit lies on the ground and dries, the beak coils up at its base and leans on its straight terminal portion, which sticks out at an angle from the axis of the coil. The tip of the dispersal unit is thus posed at an angle to the soil surface. As humidity rises, the beak uncoils, and leaning on its distal end, screws the seed-bearing tip of the dispersal unit into the soil. Upon redrying, the coiling process is repeated, but the anchoring hairs at the tip prevent it from withdrawing from its depth, with the result that the beak is pulled into the soil. This is repeated with each diurnal cycle of atmospheric humidity. These alternations in humidity become attenuated with depth in the soil, and the motions of the beak become gradually weaker as a result. The burrowing will come to an end when these motions are too weak to overcome the mechanical impedance of the soil to downward penetration. This can result in very precise positioning. In *Gymnarrhena micrantha,* which has already been dealt with in a previous section, the inflorescences which give rise to the subterranean fruits are originally formed at the soil surface, but are subsequently withdrawn to a specific depth inside the subterranean tissues of the plant by a contractile mechanism (D. Koller and Roth-Bejerano, unpublished results). This depth is presumably optimal for the subsequent germination of these seeds, since they germinate at this depth within the dead tissues of the mother plant, forming colonies. The resulting seedlings have a higher probability of finding the environmental necessities for survival where the previous generation has successfully completed its own life cycle. The fruits from the aerial inflorescences are liberated by a complex series of hygroscopic movements, requiring several cycles of moistening and drying. The persistent pappus, which is unfurled in this process, aids the dispersal by wind and also prevents the seeds from being buried when they finally come to rest in soil crevices. If the environment is suitable for germination, the seedlings form nuclei for new colonies by means of the subterranean fruits (Koller and Roth, 1964).

For plants of arid regions, the essential environmental factor is probably rain. Presence of water-leachable inhibitors of germination in the dispersal unit presumably serve as reliable indicators of the recharge of soil moisture. Somewhat similar results are obtained by plants which disperse their seeds only in response to rainfall. This is the case, for instance, in the desert plant *Blepharis persica* (Gutterman *et al.,* 1967). In the desert dwart-annual *Asteriscus pygmaeus,* the inflorescences are formed at the soil surface. Upon maturation, they become entirely enclosed by the hygroscopic infolding of the drying, lignified bases of the bracts. The latter unfold and expose the achenes when the atmospheric humidity is high, and particularly when they are wet by dew, folding up again when redried. However, the achenes are detached from the receptacle only by the impact of raindrops. Only the peripheral achenes are liberated each time, and these are nondormant. The remainder are both firmly attached and dormant, but their newly peripheral rank will gradually lose their firm attachment and dormancy (possibly during the repeated cycles of atmospheric moisture) and be prepared for dispersal and immediate germination in the next effective rain (D. Koller and Roth-Berjarano, un-published results).

## B. Temperature

Effects of temperature during both moist storage and dry storage on subsequent loss of dormancy have been recorded. Of the former, by far the most consistent and prevalent response has been to low (near-freezing) temperatures. Such effects have almost exclusively been reported in seeds with water-permeable coats, and therefore most probably act directly on the embryo. These have been briefly dealt with above. In addition, there are reports of effects of subfreezing temperatures (Midgeley, 1926), as well as of high temperatures (Jones, 1928), on seeds with water-imper-meable coats. In the former moisture was not essential. One hour of freezing was as effective as 60 days, 0°C as effective as − 20°C.

Loss of dormancy during dry storage has by and large been associated with relatively high temperatures. Thus, for instance, Ruge and Liedtke (1951) found that dormancy in seeds of various species of *Malva* was markedly reduced by relatively brief exposures to fairly high temperatures (60°–80°C) in a well-ventilated oven. It is not known whether such temperatures are ever encountered by the seeds in their natural habitats, and if they are, what would be the effects of more sustained and repeated exposures. In several species of desert plants from Arizona (*Eriophyllum wallacei, Gerea canescens, Salvia columbariae, Lepidium lasiocarpum, Sisymbrium altissimum, Streptanthus arizonicus*) exposure of the seeds to a temperature of 50°C for 1–5 weeks induced high germinability, while storage at 20°C was much less effective (Capon and Van Asdall, 1967). Seeds of a considerable number of species of Gramineae improve their germination as a result of desiccation, and this aspect will be dealt with in a subsequent section. One such species is *Panicum turgidum* (Koller and Roth, 1963). In testing the effects of temperature, it was found that the desiccation was much more effective at higher temperatures. Thus germination of controls after normal laboratory storage (without drying) was 49%, while grains stored over granulated $CaCl_2$ for 48 days ger-minated 65 and 88% under the same conditions when temperature during storage was 5° and 30°C, respectively. These results could arise from direct promotive effects of temperature, or from improved efficiency of drying. In studies on the effects of temperature during dry storage on loss of dormancy in rice (Fig. 1.17), Roberts (1965) found a negative linear relationship between storage temperature, $t$, and the logarithm of the mean dormancy period, $d$, so that $d = K_d − C_d t$ ($K_d$ and $C_d$ being con-stants). This linear relationship was valid between 27° and 47°C, in four varieties of *Oryza sativa* and two of *O. glaberrima*. In five of these vari-

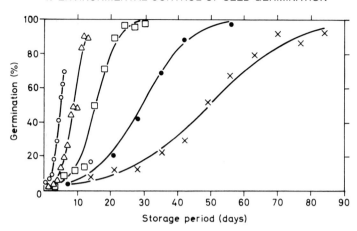

FIG. 1.17. Effect of duration and temperature of storage on subsequent germination of dormant seed of rice (*Oryza sativa* L. var. Lead 35), at 32°C. From Roberts (1965).

eties the temperature coefficient for loss of dormancy was 3.13, and in the remaining one it was 2.54. No interaction was found between storage temperature and germination temperature. Roberts suggests that the high temperature coefficient indicates a process with a high activation energy. Because the temperature coefficient does not fall off even at high temperatures, the process may not be enzymic, but this departure from normal enzymic reactions may possibly be ascribed to protection of the enzyme against heat denaturation by the high ratio of protein to water in the dehydrated seeds. It is a well-known fact that dry seeds are by and large highly tolerant to heat. It is not known to what extent the higher temperatures contributed toward the more efficient desiccation of the grains.

In a series of studies on the effects of temperature on seed coat impermeability to water in legumes, Quinlivan (1961, 1965) and Quinlivan and Millington (1962) showed a distinct effect of diurnal temperature fluctuations on increase in permeability. Dry seeds stored at constant temperatures during 5 months increased in permeability more when the storage was at a high temperature (60°C) than at a low one (15.5°C), but diurnally alternating temperatures (15.5°/60°C) were even more effective. In several cases the increase in permeability by the alternating temperature over the high constant one was considerable. In others it was quite small. From the point of view of environmental control under natural conditions it is highly significant that diurnal temperature alternations within the range which occurs in summer at the soil surface were at least equally as effective, and in some of the cases much more effective, than the most effective constant temperature, since the latter is not encountered in nature. Indeed, the decline in impermeability with time in seeds of several varieties of *Trifolium subterraneum* exposed to field conditions could be matched by seeds stored in a regime of a suitably alternating temperature. By studying pods (burrs) left under natural conditions during summer in two regions differing in the magnitude of the diurnal temperature alternation, and by comparing pods in plots with and without the dry topgrowth, it could be shown that (*a*) hot summer environments with wide soil surface temperature alternations were conducive to a more rapid loss of impermeability than cooler environments with lesser alternations, and (*b*) absence of the dry topgrowth during summer, for instance as a result of grazing, increased the rate of loss of impermeability by increasing the magnitude of the temperature alternations. Further studies (Quinlivan, 1966) showed that when the low temperature part of the cycle was 15°C, impermeability decreased progressively with increase of the temperature above 45°C in the remaining part of the cycle. Conversely, when the high temperature part of the cycle was 45°C, the reduction of impermeability was nearly equal with the remaining part of the cycle at 15° or 30°C. Seeds of *Lupinus varius* responded only when the high temperature part was higher than 45°C. It seems, therefore, that the effectiveness of the alternations is largely dependent on the level of the high temperature in the cycle. The loss of impermeability in this species as a result of such treatment occurred by fracture of the coat at the strophiole (Quinlivan, 1968).

## C. Radiation

It is well known that seeds become capable of responding to light only after they have been allowed to imbibe for a certain length of time. This has given rise to the concept that the dry seed is insensitive to light. Evenari and Neumann (1953b) questioned this concept and carried out experiments to test the possibility that the dry seeds were also receptive to light. Dry seeds of the light-sensitive lettuce variety Grand Rapids were stored under identical temperature conditions, in light and in darkness, at two relative humidities. After storage, germination was tested in darkness at 26°C. Seeds stored in 15% relative humidity in light and darkness reached 21 and 25% germination, respectively, while those which were stored at 60–70% relative humidity reached 83 and 41% respectively. The higher hydration of the seeds stored at the higher atmospheric humidity evidently sufficed to sensitize them to the presence of light during storage. Germination of dark-stored seeds was also higher when the humidity in storage was high. Roth-Bejerano *et al.* (1966) have shown that lettuce seeds, which normally require light for germination at supraoptimal temperatures, could nevertheless germinate in darkness at these temperatures following incubation at a low temperature, probably as a result of the transformation of preexisting phytochrome into a thermostable product during cold treatment. The same explanation may probablyapply to the results of Evenari and Neumann (1963b). In high humidity storage, the hydration which the seed attains suffices to allow only a very slow metabolic activity. In this state, the seed therefore appears to be insensitive to light (of short duration), in comparison with the fully imbibed state. In dark storage, the hydration is too low for germination to take place, but is sufficiently high for the thermostabilization of phytochrome to proceed, although at a slow rate, just as it can in fully imbibed seeds at near-freezing temperature. The whole subject merits additional study, on theoretical as well as practical grounds. It may, for instance, explain why certain seeds lose their requirement for light during storage. The requirement for light may also be lost if the imbibed seed is exposed to the promotive action of light and is then prevented from germinating by being immediately dehydrated again (Kincaid, 1935; R. C. Thompson, 1938; Koller and Negbi, 1959), but in other species the promotive action of light is lost upon immediate dehydration (Koller *et al.,* 1964). Light sensitive seeds may also lose their responsiveness to light by entering a state of secondary dormancy if incubated for sufficiently long periods without having their requirement for light satisfied. This is obviously not due to thermal inactivation of their active phytochrome, since they can no longer be promoted by light. In lettuce, the secondary dormancy can be relieved by incubation at low temperature, but the mechanism is not likely to be the same as described by Roth-Bejerano *et al.* (1966). In contrast to lettuce, the light requiring fractions of the seed population in *Oryzopsis miliacea* (Koller and Negbi, 1959) and *Amaranthus retroflexus* (Kadman-Zahavi, 1960) retain their original sensitivity to light even during prolonged incubation in darkness.

The fact that light-sensitive seeds start to attain sensitivity to light rapidly only after the start of imbibition has led to the belief that the imbibed state of the receptor sites within the seeds is all that is required for light sensitivity. In that case, dehydration should fully reverse the effects of imbibition. However, at least in *Artemisia monosperma* it does not. Responsiveness to light increased with time during dark incubation of the imbibed seeds, up to the sixth to eighth days (depending on the temperature). When incubation was interrupted by as many as three desiccation periods, the final sensitivity was equivalent to that of controls which had spent an equal length of time in the imbibed state. This indicated that attainment of sensitivity to light may involve a gradual build-up of the light-sensitive mechanism, not its mere rehydration. The sensitivity of such seeds to light would thus be influenced by the cumulative length of time they had been imbibed (Koller *et al.,* 1964).

### D. *Water Relations*

One of the commonest environmental factors which acts on the mature seed and affects its subsequent germination responses is atmospheric humidity. In general, dry storage is effective in increasing the germinability. The lower seed of *Xanthium pennsylvanicum* is dormant when freshly harvested. This dormancy was largely lost after about 6 months in normal dry storage, but not in storage at − 15°C. Despite the difference in dormancy, there was no difference in the growth-inhibitor content. Nor was any change noticed in the permeability of the seed coat toward the endogenous inhibitor even after 1 year of dry storage. On the other hand, within 7 days of dry storage the dormant lower seed as well as the upper seed became responsive to an atmosphere of pure oxygen, in which they germinated when imbibed. Wareing and Foda (1957) conclude that increase in permeability of the seed coats toward oxygen occurred as a result of desiccation, which permitted oxidation of the endogenous inhibitor. Studies on the changes involved in loss of dormancy in wheat grains during dry storage (Wellington, 1956; Wellington and Durham, 1961) were interpreted as indicating that desiccation was having a dual effect. On the one hand, the tensile resistance of the coats was being reduced by the effect of drying on the structure of their colloids and on the other hand the water-absorbing potential of the embryo was increased by the effects of desiccation on conversion of storage starch to sugar.

Atmospheric humidity may have far-reaching effects on seed-coat impermeability to water in seeds of certain species belonging to the *Papilionaceae* (Hyde, 1954). In the mature seeds of *Trifolium repens, T. pratense* and *Lupinus arboreus* (moisture content ⩽ 14%) there is only a single pathway by which vapor can be exchanged between the interior of the seed and the ambient air. This is an opening in the region of the hilum with special anatomic features which enable it to move hygroscopically and very rapidly, opening widely in dry air and closing in moist air. In this way it normally acts as a one-way valve which is open only under conditions which lead to loss of moisture to the outside. However, if atmospheric humidity is just below that which leads to total closure, or better still, if it is increasing at an almost imperceptible rate, water vapor is gradually taken up by the colloids enclosed inside the coat. These gradually imbibe and swell and eventually distend the coat to such an extent that the hilar valve becomes nonoperational. In this state the seed is fully permeable to liquid water. However, if the seed is redried, the valve again becomes operational and the seed regains its impermeability.

The seed thus adjusts its capacity to germinate according to the atmospheric humidity which has prevailed in its vicinity for some time past. This could be a remarkable adaptation to control germination to rains which follow an increasingly humid season.

In studies on the effects of humidity during storage on the kinetics of water uptake by seeds of annual legumes (D. Koller, unpublished results), marked differences were found between *Medicago laciniata* on the one hand and two species of *Lathyrus* on the other. The seeds were stored for 2 months in atmospheres of constant relative humidities ranging between 13.0 and 83.5% and the kinetics of their imbibition were analyzed from the time course of cumulative percentages of imbibed seeds. In *Medicago* the final percentage of imbibed seeds increased progressively with increase in humidity of storage, but did not exceed 55. Other kinetic parameters were unaffected by humidity during storage. The lag phase was equally short in all cases, and final percentages were reached equally rapidly. In *Lathyrus,* on the other hand, seeds from all storage humidities were nearly fully permeable, but the lag phase was progressively shorter and rates of imbibition were progressively higher as humidity during storage was higher. There were quantitative differences between the two species which were tested. In *L. aphaca* the lag phase varied from 7.6 to 0.1 days and the rate remained practically unchanged (12–15% per day) up to a relative humidity of 80.5% and then rose abruptly (to 91% per day). In *L. hierosolymitanus* the lag phase varied from 3.9 to 0.1 days and the rates started to increase appreciably as relative humidity rose above 49% (11–14% per day), reaching 76–78% per day at humidities of 80.5–83.0%. The ecological implications of differences in kinetics of imbibition (and therefore also of germination) were discussed in a previous section. From a functional viewpoint, it is quite clear that in *Medicago* the pathway for entry of liquid water is either fully open or fully closed (intermediate positions were limited to a very narrow range of humidity). The humidity during storage affects the proportion of seeds in each class. In *Lathyrus,* on the other hand, the humidity during storage appears to affect only the hydraulic conductivity of the seeds, possibly by changing the moisture content of their coats.

Studies on seed coat impermeability in *Lupinus varius* (Quinlivan, 1968) showed that if the moisture content of the seeds is 10% or higher, their impermeability gradually disappears during exposure to humid air at a rate which is directly related to the seed moisture content. Penetration of moisture is not confined to any specific region of the seed coat. On the other hand, if the seed moisture content falls to 8.5% or lower, the seeds lose all capacity to become permeable during humid storage, but respond to diurnal temperature alternations between 15°

and 65°C. Penetration of moisture in seeds which had thus become permeable is confined to the fracture at the strophiole. Permeability of seeds with a higher moisture content ($\geqslant$ 10%) is completely unaffected by such alternations. Under field conditions the germination pattern would depend on the degree of desiccation which the seeds had reached, until their moisture content fell below 10%, after which it would start depending on the range of diurnal temperature alternations (Fig. 1.18).

### E. Biotic Influences

The biotic environment is not without effects on seed germination. Most of these effects are mediated by changes in other components of the seed's environment, which are brought about by presence of the biotic factors. An exception is provided by the direct action of microorganisms on the physical and possibly also on the chemical properties of various coats which control germination in one way or another. Microorganisms which attack cell-wall constituents presumably play a major role in this. However, the problem does not appear to have been systematically studied.

Plants affect germination of seeds in their vicinity in several indirect ways, such as modifying the light regime quantitatively and qualitatively, adding organic substances to the liquid or gas phase, changing the proportions of the respiratory gases, etc. These aspects have already been dealt with above.

There are several ways in which animals may influence germination. Seed-harvesting animals (mostly insects, but also birds and rodents) usually collect more than they eventually consume, and store these seeds in specific hiding places. These hiding places usually have rather specialized climates, which may modify the eventual germination response. Other animals are passive agents in propagation of exozoochoric dispersal units and may land the seeds in specialized environments which could affect their subsequent germination. On the other hand, herbivores ingest dispersal units, and in the process may rid their seeds of inhibitors, or modify permeability of their coats in passage through the alimentary tract, possibly aided by the action of the microflora in it (Müller, 1934; Burton, 1948).

The part which man plays in affecting subsequent seed germination by changing the environment is probably the greatest of all. The effects of chemical pollution may be direct, on the seeds themselves, or indirect, by changing the biotic environment. Clearing away vegetation for construction, lumber, fuel, etc., also changes the physical environment. Cultivation adds the element of periodic soil disturbance, periodic harvesting, and removal of the vegetation mulch, and drastic changes in the

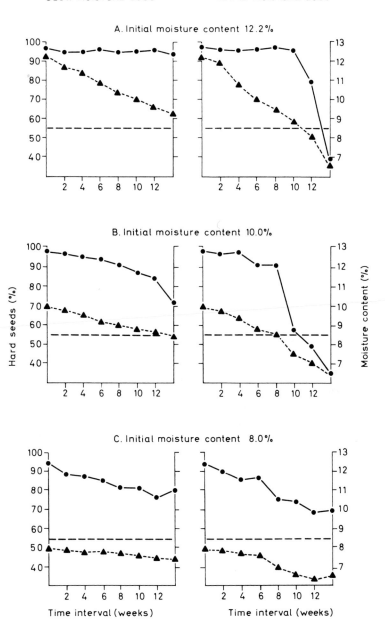

SLOW MOISTURE LOSS        RAPID MOISTURE LOSS

A. Initial moisture content 12.2%

B. Initial moisture content 10.0%

Hard seeds (%)

Moisture content (%)

C. Initial moisture content 8.0%

Time interval (weeks)        Time interval (weeks)

FIG. 1.18. Interaction of seed moisture content and daily temperature fluctuations (15°/65°C) on the impermeability of lupin seeds. The 8.5% level of moisture contents indicated by horizontal line. (●), Hard seeds; (Δ), moisture contents. From Quinlivan (1968).

soil moisture regime (Went *et al.,* 1952), all of which have consequences which have been dealt with above in greater or lesser detail. In general, however, too little attention has thus far been paid to effects of the biotic factor on seed germination.

## F. Fire

The occurrence of fires which burn off the natural vegetation, frequently results in a flush of germination of species whose seedlings could rarely be found prior to the fire. The exact cause of the phenomenon has not received a great deal of attention, although its role in the plant succession which revegetates the denuded area is frequently quite vital for soil conservation, watershed management, and flood control. It is safe to assume that for at least some of the species the effect is indirect, by the changes which occur in the microenvironment as a result of removal of the vegetation cover. This causes quantitative changes in the radiation balance, as well as qualitative and quantitative changes in the light regime, which may promote germination of species with a requirement for light, for diurnally fluctuating temperatures, or for both, or of those whose air-dried seeds become permeable to water as a result of exposure to the wider amplitude of temperature alternations of the bare soil surface. At the same time, fires remove the litter and thus eliminate one of the sources of germination inhibitors (Went *et al.,* 1952). In addition to these indirect effects, fire may have direct ones on the seed itself. One study which concerned itself with such a case was made by Stone and Juhren (1951) on germination of *Rhus ovata* which, in the chaparral vegetation of southern California, occurs almost exclusively after a burn. The direct role of fire in inducing this germination was ascertained by controlled burning of excelsior over sand in which seeds were buried at a depth of one-quarter inch. Germination after burning reached 38% as against 1% in the unburned control. The maximum temperature at seed depth was 70°–80°C. Exposure of the dry seeds to controlled temperatures for different duration resulted in about 25% germinable seeds after exposure to 80°C for 5–60 minutes, and about 33% after exposure to 100°C for 5–10 minutes. The effect of heat was traced to the formation of cracks in the underlying layer of the seed coat, along the edge over the micropyle, with a resulting loss in impermeability to water. However, as the soil surface temperatures in areas cleared of vegetation frequently reached 60°C for several hours each day, the authors feel that this indirect effect and not the fire itself could be responsible for the phenomenon. Experiments were not done to determine whether or not the mere removal of vegetation, without the fire, would also lead to germination of *Rhus*.

## V. Prematuration Conditioning

One of the most disconcerting facts which seed physiologists are faced with is the high degree of physiological variability among seeds. From what has already been said it is quite clear that there are other factors, in addition to the inherent genetic variability, that contribute to this increase. Some of the variation results from unintentional differences in storage conditions, some from changes which occur during storage, even under strictly controlled conditions and some from the degree of maturity at which the seeds are harvested, even though the date of harvest may be the same. Even when all these sources of variation are accounted for, there remains a residual variability which by elimination has been attributed to the year-to-year variations in climate during seed formation and ripening, and to the more permanent differences in climate between different habitats where the seeds had been produced (Quinlivan, 1965). Quinlivan (1966) compared the effects of diurnally alternating temperatures on the rate of loss of seed coat impermeability in *Trifolium subterraneum* and *Lupinus varius* from two sites which differed in the length of their growing season. The rate was consistently and significantly higher in seeds which had been produced in the site with the shorter growing season. Midgeley (1926) concluded that availability of soil moisture influenced the degree of seed coat impermeability in *Vicia villosa*, by controlling the amount of growth (i.e., the length of the growing season).

Studies with the annual legume *Ononis sicula* (Evenari *et al.,* 1966) showed that seeds which had been produced under short-day conditions had a much higher average conductivity for liquid water than those produced under long days. After two years storage, the difference was no longer so much in the conductivity as in the proportion of seeds which had become completely impermeable, which was much greater in the seeds grown in long photoperiods. At any time, a brief desiccation (7 days over $CaCl_2$) reduced the fraction of permeable seeds to zero. An analysis of the phenomenon showed that its cause might lie in a differential effect of photoperiod on the maturation of the pod and of its seeds. Under short photoperiods the pod ripens and shatters before the seeds have reached full maturity. This is indicated by the relative abundance of greenish-yellow seeds in the population, whose conductivity was much higher than that of the yellow seeds. In long photoperiods the greenish-yellow fraction was very small, indicating that the pod had reached full maturity at about the same time as the seed. The closer the seed is to maturity the lower is its moisture content, and the coats are probably the first to be affected. During the early stages of desiccation (and maturation) the overall conductivity of the coats is lowered until they lose permeability altogether. At the same time the desiccation of the interior of the seed leads to its shrinkage until the hilar valve becomes operational. It would be interesting to know if the fully impermeable seeds would lose their impermeability in response to large diurnal temperature alternations as do those of *Lupinus varius* (Quinlivan, 1968). Subsequent studies (Gutterman, 1969) established the fact that the photoperiodic effects were exerted during a very late stage in pod development after it had grown to its full size but while it was still green. Photoperiodic effects on the properties of the seed coat were also observed in *Chenopodium amaranticolor* (Lona, 1947). Plants which are induced to flower in short-days and then allowed to form seeds in long or short days produce readily germinable seeds only in the latter. Seeds produced under long-day conditions have much thicker coats, which inhibit germination apparently by their mechanical resistance to embryo expansion, not by impermeability to water. Seed weight of *Chenopodium polyspermum* was progressively reduced from 63 to 28 mg per 100 seeds as the photoperiod under which they were produced was increased from 10 to 24 hours. Seeds which were produced on plants in long days germinated very poorly in darkness at 25° and 30°C, and were promoted by light only at 30°C. In contrast, plants grown in short days produced seeds which germinated fully at 30°C in both light and darkness, and were promoted by light at 25°C (Jacques, 1957). Studies with *Portulaca oleracea* (Gutterman, 1969) showed that germination is progressively delayed and slowed as the photoperiod under which the seeds had matured is lengthened. The capacity to germinate at suboptimal temperatures is also progressively reduced with increasing photoperiod. The temperature-response of the seeds is also dependent on the photoperiod under which they had matured. At 26°C in light, only those from 8-hour photoperiods showed any germination. Upon transfer to 35°C (after 72 hours) germination was observed also in seeds from 11-hour photoperiods, while seeds from 13-hour photoperiods germinated fully only upon transfer to 42°C (after 168 hours). The photoperiodic effects are to a large extent exerted in the last stages of seed maturation, but are more pronounced if the developing flowers are also subjected to the same photoperiodic treatment. The difference between the photoperiodic preconditioning is accentuated under limiting temperature conditions, for instance in a thermoperiodic regime when the daily exposure to the optimal temperature (40°C) was short. Takahashi (1960) reports that the lag phase in germination of rice seed was longer when the seeds were derived from plants grown in natural day lengths than in 8-hour day lengths.

The temperature regime to which the mother plant is exposed may also affect the germination response of the seed. Von Abrams and Hand (1956) found a negative correlation between the dormancy (cold requirement) of seeds of hybrid roses and the level of temperature that prevailed during the 30–60 days which preceded their harvest in that particular year. The role of temperature was then directly confirmed by experiments. Plants which were placed in a warm greenhouse during the period of seed maturation produced seeds with a lower level of dormancy than those which remained outdoors. J. F. Harrington and Thompson (1952) found a direct relationship between the level of dark germination in lettuce seed at supraoptimal temperatures (26°C) and the air temperature in the field during 10–30 days before harvest. The effects of controlled environmental conditions during seed ripening (after the appearance of flower buds but before anthesis) on subsequent germination responses were studied in lettuce (Koller, 1962). The tolerance of the seeds for high germination temperatures was markedly increased when their formation and maturation took place in regimes of higher temperature and/or longer photoperiods. The response to light remained unchanged, but temperature dependence was quantitatively altered. There were indications that other factors which determine the length of the maturation period, such as the intensity and possibly spectral energy distribution of light, might also affect subsequent germination. A heat stress during the maturation of wheat grains had marked aftereffects on their germinability. When applied 7–10 days after awn emergence, germinability was reduced, probably as a result of injury. When applied 3 weeks after awn emergence, germinability was enhanced, probably by reduction of the thickness of the coats, which increased their conductance to water, and by decrease in the concentration of an endogenous inhibitor (R. A. Khan and Laude, 1969).

## REFERENCES

Aamodt, O. S. (1935). Germination of Russian pigweed in ice and on frozen soil. *Sci. Agr.* **15**, 507.

Al-Naib, F. A., and Rice, E. L. (1971). Allelopathic effects of *Platanus occidentalis*. *Bull. Torrey Bot. Club* **98**, 75.

Arndt, W. (1965). The impedance of soil seals and the forces of emerging seedlings. *Aust. J. Agr. Res.* **3**, 55.

Asakawa, S. (1957a). Studies on the delayed germination of *Fraxinus mandschurica* var. *japonica* seeds. (5) Effect of the compound stratification on germination. *Ringyo Shikenjo Kenkyu Hokoku* **95**, 71.

Asakawa, S. (1957b). Studies on the delayed germination of *Fraxinus mandschurica* var. *japonica* seeds. (7) Thermoperiodicity in germination. *Ringyo Shikenjo Kenkyu Hokoku* **103**, 25.

Ballard, L. A. T. (1958). Studies of dormancy in the seeds of subterranean clover (*Trifolium subterraneum* L.). I. Breaking of dormancy by carbon dioxide and by activated carbon. *Aust. J. Biol. Sci.* **11**, 264.

Ballard, L. A. T., and Grant Lipp, A. E. (1967). Seed dormancy: Breaking by uncouplers and inhibitors of oxidative phosphorylation. *Science* **156**, 398.

Ballard, L. A. T., and Grant Lipp. A. E. (1969). Studies of dormancy in the seeds of subterranean clover (*Trifolium subterraneum* L.). III. Dormancy breaking by low concentrations of oxygen. *Aust. J. Biol. Sci.* **22**, 279.

Barley, K. P., Farrell, D. A., and Greacen, E. L. (1965). The influence of soil strength on the penetration of a loam by plant roots. *Aust. J. Soil Res.* **3**, 69.

Barton, L. V. (1965). Seed dormancy: General survey of dormancy types in seeds. *In* "Handbuch der Pflanzenphysiologie" (W. Ruhland, ed.), Vol. 15, Part 2, p. 699. Springer-Verlag, Berlin and New York.

Barton, L. V., and Crocker, W. (1948). "Twenty Years of Seed Research." Faber & Faber, London.

Bibbey, R. O. (1948). Physiological studies on weed seed germination. *Plant Physiol.* **23**, 467.

Binet, P. (1964). Action de la température et de la salinité sur la germination des graines de *Plantago maritima* L. *Bull. Soc. Bot. Fr.* **111**, 407.

Black, M., and Naylor, J. M. (1959). Prevention of onset of seed dormancy by gibberellic acid. *Nature (London)* **184**, 468.

Black, M., and Wareing, P. F. (1955). Growth studies in woody species. VII. Photoperiodic control of germination in *Betula pubescens* Ehrh. *Physiol. Plant.* **8**, 300.

Black, M., and Wareing, P. F. (1960). Photoperiodism in the light-inhibited seed of *Nemophila insignis*. *J. Exp. Bot.* **11**, 28.

Bonner, F. T. (1968). Water uptake and germination of red oak acorns. *Bot. Gaz.* **129**, 83.

Boorman, L. A. (1968). Some aspects of the reproductive biology of *Limonium vulgare* Mill. and *Limonium humile* Mill. *Ann. Bot. (London)* [N.S.] **32**, 803.

Borris, H. (1936). Über das Wesen der Keimungsförderenden Wirkung der Erde. *Ber. Deut. Bot. Ges.* **54**, 472.

Borthwick, H. A., and Cathey, H. M. (1962). Role of phytochrome in control of flowering of *Chrysanthemum*. *Bot. Gaz.* **123**, 155.

Borthwick, H. A., Hendricks, S. B., Toole, E. H., and Toole, V. K. (1954). Action of light on lettuce seed germination. *Bot. Gaz.* **115**, 205.

Borthwick, H. A., Hendricks, S. B., Schneider, M. J., Taylorson, R. B., and Toole, V. K. (1969). The high-energy light action controlling plant response and development. *Proc. Nat. Acad. Sci. U.S.* **64**, 479.

Briggs, W. M., and Fork, D. C. (1969). Long-lived intermediates in phytochrome transformation. I. *In vitro* studies. *Plant Physiol.* **44**, 1081.

Brown, R. (1965). The germination of angiospermous parasite seeds. *In* "Handbuch der Pflanzenphysiologie" (W. Ruhland, ed.), Vol. 15, Part 2, p. 925. Springer-Verlag, Berlin and New York.

Bünsow, R., and von Bredow, K. (1958). Wirkung von Licht und Gibberellin auf die Samenkeimung der Kurztagpflanze *Kalanchoe blossfeldiana*. *Biol. Zentralbl.* **77**, 132.

Burton, G. W. (1948). Recovery and viability of seeds of certain southern grasses and lespedeza passed through the bovine digestive tract. *J. Agr. Res.* **76**, 95.

Capon, B., and Van Asdall, W. (1967). Heat pre-treatment as a means of increasing germination of desert annual seeds. *Ecology* **48**, 305.

Chen, S. S. C., and Thimann, K. V. (1966). Nature of seed dormancy in *Phacelia tanacetifolia*. *Science* **153**, 1537.

Cohen, D. (1958). The mechanism of germination stimulation by alternating temperatures. *Bull. Res. Counc. Isr., Sect. D* **6**, 111.

Collis-George, N., and Williams, J. (1968). Comparison of the effects of soil matric potential and isotropic effective stress on the germination of *Lactuca sativa. Aust. J. Soil Res.* **6,** 179.

Crocker, W. (1928). Storage, after-ripening and germination of apple seeds. *Amer. J. Bot.* **15,** 625.

Crocker, W., and Barton, L. V. (1957). "Physiology of Seeds." Chronica Botanica, Waltham, Massachusetts.

Crocker, W., Thornton, N. C., and Schroeder, E. M. (1946). Internal pressure necessary to break shells of nuts and the role of the shells in delayed germination. *Contrib. Boyce Thompson Inst.* **14,** 173.

Cumming, B. G. (1963). The dependence of germination on photoperiod, light quality and temperature, in *Chenopodium* spp. *Can. J. Bot.* **14,** 1211.

Datta, S. C., and Chakravarty, H. L. (1962). Germination studies on the seeds of *Citrullus. Indian Agr.* **6,** 220.

Davis, W. E., and Rose, R. C. (1912). The effects of external conditions upon the after-ripening of the seeds of *Crataegus mollis. Bot. Gaz.* **54,** 49.

Edwards, T. I. (1932). Temperature relations of seed germination. *Quart. Rev. Biol.* **7,** 428.

Edwards, T. I. (1933). The germination and growth of *Peltandra virginica* in absence of oxygen. *Bull. Torrey Bot. Club* **60,** 573.

Esashi, Y., and Leopold, C. A. (1968). Physical forces in dormancy and germination of *Xanthium* seeds. *Plant Physiol.* **43,** 871.

Esashi, Y., and Leopold, C. A. (1969). Dormancy regulation in subterranean clover seeds by ethylene. *Plant Physiol.* **44,** 1470.

Evenari, M. (1949). Germination inhibitors. *Bot. Rev.* **15,** 153.

Evenari, M. (1952). The germination of lettuce seeds. I. Light, temperature and coumarin as germination factors. *Palestine J. Bot., Jerusalem Ser.* **5,** 138.

Evenari, M. (1956). Seed germination. *Radiat. Biol.* **3,** 519–549.

Evenari, M. (1961a). Chemical influences of other plants (allelopathy). *In* "Handbuch der Pflanzenphysiologie" (W. Ruhland, ed.), Vol. 16, p. 691. Springer-Verlag, Berlin and New York.

Evenari, M. (1961b). A survey of the work done in seed physiology by the Department of Botany, Hebrew University, Jerusalem (Israel). *Proc. Int. Seed Test. Ass.* **26,** 597.

Evenari, M. (1965). Light and seed dormancy. *In* "Handbuch der Pflanzenphysiologie" (W. Ruhland, ed.), Vol. 15, Part 2, p. 804. Springer-Verlag, Berlin and New York.

Evenari, M., and Neumann, G. (1952). The germination of lettuce seed. II. The influence of fruit coat, seed coat and endosperm upon germination. *Bull. Res. Counc. Isr., Sect. D* **2,** 75.

Evenari, M., and Neumann, G. (1953a). The germination of lettuce seeds. III. The effect of light on germination. *Bull. Res. Counc. Isr., Sect. D* **3,** 136.

Evenari, M., and Neumann, G. (1953b). The germination of lettuce seeds. IV. The influence of relative humidity of the air on light effect and germination. *Palestine J. Bot., Jerusalem Ser.* **6,** 96.

Evenari, M., Konis, E., and Zirkin, D. (1948). On the germination of some rosaceous seeds. II. The germination of *Kerassi* seeds. *Palestine J. Bot., Jerusalem Ser.* **4,** 166.

Evenari, M., Koller, D., and Gutterman, Y. (1966). Effects of the environment of the mother plant on germination by control of seed coat impermeability to water in *Ononis sicula* Guss. *Aust. J. Biol. Sci.* **19,** 1007.

Flemion, F. (1933). Dwarf seedlings from non-after-ripened embryos of *Rhodotypos kerriodies. Contrib. Boyce Thompson Inst.* **5,** 161.

Flemion, F. (1934). Dwarf seedlings from non-after-ripened embryos of peach, apple and hawthorne. *Contrib. Boyce Thompson Inst.* **6,** 205.

Flemion, F. (1959). Effects of temperature, light and gibberellic acid on stem elongation and leaf development in physiologically dwarfed seedlings of peach and *Rhodotypos*. *Contrib. Boyce Thompson Inst.* **20**, 57.

Funke, G. L. (1943). The influence of *Artemisia absinthium* on neighboring plants. *Blumea* **5**, 281.

Galil, J. (1965). Vegetative dispersal of *Allium ampeloprasum* L. II. Sprouting of bulblets. *Isr. J. Bot.* **14**, 184.

Giersbach, J., and Crocker, W. (1932). Germination and storage of wild plum seeds. *Contrib. Boyce Thompson Inst.* **4**, 39.

Grant Lipp, A. E., and Ballard, L. A. T. (1959). The breaking of dormancy of some legumes by carbon dioxide. *Aust. J. Agr. Res.* **10**, 495.

Griswold, S. G. M. (1936). Effect of alternate moistening and drying on germination of seeds of western range plants. *Bot. Gaz.* **98**, 243.

Gutterman, Y. (1969). "The Photoperiodic Response of Some Plants and the Effect of the Environment of the Mother Plant on the Germination of Their Seeds." Ph.D. Thesis, Hebrew University, Jerusalem (in Hebrew, English summary).

Gutterman, Y., Witztum, A., and Evenari, M. (1967). Seed dispersal and germination in *Blepharis persica* (Burm.) Kuntze. *Isr. J. Bot.* **16**, 213.

Haasis, F. W. (1928). Germinative energy of lots of coniferous tree seed, as related to incubation temperature and to duration of incubation. *Plant Physiol.* **3**, 365.

Hadas, A. (1970). Factors affecting seed germination under soil moisture stress. *Isr. J. Agr. Res.* **20**, 3.

Hammouda, M. A., and Bakr, Z. Y. (1969). Some aspects of germination of desert seeds. *Phyton (Horn, Austria)* **13**, 183.

Harper, J. L., and Benton, R. A. (1966). The behaviour of seeds in soil. II. The germination of seeds on the surface of a water supplying substrate. *J. Ecol.* **54**, 151.

Harper, J. L., and McNaughton, I. H. (1960). The inheritance of dormancy in inter- and intraspecific hybrids of *Papaver. Heredity* **15**, 315.

Harrington, G. T. (1923). Use of alternating temperatures in the germination of seeds. *J. Agr. Res.* **23**, 295.

Harrington, J. F., and Thompson, R. C. (1952). Effect of variety and area of production on subsequent germination of lettuce seed at high temperature. *Proc. Amer. Soc. Hort. Sci.* **59**, 445.

Hart, J. W., and Berrie, A. M. M. (1966). The germination of *Avena fatua* under different gaseous environments. *Physiol. Plant.* **19**, 1020.

Hartmann, K. M. (1966). A general hypothesis to interpret 'high energy phenomena' of photomorphogenesis on the basis of phytochrome. *Photochem. Photobiol.* **5**, 349.

Haut, I. C. (1932). The influence of drying on the after-ripening and germination of fruit-tree seeds. *Proc. Amer. Soc. Hort. Sci.* **29**, 371.

Haut, I. C. (1938). Physiological studies on after-ripening and germination of fruit-tree seeds. *Md. Agr. Exp. Sta., Misc. Publ.* **420**, 1.

Hendricks, S. B., Toole, V. K., and Borthwick, H. A. (1968). Opposing actions of light in seed germination of *Poa pratensis* and *Amaranthus arenicola. Plant Physiol.* **43**, 2023.

Heslop-Harrison, J. (1963). Sex expression in flowering plants. *Brookhaven Symp. Biol.* **16**, 109–125.

Heydecker, W., and Orphanos, P. I. (1968). The effects of excessive moisture on the germination of *Spinacia oleracea* L. *Planta* **83**, 237.

Hillman, W. S. (1967). The physiology of phytochrome. *Annu. Rev. Plant Physiol.* **18**, 301.

Hyde, E. O. C. (1954). The function of the hilum in some Papilionaceae in relation to the ripening of the seed and the permeability of the testa. *Ann. Bot. (London)* [N.S.] **18,** 241.

Isikawa, S. (1954). Light sensitivity against the germination. I. "Photoperiodism" of seeds. *Bot. Mag.* **67,** 51.

Isikawa, S. (1962). Light sensitivity against the germination. III. Studies on various partial processes in light sensitive seeds. *Jap. J. Bot.* **18,** 105.

Isikawa, S., and Yokohama, Y. (1962). Effect of "intermittent irradiation" on the germination of *Epilobium* and *Hypericum* seeds. *Bot. Mag.* **75,** 127.

Isikawa, S., Fujii, T., and Yokohama, Y. (1961). Photoperiodic control of the germination of *Eragrostis* seeds. *Bot. Mag.* **74,** 14.

Ives, S. A. (1923). Maturation and germination of seeds of *Ilex opaca. Bot. Gaz.* **75,** 60.

Jacques, R. (1957). Quelques données sur le photopériodisme de *Chenopodium polyspermum* L.: Influence sur la germination des graines. *U.S.I.B. Publ.* **34B,** 125.

Jones, J. P. (1928). A physiological study of dormancy in vetch seeds. *Cornell Univ., Agr. Exp. Sta., Mem.* **120,** 1.

Joseph, H. C. (1929). Germination and viability of birch seeds. *Contrib. Boyce Thompson Inst.* **2,** 17.

Juhren, M., Went, F. W., and Phillips, E. (1956). Ecology of desert plants. IV. Combined field and laboratory work on germination of annuals in the Joshua Tree National Monument, California. *Ecology* **37,** 318.

Kadman-Zahavi, A. (1955). The effect of light and temperature on the germination of *Amaranthus blitoides* seeds. *Bull. Res. Counc. Isr., Sect. D* **4,** 370.

Kadman-Zahavi, A. (1960). Effects of short and continuous illuminations on the germination of *Amaranthus retroflexus* seeds. *Bull. Res. Counc. Isr., Sect. D* **9,** 1.

Kahn, A. (1960). An analysis of "dark-osmotic inhibition" of germination of lettuce seed. *Plant Physiol.* **35,** 1.

Khan, R. A., and Laude, H. M. (1969). Influence of heat stress during seed maturation on germinability of barley seed at harvest. *Crop Sci.* **9,** 55.

Kincaid, R. R. (1935). The effects of certain environmental factors on germination of Florida cigar wrapper tobacco seeds. *Fla. Agr. Exp. Sta., Bull.* **277,** 1.

King, L. J. (1952). Germination and chemical control of the giant foxtail grass. *Contrib. Boyce Thompson Inst.* **16,** 469.

Köves, E., and Varga, M. (1959). Comparative examination of water- and ether-soluble inhibiting substances in dry fruits. *Phyton (Buenos Aires)* **12,** 93.

Koller, D. (1956). Germination-regulating mechanisms in some desert seeds. III. *Calligonum comosum* L'Her. *Ecology* **37,** 430.

Koller, D. (1957). Germination-regulating mechanisms in some desert seeds. IV. *Atriplex dimorphostegia* Kar. et Kir. *Ecology* **38,** 1.

Koller, D. (1962). Pre-conditioning of germination in lettuce at time of fruit ripening. *Amer. J. Bot.* **49,** 841.

Koller, D. (1969). The physiology of dormancy and survival of plants in desert environments. *Symp. Soc. Exp. Biol.* **23,** 449.

Koller, D. (1971). Analysis of the dual action of white light on germination of *Atriplex dimorphostegia* (Chenopodiaceae). *Isr. J. Bot.* **19,** 499.

Koller, D., and Negbi, M. (1959). The regulation of germination in *Oryzopsis miliacea. Ecology* **40,** 20.

Koller, D., and Roth, N. (1963). Germination-regulating mechanisms in some desert seeds. VII. *Panicum turgidum* (Gramineae). *Isr. J. Bot.* **12,** 64.

Koller, D., and Roth, N. (1964). Studies on the ecological and physiological significance of amphicarpy in *Gymmarrhena micrantha* (Compositae). *Amer. J. Bot.* **51,** 26.

Koller, D., Mayer, A. M., Poljakoff-Mayber, A., and Klein, S. (1962). Seed germination. *Annu. Rev. Plant Physiol.* **13,** 437.

Koller, D., Poljakoff-Mayber, A., Diskin, T., and Berg, A. (1963). The regulation of germination in *Citrullus colocynthis* (L.) Schrad. *Amer. J. Bot.* **50,** 597.

Koller, D., Sachs, M., and Negbi, M. (1964). Germination-regulating mechanisms in some desert seeds. VIII. *Artemisia monosperma. Plant Cell Physiol.* **5,** 85.

Lang, A. (1965). Effects of some internal and external conditions on seed germination. *In* "Handbuch der Pflanzenphysiologie" (W. Ruhland, ed.), Vol. 15, Part 2, p. 848. Springer-Verlag, Berlin and New York.

Lauer, E. (1953). Über die Keimtemperatur von Ackerunkräutern und deren Einfluss auf die Zusamensetzung von Unkrautgesellschaften. *Flora (Jena)* **140,** 551.

Lee, I. K., and Monsi, M. (1963). Ecological studies on *Pinus densiflora* forest. 1. Effects of plant substances on floristic composition of the undergrowth. *Bot. Mag.* **76,** 400.

Lehmann, E. (1918). Über die minimale Belichtungskeit welche die Keimung der Samen von *Lythrum salicaria* auslöst. *Ber. Deut. Bot. Ges.* **35,** 157.

Lodhi, M. A. K., and Rice, E. L. (1971). Allelopathic effects of *Celtis laevigata. Bull. Torrey Bot. Club* **98,** 83.

Lona, F. (1947). L'influenza delle condizioni ambientali, durante l'embriogenesi, sulla caratteristiche del seme e della pianta che ne deriva. *Pubbl. 1st. Bot. Univ. Milano, Lavori Bot. (G. Gola Anniv. Vol.)* pp. 277–316.

Lubke, M. A., and Cavers, P. B. (1969). The germination ecology of *Saponaria officinalis* from riverside gravel banks. *Can. J. Bot.* **47,** 529.

McDonough, W. T. (1967). Dormant and non-dormant seed. Similar germination responses when osmotically inhibited. *Nature (London)* **214,** 1147.

Mall, L. P. (1954). Germination of seeds of three common weeds of dry phase of low-lying lands. *Proc. Nat. Acad. Sci., India, Sect. B* **24,** 197.

Mancinelli, A. L., and Borthwick, H. A. (1964). Photocontrol of germination and phytochrome reaction in dark germinating seeds of *Lactuca sativa* L. *Ann. Bot. (Rome)* **28,** 9.

Mancinelli, A. L., Borthwick, H. A., and Hendricks, S. B. (1966). Phytochrome action of tomato seed germination. *Bot. Gaz.* **127,** 1.

Manohar, M. S., and Heydecker, W. (1964a). Effects of water potential on germination of pea seeds. *Nature (London)* **202,** 22.

Manohar, M. S., and Heydecker, W. (1964b). Water requirements for seed germination. *Nottingham Univ. Sch. Agr. Rep.* pp. 55–62.

Meneghini, M., Vicente, M., and Noronha, A. B. (1968). Effect of temperature on dark germination of *Rumex obtusifolius* L. seeds. A tentative physico-chemical model. *Arq. Inst. Biol. (Sao Paulo)* **35,** 33.

Midgeley, A. R. (1926). Effect of alternate freezing and thawing on impermeability of alfalfa and dodder seeds. *J. Amer. Soc. Agron.* **18,** 1087.

Milthorpe, F. L. (1950). Changes in the drought resistance of wheat seedlings during germination. *Ann. Bot. (London)* [N.S.] **14,** 79.

Morinaga, T. I. (1926a). Germination of seeds under water. *Amer. J. Bot.* **13,** 126.

Morinaga, T. I. (1926b). Effect of alternating temperatures upon the germination of seeds. *Amer. J. Bot.* **13,** 141.

Morinaga, T. I. (1926c). The favorable effects of reduced oxygen supply upon the germination of certain seeds. *Amer. J. Bot.* **13,** 159.

Morley, F. H. W. (1958). The inheritance and ecological significance of seed dormancy in subterranean clover (*Trifolium subterraneum* L.) *Aust. J. Biol. Sci.* **11,** 261.

Müller, P. (1934). Beitrag zur Keimverbreitungsbiologie der Endozoochoren. *Ber. Schweiz. Bot. Ges.* **43,** 241.

Nagao, M., Esashi, Y., Tanaka, T., Kumagai, T., and Fukumoto, S. (1959). Effects of photoperiod and gibberellin on germination of seeds of *Begonia evansiana* Andr. *Plant Cell Physiol.* **1**, 39.

Negbi, M., and Evenari, M. (1961). The means of survival of some desert summer annuals. *In* "Plant-Water Relationships in Arid and Semi-Arid Conditions," Proc. Madrid Symp., pp. 249–259. UNESCO, Paris.

Negbi, M., Rushkin, E., and Koller, D. (1966). Dynamic aspects of water-relations in germination of *Hirschfeldia incana* seeds. *Plant Cell Physiol.* **7**, 363.

Ogawara, K., and Ono, K. (1956). The germination-inducing action of heat and the removal of heat-injury by the illumination in the light-favored tobacco seed. *Okayama Univ. Sch. Educ. Bull.* **2**, 50.

Ohga, I. (1926). The germination of century-old and recently harvested Indian lotus fruits, with special reference to the effects of oxygen supply. *Contrib. Boyce Thompson Inst.* **1**, 289.

Pelton, J. (1956). A study of seed dormancy in eighteen species of high altitude Colorado plants. *Butler Univ. Bot. Stud.* **13**, 74.

Pfeiffer, N. E. (1934). Morphology of the seed of *Symphoriocarpos racemosus* and the relation of fungal invasion of the coat to germination capacity. *Contrib. Boyce Thompson Inst.* **6**, 103.

Pittendrigh, C. S., and Bruce, V. M. (1959). Daily rhythms as coupled oscillator systems and their relation to thermoperiodism and photoperiodism. *In* "Photoperiodism and Related Phenomena in Plants and Animals," Publ. No. 55, pp. 475–505. Amer. Ass. Advance. Sci., Washington, D.C.

Quinlivan, B. J. (1961). The effect of constant and fluctuating temperatures on the permeability of the hard seeds of some legume species. *Aust. J. Agr. Res.* **12**, 1009.

Quinlivan, B. J. (1965). The influence of the growing season and the following dry season on the hardseededness of subterranean clover in different environments. *Aust. J. Agr. Res.* **16**, 277.

Quinlivan, B. J. (1966). The relationship between temperature fluctuations and the softening of hard seeds of some legume species. *Aust. J. Agr. Res.* **17**, 625.

Quinlivan, B. J. (1968). The softening of hard seeds of sand-plain lupine (*Lupinus varius* L.). *Aust. J. Agr. Res.* **19**, 507.

Quinlivan, B. J., and Millington, A. J. (1962). The effect of a mediterranean summer environment on the permeability of hard seeds of subterranean clover. *Aust. J. Agr. Res.* **13**, 378.

Roberts, E. H. (1965). Dormancy in rice seeds. IV. Varietal responses to storage and germination temperatures. *J. Exp. Bot.* **16**, 341.

Rollin, P. (1958). Action de la lumière sur la germination des graines de *Phacelia tanacetifolia*. *Rev. Gen. Bot.* **66**, 440.

Rollin, P. (1959). Mise en évidence de deux dormances chez les akènes de *Bidens radiatus*. *Rev. Gen. Bot.* **66**, 636.

Rollin, P. (1968). La photosensibilité des graines. *Bull. Soc. Fr. Physiol. Veg.* **14**, 47.

Rollin, P., and Maignan, G. (1966). La nécessité du phytochrome $P_{rl}(=P_{730})$ pour la germinades akènes de *Lactuca sativa* L. Variété "Reine de Mai." *C. R. Acad. Sci.* **263**, 756.

Roth-Bejerano, N., Koller, D., and Negbi, M. (1966). Mediation of phytochrome in the inductive action of low temperature on dark germination of lettuce seed at supra-optimal · temperature. *Plant Physiol.* **41**, 962.

Roth-Bejerano, N., Koller, D., and Negbi, M. (1971). Photocontrol of germination in *Hyoscyamus desertorum*, a kinetic analysis. *Israel J. Botany* **20**, 28.

Ruge, U., and Liedtke, D. (1951). Zur periodischen Keimbereitschaft einiger Malven-Arten. *Ber. Deut. Bot. Ges.* **64**, 141.

Sachs, J. (1860). Physiologische Untersuchungen über der Abhängigkeit der Keimung von der Temperatur. *Jahrb. Wiss. Bot.* **2,** 338.

Sauer, J., and Struik, G. (1964). A possible ecological relation between soil disturbance, light-flash, and seed germination. *Ecology* **45,** 884.

Scheibe, J., and Lang, A. (1965). Lettuce seed germination: Evidence for a reversible light-induced increase in growth potential and for phytochrome mediation of the low temperature effect. *Plant Physiol.* **40,** 485.

Schroeder, E. M., and Barton, L. V. (1939). Germination and growth of some rock garden plants. *Contrib. Boyce Thompson Inst.* **10,** 235.

Sedgley, R. H. (1963). The importance of liquid-seed contact during the germination of *Medicago tribuloides* Desr. *Aust. J. Agr. Res.* **14,** 646.

Shur, I. (1965). "Physiology of Germination of *Citrullus colocynthis* Seeds." M.Sc. Thesis, Hebrew University, Jerusalem (in Hebrew).

Sifton, H. B. (1927). On the germination of the seed of *Spinacia oleracea* at low temperatures. *Ann. Bot. (London)* **41,** 557.

Simpson, G. M. (1966). A study of germination in the seed of wild rice (*Zizania aquatica*). *Can. J. Bot.* **44,** 1.

Soriano, A. (1953). Estudios sobre germinacion. I. *Rev. Invest. Agr. (Buenos Aires)* **7,** 315.

Stearns, F., and Olson, J. (1958). Interactions of photoperiod and temperature affecting seed germination in *Tsuga canadensis. Amer. J. Bot.* **45,** 53.

Steinbauer, G. P. (1937). Dormancy and germination of *Fraxinus* seeds. *Plant Physiol.* **12,** 813.

Steinbauer, G. P., and Grigsby, B. (1960). Dormancy and germination of the docks (*Rumex* spp.). *Proc. Ass. Off. Seed Anal.* **50,** 112.

Stokes, P. (1953). The stimulation of growth by low temperature in embryos of *Heracleum sphondylium* L. *J. Exp. Bot.* **4,** 222.

Stokes, P. (1965). Temperature and seed dormancy. *In* "Handbuch der Pflanzenphysiologie" (W. Ruhland, ed.), Vol. 15, Part 2, p. 746. Springer-Verlag, Berlin and New York.

Stone, E. C., and Juhren, G. (1951). The effect of fire on the germination of *Rhus ovata* Wats. *Amer. J. Bot.* **38,** 368.

Takahashi, N. (1960). The pattern of water absorption in rice seed during germination. I. The relationship between the pattern and the germination velocity. *Proc. Crop Sci. Soc. Jap.* **29,** 1 (in Japanese, English summary).

Takahashi, N., and Oka, H. (1957). A preliminary note on moist storage of dormant wild rice seeds. *Annu. Rep. Nat. Inst. Genet. (Jap.)* **8,** 42.

Tang, W. T., and Chiang, S. M. (1955). Studies on the dormancy of rice seed. *Nat. Taiwan Univ. Coll. Agr. Mem.* **4,** 1 (in Chinese, English summary).

Taylor, G. B., and Rossiter, R. C. (1967). Germination response to leaching in dormant seed of *Trifolium subterraneum* L. *Nature (London)* **216,** 389.

Taylorson, R. B., and Borthwick, H. A. (1969). Light filtration by foliar canopies: Significance for light controlled weed seed germination. *Weed Sci.* **17,** 48.

Thompson, P. A. (1969). Germination of *Lycopus europaeous* L. in response to fluctuating temperatures and light. *J. Exp. Bot.* **20,** 1.

Thompson, R. C. (1938). Dormancy in lettuce seed and some factors influencing its germination. *U.S. Dep. Agr., Tech. Bull.* **655,** 1.

Thornton, N. C. (1935). Factors influencing germination and development of dormancy in cocklebur seeds. *Contrib. Boyce Thompson Inst.* **7,** 477.

Thornton, N. C. (1936). Carbon dioxide. IX. Germination of lettuce seeds in high temperatures in both light and darkness. *Contrib. Boyce Thompson Inst.* **8,** 25.

Toole, E. H., Toole, V. K., Borthwick, H. A., and Hendricks, S. B. (1955). Interaction of temperature and light in germination of seeds. *Plant Physiol.* **30**, 473.

Toole, E. H., Hendricks, S. B., Borthwick, H. A., and Toole, V. K. (1956). Physiology of seed germination. *Annu. Rev. Plant Physiol.* **7**, 299.

Ungar, I. A. (1962). Influence of salinity on seed germination in succulent halophytes. *Ecology* **43**, 763.

Vaartaja, O. (1956). Photoperiodic responses of seed of certain trees. *Can. J. Bot.* **34**, 377.

Van Wijk, W. R., and De Vries, D. A. (1963). Periodic temperature variations in a homogeneous soil. *In* "Physics of Plant Environment" (W. R. Van Wijk, ed.), p. 102. Wiley, New York.

Varga, M. (1957a). Examination of growth-inhibiting substances separated by paper chromatography in fleshy fruits. III. Changes in concentration of growth-inhibiting substances as a function of the ripening. *Acta Biol. (Szeged)* **3**, 225.

Varga, M. (1957b). Examination of growth-inhibiting substances separated by paper chromatography in fleshy fruits. IV. Paper chromatographic analysis of lemon juice containing germinated seeds. *Acta Biol. (Szeged)* **3**, 233.

Villiers, T. A., and Wareing, P. F. (1965). The possible role of low temperature in breaking the dormancy of seeds of *Fraxinus excelsior. J. Exp. Bot.* **16**, 519.

Von Abrams, G. J., and Hand, M. E. (1956). Seed dormancy in *Rosa* as a function of climate. *Amer. J. Bot.* **43**, 7.

Wagner, E., and Mohr, H. (1966). Kinetic studies to interpret 'high energy phenomena' of photomorphogenesis on the basis of phytochrome. *Photochem. Photobiol.* **5**, 397.

Waisel, Y. (1958). Germination behaviour of some halophytes. *Bull. Res. Counc. Isr., Sect. D* **6**, 187.

Wareing, P. F., and Foda, H. A. (1957). Growth inhibitors and dormancy in *Xanthium* seeds. *Physiol. Plant.* **10**, 266.

Warington, K. (1936). The effect of constant and fluctuating temperatures on the germination of weed seeds in arable soils. *J. Ecol.* **24**, 185.

Wellington, P. S. (1956). Studies on the germination of cereals. 2. Factors determining the germination of wheat grains during maturation. *Ann. Bot. (London)* [N.S.] **20**, 481.

Wellington, P. S., and Durham, V. M. (1961). Studies on the germination of cereals. 3. The effects of the covering layers on the uptake of water by the embryo of the wheat grain. *Ann. Bot. (London)* [N.S.] **25**, 185.

Wells, P. V. (1959). Ecological significance of red-light sensitivity in germination of tobacco seed. *Science* **129**, 41.

Went, F. W. (1948). Ecology of desert plants. I. Observations on the germination in the Joshua Tree National Monument, California. *Ecology* **29**, 242.

Went, F. W. (1949). Ecology of desert plants. II. The effect of rain and temperature on germination and growth. *Ecology* **30**, 1.

Went, F. W. (1957). "Experimental Control of Plant Growth." Chronica Botanica, Waltham, Massachusetts.

Went, F. W., and Westergaard, M. (1949). Ecology of desert plants. III. Development of plants in the Death Valley National Monument, California. *Ecology* **30**, 26.

Went, F. W., Juhren, G., and Juhren, M. C. (1952). Fire and biotic factors affecting germination. *Ecology* **33**, 351.

Wesson, G., and Wareing, P. F. (1969a). The role of light in the germination of naturally occurring populations of buried weed seeds. *J. Exp. Bot.* **20**, 402.

Wesson, G., and Wareing, P. F. (1969b). The induction of light sensitivity in weed seeds by burial. *J. Exp. Bot.* **20**, 414.

Witztum, A., Gutterman, Y., and Evenari, M. (1969). Integumentary mucilage as an oxygen barrier during germination of *Blepharis persica* (Burm.) Kuntze. *Bot. Gaz.* **130**, 238.

# 2

## METABOLISM OF GERMINATING SEEDS

### Te May Ching

## I.   Introduction

The metabolism of germinating seeds is amphibolic; that is, it is both catabolic in the sense of degrading reserve compounds to provide energy and raw materials for the early growth of the seedling, and anabolic in the sense of producing machinery for protein synthesis and biogenesis of various organelles needed for the catabolic activity as well as the true anabolic synthesis of new cells and tissues. Because most seeds are partially differentiated, major catabolic activities are localized in storage tissues, such as the endosperm of seeds of both monocotyledons and dicotyledons, megagametophytes of gymnospermous seeds, and cotyledons of dicotyledonous seeds. True anabolism usually takes place in the embryo or embryo axis. This unique division of labor reduces metabolic complexity within a single organ or organelle to such a degree that unidirectional degradative pathways and synthesis of specific enzymes and organelles can be studied with relative ease and certainty.

Much information on seed germination has been accumulated in recent years. This chapter summarizes the relevant metabolic studies dealing with macromolecules, soluble phosphorus compounds, lipids, carbohydrates, proteins, and amino acids, not as ends in themselves, but as beginnings of things to come. The current concepts of developmental biology, gene action, and metabolic regulation will be introduced as a background for a better understanding of the experimental approaches and conclusions reached by different investigators in studies discussed in the later part of the chapter. The major metabolic events are not the same in seeds of different species because of their morphological structure, developmental determination, chemical nature of their reserves, physiological maturity, and germination conditions. Since a comprehensive coverage of all species is difficult, this chapter deals primarily with the metabolism of mature, common, fatty, starchy, and proteinaceous seeds germinating under normal conditions. The physiology and metabolism of seeds with specific dormancy (Chapter 3) and anomalous storage history (Chapter 4) and that of seeds which are germinated in specific environments (Chapter 1) or under abnormal conditions (Chapter 5) are discussed separately in this volume.

## II. Present State of Knowledge

### A. Concept of Development in Relation to Seed Germination

Seed germination is one phase of the developmental process from fertilized egg cell to mature plant. In this phase a partially differentiated embryo resumes its course of development after a period of quiescence. The whole developmental course of an organism is genetically programmed and environmentally modulated. Each developmental phase exhibits a characteristic pattern of metabolism evidenced by enzyme activities that differ in kind, rate, and location. Our current concept is that the kind of enzymes produced by gene action results from external or endogenous signal(s) that turns on particular portions of the deoxyribonucleic acids (DNA). Once the portions of DNA are turned on, complementary messenger ribonucleic acids (mRNA) are copied. The specific codons (base triplets) in the mRNA determine the primary structure or the sequence of amino acids in enzyme molecules. The transcribed mRNA at one stage may be translated directly or in succeeding stages to proteins which may become active enzymes or require further activation. The rates of enzyme activities are controlled by the quantity of the active enzyme, substrate(s), cofactor(s), coenzyme(s), the presence of inhibitor(s) or stimulator(s), the physical and chemical microenvironment including temperature, light, and gaseous phase, if involved, pH, hydration, ionic strength, etc., and lastly the outlet for the reaction products. Although the factors determining the location of a specific enzyme are not yet well understood, some generalizations can be made. At the organelle level, the assemblage of functional enzymes is partially directed by organellar DNA, such as in chloroplasts and mitochondria (Criddle, 1969; Nass, 1969). At the cell level, hormones, nutrients, and some specific environmental (including microchemical) stimuli are required to change the phase of differentiation, and cell division usually occurs prior to the phase change (Bopp, 1968; Halperin, 1969; Ebert, 1968). Perhaps this phase change is an overriding mechanism whereby a large block of genes is switched on and other groups are turned off. Each daughter cell may have different or similar patterns of circuitry which in turn determine the location of various enzymes to be produced. Organs are then derived from differentiated cells characteristic of a species, and the mature plant consists of an orderly collection of organs such as root, shoot, stem, and leaves.

The successive steps of information transfer, with each step depending on the preceding one, should be emphasized in overall plant development. Developmental processes are determined not only according to the genetic information, but are regulated also by the metabolic information which is collected in successive steps.

Even though both seeds and eggs of animals are "autotrophic" and protected by seed coats, jelly layers, or shells, most seeds are partially differentiated to multicellular storage organs and embryo or embryo axis, whereas eggs usually are unicellular with ultrastructural differentiation (in sea urchin eggs, Harris, 1969). The closest counterparts in the animal kingdom to seeds are the eggs of brine shrimp *(Artemia salina)* which contain multicellular embryos at the gastrula stage. Their embryo development is often interrupted at this point by a period of severe desiccation and the encysted gastrulae become dormant but retain viability for a long time. Upon hydration under suitable conditions of oxygen tension and temperature, the dormant eggs resume active metabolism and develop to form nauplius larva without further cell division (Clegg and Golub, 1969).

In seeds, each embryonic part has a determined metabolic pattern of its own as well as a blueprint of further development, but the mature storage organ usually has reached a committed phase of development. Therefore, the cells of endosperm, megagametophyte, or cotyledons usually are unable to divide and change phase. Some degree of autonomy, however, still operates in these storage cells. The most limited one is the endosperm of Gramineae where only the cells in the aleurone layer are viable and can be induced to synthesize enzymes for degradation of reserves. The majority of center cells of the endosperm have dissociated nuclei and are devoid of mitochondria (MacLeod, 1969) and thus cannot provide energy requirements for synthesis. The triploid endosperm of dicotyledonous seed, e.g., castor bean *(Ricinus communis)* and the haploid megagametophyte of gymnosperms possess semiautonomy in their ability to produce new machinery for protein synthesis (i.e., ribosome, mRNA, and tRNA) and additional mitochondria and glyoxysomes for energy supply and for manufacturing translocatable substrates (Marré, 1967; Gerhardt and Beevers, 1970; Ching, 1970b). The cells in cotyledons of dicots are more versatile in perceiving external light stimuli and differentiating proplastids into chloroplasts, in addition to synthesizing proteins and developing mitochondria, etc.

The metabolism of seed germination, therefore, will vary with different species of plants, owing to their genetically determined diversity in seed morphology, chemical reserves, and special requirements of signals for changing phase or of stimuli for derepressing genes.

### B. Gene Action and Metabolic Modulation

1. GENE ACTION: TRANSCRIPTION

The cells of an organism generally contain identical genomes as precisely sequenced bases grouped in linear fashion in DNA molecules. Genes are functional units of heredity, each of which is composed of a region of the DNA strand. Only when the genes are turned on or derepressed, can they be transcribed. Contiguous genes, particularly those coding for enzymes of one metabolic pathway, often are transcribed in one block, and noncontiguous genes may be coordinated by a series of mechanisms (Britton and Davidson, 1969). Even though the precise mechanism is obscure at the present, so-called "turning on" is believed to involve the uncovering of genes by removing histones, modifying histones, or deactivating repressors. Studies of such repressors suggest that they are composed of a small nuclear RNA (about forty nucleotides in length), an acidic chromosomal protein and a histone (basic nucleoprotein) (Bekhor *et al.*, 1969). Dormant DNA is enveloped by nucleoproteins or repressors (Fig. 2.1) while active DNA has 0.3–23% of its bases being transcribed at one time (Bonner, 1965; D. Chen *et al.*, 1968a). The DNA in embryonic pea *(Pisum sativum)* cotyledons might have template

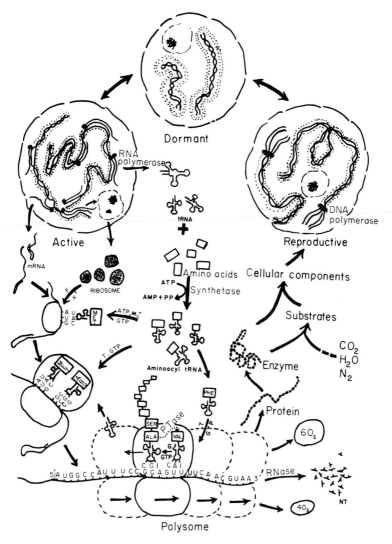

FIG. 2.1. Diagrammatic illustration of gene action and protein synthesis in higher plants. (GTP, guanosine triphosphate; AMP, adenosine monophosphate; PP, pyrophosphate; T, transfer factor; G, translocation factor; PTase, peptide transferase; NT, nucleotides.)

activity as high as 32% (Fambrough *et al.,* 1968). The process of transcription is conditioned by the presence and the activity of RNA polymerase and the proper quantity and proportion of the four major and some
special ribonucleotides needed for synthesis of three kinds of ribonucleic acids. The most frequently copied kind is the mRNA to yield a
group of medium sized polynucleotides (hundreds of bases in length).
Polycistronic mRNA may be the unit of transcription for cell economy
and efficient cellular control, but no evidence for this has yet been
found in plant materials (Loening, 1968). The longest RNA strands are
the ribosomal ribonucleic acids (rRNA). Each cytoplasmic ribosome consists of one molecule each of 25 S RNA (S, sedimentation coefficient in
Svedberg Units), molecular weight 1.3 million, 18 S RNA, molecular
weight 0.7 million, and 5 S RNA, and 15 to 20 electrophoretically distinct protein components in two subunits (Loening, 1968). The smallest
RNA strands are the transfer ribonucleic acids (tRNA) usually made by
70–80 nucleotides arranged in a clover-leaf configuration for maximum
base pairing and stability. Transfer RNA's from all species have an
average molecular weight of 25,000 with a high content of guanosine and
cytosine and some unusual and methylated bases (Loening, 1968). As
each RNA is a complementary replica of the cistronic DNA, it contains a
precise nucleotide sequence and a particular base ratio. The sequence in
mRNA carries the message in base triplets (codons) to code for the primary structure of a protein (enzyme), while a particular sequence in
tRNA characterizes its specificity in forming a particular aminoacyl-
tRNA complex for protein synthesis as well as its ability to recognize
a messenger codon. The significance of the rRNA sequence is unknown
at present.

The relative amounts of these RNA's produced at one time may be
shown by the percent of DNA hybridized in germinating wheat embryos.
Forty-eight-hour germinated wheat embryos produced 2.30% of their
DNA base length as mRNA's, 0.56% as rRNA's, and 0.48% as tRNA's
(D. Chen *et al.,* 1968a). A low synthetic rate of rRNA and tRNA for
turnover-type maintenance probably is common in nondividing cells, but
a high rate should be observed in dividing cells. However, a rapid rate of
mRNA synthesis must be maintained in order to keep pace with active
developmental processes.

2. PROTEIN SYNTHESIS: TRANSLATION

The major function of the three kinds of RNA is for protein synthesis
as diagrammatically illustrated in Fig. 2.1. The mechanics of protein
synthesis involve first the activation of each amino acid composing the
polypeptide chain by its respective aminoacyl-tRNA synthetase (EC

6.1.1.) in the presence of adenosine triphosphate (ATP) to form amino-acyl-adenylate and pyrophosphate. With the aid of its specific tRNA, the aminoacyl-adenylate is changed to aminoacyl-tRNA which is then bound on a ribosome that has been attached to mRNA. The binding reaction requires guanosine triphosphate (GTP), $Mg^{2+}$ and a soluble protein designated as transfer factor T. The next step is the translocation of aminoacyl-tRNA from the ribosomal acceptor site to a donor site. GTP and a ribosomal protein factor G are needed for the translocation. After another aminoacyl-tRNA is bound to the acceptor site, ribosomal peptidyl transferase transfers the donor amino acid on to the acceptor amino acid. Simultaneously the freed tRNA is ejected from the ribosome and the acceptor peptidyl-tRNA is translocated to the donor site. The cycle of elongation starts again with a new aminoacyl-tRNA, and a peptide is formed by repeating the cycle. The sequence of amino acids in a peptide is coded in mRNA, and decoding is accomplished by matching anticodon in tRNA with codon in mRNA. Each ribosome moves along the mRNA and forms one copy of a specified polypeptide. Several ribosomes may be attached concurrently to a messenger RNA to form a polysome (Lipmann, 1969).

After sufficient number of polypeptide copies are translated from one mRNA, the mRNA is hydrolyzed by ribonuclease to nucleotides and the message is thus eliminated. The half-life of mRNA's varies from minutes in bacteria, hours in active plant cells (Cherry, 1967), to long-lived in dormant organisms which have mRNA that may last as long as the life span of a seed. The longevity of some legume seeds is reported to be several hundred years (Barton, 1961).

In bacteria, the initiation of protein synthesis involves the binding of the 30 S ribosomal subunit and an initiator tRNA (formylmethionine-tRNA) to a specific initiator codon, usually AUG or GUG, on the 5' terminus of a mRNA, and the subsequent adding of a 50 S subunit with the aid of GTP and soluble protein factors. The termination of protein synthesis in bacteria is signaled by codons, usually UAA, UAG, or UGA, on the 3' terminus of a mRNA (Scolnick *et al.,* 1968). The only non-bacterial initiation system thus far reported is that of the wheat embryo (Jerez *et al.,* 1969; Marcus, 1970a,b). In common with the bacterial system, three initiation factors are found. However, in contrast to the latter system, ATP is required in addition to GTP and met-tRNA (nonformy-lated) is the initiation component (Marcus, 1970b). In reticulocytes of rabbits, evidence presented suggests the sequential binding of subunits onto mRNA in a similar manner as the bacterial system (Hoerz and McCarty, 1969). A new nuclease inhibitor, diethyl pyrocarbonate, may be useful in isolating enough undegraded endogenus mRNA for these types of studies (Fedorcsak *et al.,* 1969; Weeks and Marcus, 1969). At present heterologous tobacco mosaic virus RNA, polyuridylic acid (poly U), bacterial RNA, and yeast RNA are used for the template.

As regards the aminoacyl transfer reaction, two factors have been isolated from rice embryos (App, 1969). Both were required for *in vitro* synthesis of polyphenylalanine with poly U as the template, and they were present in crude supernatant and on crude ribosomes. Factor I could be removed by deoxycholate (DOC, 0.6%) from crude ribosomes, while Factor II could be extracted from DOC-washed ribosomes by 0.5 *M* KCl.

In the postribosomal supernatant of wheat germ homogenate, two soluble aminoacyl transfer factors were isolated. One of the factors catalyzes a GTP and polyuridylic acid-dependent binding of phenylalanyl-tRNA to ribosomes while the second factor facilitates the formation of peptidyl puromycin from nonenzymically bound phenylalanyl-tRNA (Legocki and Marcus, 1971).

## 3. FATE OF SYNTHESIZED RNA AND PROTEINS: ASSEMBLY

The synthesized rRNA of eukaryotic cells associates with proteins in the nucleolus and forms ribosomes which are then discharged into the cytoplasm through nuclear pores. The mRNA may be long lived and packaged with proteins as the informosomes in the cytoplasm of developing sea urchin eggs (Spirin, 1966). While different maturation processes of newly synthesized RNA's in animal cells have been reported (Burdon, 1971), only limited investigation was conducted on plant materials (Chakravorty, 1969b). The maturation processes, however, were similar to those of animal cells.

The newly synthesized protein spontaneously assumes a three-dimensional configuration that is caused by interaction among intramolecular amino acid sequences (Lipmann, 1969). Therefore, the basic unit configuration (monomer) is genetically specified. Actual catalytic enzymes, however, may be polymers of several identical or different monomers connected or modified by divalent cations, metabolites, or coenzymes. Good examples are common in biological materials: glutamine synthetase (EC 6.3.1.2) in bacteria has one to thirteen identical subunits connected by adenylic acid (AMP) (Atkinson, 1969), and tryptophan synthetase (EC 4.2.1.20) has two subunits with different amino acid sequences bound by pyridoxal phosphate (Goldberg *et al.,* 1966). While genetic regulation of multiple molecular forms of enzymes is well established (Scandalios, 1969), the state of metabolism also exerts its influence (Atkinson, 1969). One can imagine the dynamic aspects of a living cell when one learns that AMP increases from 30 nmoles/gm in imbibed lettuce seeds to 240 nmoles after 3 minutes of anaerobiosis (Pradet *et al.,* 1968).

Even though the number of enzymic proteins in cells may be as large as two to three magnitudes over the nonenzymic proteins, gene-specified proteins may also (*1*) be conserved and only be activated by a signal as lipase (EC 3.1.1.3), $\beta$-amylase (EC 2.3.1.2), and protease (EC 3.4.4) in seeds, (*2*) be stored as reserves like protein bodies in seeds, (*3*) form structural components of organelles as the structural protein of mitochondria, (*4*) assume a supporting role like acyl-carrier protein for fatty acid synthesis, (*5*) act as an inhibitor for a specific reaction as the trypsin inhibitor in soybeans, and (*6*) degrade existing enzymes and functioning structural molecules as lysomal enzymes and nucleases. What controls and coordinates the fate of each protein formed is still an enigma. However, the hundreds of proteins transcribed, activated, or modified in each developmental phase are precisely coordinated in time and space so that the developmental pattern of a species is flawlessly unfolded in a suitable environment. Under abnormal conditions, the developmental pattern deviates from the normal course by changing not only the kind and quantity of proteins and enzymes synthesized, but also by modifying their rate or nature of activity and degradation.

## 4. METABOLIC MODULATION: ACTIVITY

Metabolic modulation is the final and most obvious regulatory step in controlling the germination process. Seeds are autotrophic or self-sufficient with respect to their food and energy supply, but the major food reserve needs to be hydrolyzed and transformed in order to be transported and utilized for growth. The energy supply requires mitochondria, adenosine diphosphate (ADP), inorganic phosphate, and substrates. Readily utilizable substrates, preexisting enzymes, and organelles are present in seeds. However, they are not complete in kind, nor are they sufficient in quantity to carry out the developmental pattern programmed by the genome. Furthermore, intermittent destruction and resynthesis (turnover) of most existing enzymes and some organelles are essential in maintaining optimum activity. All these sequential developmental events rely on metabolic coordination; and the elimination or addition of any one substrate, metabolite, or enzyme, the changes in oxygen and water supply, the presence of inhibitors or promoters, or the variation of temperature will alter the course of the normal germination process.

The known controlling mechanisms of enzyme activities or the reported metabolically modulated enzyme activities in the germination process are summarized briefly here and will be discussed in more detail in later sections.

a. END-PRODUCT INHIBITION. Glucose, sucrose, or fructose inhibits the development of isocitrate lyase (EC 4.1.3.1) in fatty seeds (Cocucci and Caldogno, 1967) and glucose 6-phosphate checks the activity of purified isocitrate lyase (Nagamachi *et al.,* 1967). Inorganic phosphate inhibits phytase (EC 3.1.3.8) in wheat scutellum (Bianchetti and Sartirani, 1967).

b. SUBSTRATE INDUCTION. The activity of hexokinase (EC 2.7.1.1) and fructokinase (EC 2.7.1.4) is doubled by incubating cotyledons of castor bean with either glucose or fructose (Marré *et al.,* 1965a). A proportional increase of nitrate reductase (EC 1.6.6.1) activity is induced by nitrate concentration in radish cotyledons (Beevers *et al.,* 1965) and barley aleurone layers (Ferrari and Varner, 1969).

c. HORMONAL STIMULATION. Gibberellin (GA) stimulates the synthesis of $\alpha$-amylase (EC 3.2.1.1), protease, and ribonuclease (EC 2.7.7.16) in barley (Filner *et al.,* 1969; Galston and Davies, 1969), and the secretion of ATPase (EC 3.6.1.3), GTPase, phytase, phosphomonoesterase (EC 3.1.3), phosphodiesterase (EC 3.1.4.1), carbohydrate hydrolyzing enzymes other than amylase and peroxidase (EC 1.11.1.7) from the aleurone layer of barley (Pollard, 1969). GA also increases lipase in cottonseeds (Black and Altschul, 1965), and isocitrate lyase in hazel cotyledons (Pinfield, 1968). Protease in squash is induced by cytokinin and benzyladenine (Penner and Ashton, 1967). Auxin increases cellulase (EC 3.2.1.4) in young epicotyl of peas (Fan and Maclachlan, 1966). Indoleacetic acid might stimulate citrate synthase (EC 4.1.3.7) from bean seedlings (Sarkissian and Schmalstieg, 1969).

d. INDUCER, EFFECTOR, OR MODIFIER REGULATION. Cysteine activates preexisting $\beta$-amylase in grains (Spradlin *et al.,* 1969). High concentration of ATP inhibits phosphorylase I (EC 2.4.1.1) of corn endosperm (Tsai and Nelson, 1968). Hydroxylamine, glutamine, purine, or pyrimidine nucleotides induce lipase in the endosperm of germinating wheat (Tavener and Laidman, 1969), while hydroxylamine or glutamine plus indoleacetic acid induce lipase in the aleurone tissue of germinating wheat (Eastwood *et al.,* 1969). $ATP^{4-}$ inhibits the activity of nucleotide monophosphokinase (EC 2.7.4.6.) isolated from cotyledons of germinating cucumber seeds, whereas $MgATP^{2-}$ is the substrate of the enzyme (May and Symons, 1968).

### C. *Hysteresis in Seed Germination*

In addition to the four controlling levels mentioned above, treatment with herbicides, fumigants, insecticides, fertilizer, and irrigation practices following planting as well as seed maturing conditions, artificial drying, seed maturity, seed dormancy, conditions and duration of storage, influence the metabolism of germinating seed in various ways.

Because of the impact on future crop production and nutritional supply of the ever-increasing world population, some examples of hysteresis are worthy of mention. One is the increase of protein content and yield of seeds that could be achieved by spraying plants with simazine or atrizine (herbicides) or nitrogen fertilizer. The improved quantity and quality are transmitted to the offspring (Ries *et al.*, 1967; Schweizer and Ries, 1969). Another example is the induction of rapid germination and growth and increased crop production by repeated hydration and dehydration of the seeds prior to planting (Austin *et al.*, 1969). The possibility of the uptake of heterologous DNA's by barley seedlings and further incorporation of some of the exogenous DNA into endogenous DNA (Ledoux and Huart, 1969) indicate that someday desirable genes (DNA sections) might be incorporated into adapted crops for high yield, disease resistance, etc. Some adverse effect may also result from modern agricultural practice, such as the case of overdrying of seeds for the purpose of safe storage which, unfortunately, changes normal growth to an attenuated one (Klingmüller and Lane, 1960; Nutile, 1964).

### III.   Basal Metabolism

There are basal metabolic activities common to all seeds during germination and some specific activities are limited to certain seeds because of their structure or the nature of their food reserve. The former will be discussed in this section and the latter in the following sections.

### A.   *Awakening Step*

The importance of the awakening step from a dry dormant seed to an active metabolic state has gained recognition in recent years (Woodstock and Pollock, 1965). This step usually lasts from minutes to several hours at an optimum temperature in water and in the presence of oxygen. It is often called the imbibition period during which three major events are occurring. First is the rapid water uptake by biocolloids in dry seeds following the sequence of first, second, and then multiple layers of water molecules enveloping structural as well as soluble cellular constituents (Hlynka and Robinson, 1954; Jirgensons, 1962). Second is the reactivation of preexisting macromolecules and organelles, both of which are formed during seed maturation, but are subsequently stored or inactivated by the dehydration process of mature seeds (Marré, 1967). Third is the respiration resulting in adenosine triphosphate (ATP) formation which provides energy for synthesis of substrates, coenzymes, and proteins (enzymes), all of which are necessary to initiate anabolic activities. The anabolic activities consist of the synthesis of various substrates, DNA, rRNA, mRNA, tRNA, more enzyme proteins, structural proteins and lipids for chromosomes, nucleoli, ribosomes, mitochondria, chloroplasts, dictyosomes, microsomes, peroxisomes, membranes, and cell wall materials for extension of existing cells and formation of new cells in the embryo axis. In storage tissue, similar activities are occurring except for nuclear synthesis. Turnover type incorporation of DNA precursors, however, has often been observed in storage tissue (Marcus and Feeley, 1962; Cherry, 1967; Sparvoli, 1969).

## 1. IMBIBITION

The rate and quantity of water imbibed by seeds vary with the nature of the seed coat, chemical composition of the seed and size, and imbibition temperature when water is not limiting. The penetration of water into large storage tissue is from the periphery to interior of the tissue, so the epidermal and vascular cells of pea and bean *(Phaseolus vulgaris)* cotyledons (Kollöffel, 1969), aleurone cells of cereals, epidermal cells of castor bean endosperm, and coniferous megagametophyte usually show the earliest signs of respiratory dehydrogenases. At the cellular level, the areas next to the cell wall and nucleus and the space between storage organelles become hydrated first. Tissue swelling then follows, and more water uptake occurs until the tissues have a 40–60% water content (67–150% on a dry weight basis). In embryos or excised embryo axes, water uptake is rapid and uniform and often exceeds the water content of its storage organ by up to 10%. Water uptake in seeds often ceases for several hours or days at the end of the imbibition period. It then resumes and is rapid at the time of radicle emergence until the storage tissue and growing seedling have water contents of 70–90% (Goo, 1951; Stanley, 1958; Ching, 1959, 1965; Mayer and Poljakoff-Mayber, 1963; Ingle *et al.*, 1964; Hall and Hodges, 1966; Opik, 1966; Bain and Mercer, 1966; Pradet *et al.*, 1968).

A distinct imbibition stage usually is lacking in whole wheat seed (Fig. 2.2) because it absorbs water gradually into the bulk endosperm that contains mostly insoluble starch granules. Wheat embryo, however, definitely exhibits a rapid imbibition and follows by a lag period of water uptake as shown by Marcus *et al.* (1966) in Fig. 2.3. Proteinaceous and fatty seeds, in contrast, demonstrate a rapid imbibitional water uptake, and the water absorption is temperature dependent (Fig. 2.2).

In pea and rape *(Brassica napus)* seeds, the storage organs are the cotyledons which comprise respectively 95 and 90% of the decoated embryo dry weight. Peas contain about 40% protein and 45% starch, while rape seeds are composed of roughly 40% protein and 40% lipids.

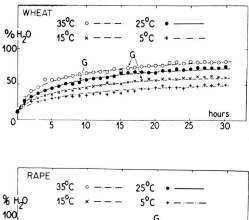

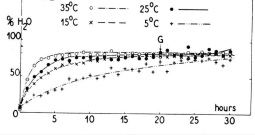

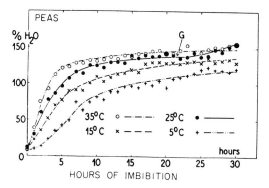

FIG. 2.2. Water uptake by seeds of wheat, rape, and peas at different temperatures (Allerup, 1958). G, germination.

Approximately 80% of the seed proteins are in storage organelles called protein bodies. The storage proteins of dry seeds usually are in crystalline form. The remaining 20% of the seed proteins are distributed in nuclei, mitochondria, proplastids, microsomes, and in cytosol (Larson and Beevers, 1965; Penner, 1965; Beevers and Guernsey, 1966; Newcomb, 1967; Ching, 1968; Larson, 1965). Seed starches usually are stored in starch granules in amyloplasts or proplastids. Lipids are found in lipid bodies (fat bodies or spherosomes). All storage organelles have membranes which usually contain 60% protein and 40% polar lipids. Storage organelles also are present in embryos or embryo axes, but these are smaller and fewer than in storage cells (Setterfield *et al.,* 1959; Ching, 1965; Yatsu, 1965; Horner and Arnott, 1966; Klein and Ben-Shaul, 1966).

During imbibition, water is probably first absorbed mainly by metabolically functional organelles, the cytosol, and outer membranes of storage organelles, considering that starch granules, reserve proteins, and fats are all water insoluble. The quantity of water taken up into localized sites in cells could provide a medium suitable for metabolic activity to proceed, even though the total water content is low for the whole seed.

## 2. REACTIVATION OF PREEXISTING MACROMOLECULES AND ORGANELLES

Contrary to general thinking, simple hydration of preexisting dry organelles (mitochondria, ribosomes, nuclei, membranes, glyoxysomes, etc.) and dry macromolecules (enzymes, tRNA, mRNA, etc.) may not reactivate their functions readily, even though under normal conditions water appears to be the only requirement for germination. The following discusses the recent significant findings related to this aspect.

a. RIBOSOMES. The functional status of ribosomes in seeds is probably related to ripening time, metabolic conditions, and the degree of drying. Marré (1967) noted that during the normal course of development in castor beans, the functional integrity of monosomes was maintained throughout the entire period of maturation until the last phase of rapid drying. At that time structural changes occurred leading to monosome inactivation. He concluded that dehydration and aging might be the cause of structural changes. Sturani (1968) supported this view by the observation that ribosomes isolated from dry castor bean endosperm were able to bind poly U, but the complex was still very low in protein-synthesizing ability.

Dehydration of wheat (*Triticum* spp.) embryos during the first 24 hours of germination did not result in deleterious effects upon rehydration, whereas drying of embryos which had germinated for periods longer than 48 hours led to irreversible growth arrest. This arrest was mainly caused by DNA breakdown and formation of false messages (D. Chen *et al.,* 1968b). Untimely dehydration during seed maturation also may cause such irreversible effects. However, mild dehydration by exposing maturing castor beans to water vapor of 98% relative humidity for 48 hours decreased both monosome and polysome content of the endosperm, and the decrease could be reversed by rehydration on wet filter paper for 48 hours (Marré, 1967). It would be of interest to examine these dehydrated immature seeds for their content, composition, and function of ribosomes in subsequent germination, since it is known that water stress could alter base ratio of RNA in corn (West, 1962) and reduce polysome formation in germinating castor beans (Sturani *et al.,* 1968).

Dehydration methods and conditions greatly modify the function of ribosomes. Freeze drying of isolated ribosomes from *Escherichia coli* in 5 m$M$ Mg$^{2+}$ and 1 m$M$ spermidine, a naturally occurring oligoamine [H$_2$N(CH$_2$)$_4$NH(CH$_2$)$_3$NH$_2$], retained the protein-synthesizing ability, while vacuum drying alone required the additional presence of a synthetic hydrophilic gel (Sephadex G-100, G-50, or Biogel) (Tamaoki *et al.,* 1969). Furthermore, quick drying under high vacuum was better than slow drying under low vacuum. Neither bovine serum albumin nor dextran could substitute for the gel. Therefore, it appears that in order to retain the function of cellular organelles during dehydration, an optimum divalent cation concentration must be maintained, a protective agent (Spermidine) must be present, and the water content of the ribosomes and smaller molecules isolated with the ribosomes should be reduced by the gel prior to vacuum drying. At the cellular level, dehydration is obviously a more delicate process if one hopes to preserve viability. A successful example is the freeze-drying method used for preserving coniferous pollen. The process involves air drying from 40–20% down to 12–8% water (fresh weight basis) prior to freeze drying to 3–5% which is necessary to retain the high germinability of the pollen (Ching and Ching, 1964). The air drying step is apparently a means to remove the free water molecules, and the bound ones are necessary in maintaining precise spatial relationship of molecular conformation of the biopolymers. If these "conformational cushion" type of water molecules are sublimed *in situ* by freeze drying, accurate conformation and unimpaired function would be insured upon rehydration. In naturally dried seeds of the temperate zone, only free water is removed and the cushion-type bound water molecules remain *in situ*. Only over-drying will cause delayed germination and reduced seedling growth (Klinmüller and Lane, 1960; Nutile, 1964). In seeds of citrus and tropical plants, even air drying is detrimental (Barton, 1961). Apparently, factors other than the physical status of water molecules are involved also. For example, the concentration of divalent cations and soluble metabolites, the activity of ribonuclease and proteases etc., should be investigated in addition to the study of ribosomes and organelles at different stages of maturation.

Activation of ribosomes for *in vitro* protein synthesis appears to be a temperature-dependent process in wheat embryos as ribosomes from embryos imbibed in water at 4°C for 30 minutes incorporated only 2.5%

of the radioactive amino acid incorporated by those imbibed at 23°C (Marcus et al., 1966). Thus, imbibition at proper temperature and in the presence of oxygen is a necessary step for ribosome activation. Hydration, however, appears not to be the reason for ribosome activation as embryos imbibed at 4°C absorbed about 90% of the water taken up by embryos at 23°C. Yet they were not able to synthesize protein. Marcus et al. (1968) have developed a successful system to activate ribosomes and to form polysomes in homogenates of the embryo by incubating the homogenate at 30°C with ATP. This activating system was not affected by actinomycin D or deoxyribonuclease, but was strongly inhibited by ribonuclease and cycloheximide (Marcus and Feeley, 1966). These findings indicate that intact, preexisting RNA's and protein synthesis are required for ribosome reactivation and polysome formation. Perhaps ATP content in the tissue is the limiting factor for the awakening step of germination.

b. Messenger RNA's. The occurrence of long-lived mRNA's which are formed at seed maturation but later direct protein synthesis at the time of germination has been well documented based on the results of inhibitors (Waters and Dure, 1966; D. Chen et al., 1968a; Ihle and Dure, 1969). The origin and functional status of one of the long-lived messages was explored in maturing cotton (*Gossypium* spp.) seeds (Ihle and Dure, 1970). The message for producing a protease at germination was transcribed as early as the middle stage of seed formation, but the message was not translatable because of the presence of an inhibitor which could be removed by washing and which could also be substituted by abscisic acid. The inhibitor was no longer functional in mature seeds, and the message became very active as the template for synthesis of the protease. Whether prolonged dry conditions of seeds will reduce the template activity of the message is not presently clear, as Ihle and Dure (1970) studied precocious germination only (i.e., directly germinating maturing seeds on an agar medium without a dry storage period).

In normally matured seeds, in addition to the lack of polysomes in dry seeds (Marcus et al., 1966; Barker and Rieber, 1967; Marré, 1967), the functional status of long-lived mRNA at the early imbibition period of germination also is definitely low. This is shown by the fact that little endogenous protein-synthesizing ability was found in ribosomes isolated from dormant seeds, but poly U or mRNA from germinating seeds stimulated amino acid incorporation (Marcus and Feeley, 1964a, 1965; Allende and Bravo, 1966; Barker and Rieber, 1967; Jachymczyk and Cherry, 1968; Sturani, 1968).

Are the long-lived mRNA's in the form of informosomes encapsulated within a protein coat as in sea urchin eggs and do they require the action of protease to release them (Spirin, 1966; Brahmachary, 1968) when seeds are exposed to germinating conditions? A positive answer is based on direct evidences listed in the following paragraphs and two common observations noted in seeds: (*1*) protease is present in many seeds (Irving and Fontaine, 1945; Engel, 1947a; Olfelt *et al.*, 1955; Young and Varner, 1959; Rossi-Fanelli *et al.*, 1965; Penner and Ashton, 1967) and (*2*) low imbibition temperature reduces early growth of the seedlings (Pollock and Toole, 1966). These observations might indicate that low temperature results in a low rate of protease activity and thus slows down the releasing mechanism of long-lived mRNA.

Direct evidence of such preformed mRNA in dry seeds was presented by Marcus and Feeley (1966). Using a combination of simultaneously activating ribosomes and incorporating labeled amino acid as an assay system, they tested the template activity of various fractions that were separated by differential centrifugation. The high template activity was found to be associated with a pellet sedimented at $5900 \times g$ as well as a RNA fraction. The solublized pellet had higher activity than the RNA fraction, and a protein fraction might be involved with the conserved mRNA (Marcus, 1969). Whether the $5900 \times g$ pellet contains informosome remains to be clarified.

Mori *et al.* (1968, 1969) recently isolated one species of heavy soluble RNA (18 S and 9 S) from the postmicrosomal supernatant of cotyledon homogenate of soybean *(Glycine max)* seeds. The most interesting features of this sRNA lies in its high template activity for protein synthesis and in its correlative increase in amount with maturation and corresponding decrease with germination. Based on the possibility of precipitating this sRNA at $200,000 \times g$ from the supernatant, the authors concluded that this heavy sRNA might exist as a nucleoprotein in cells. Whether they are the long-lived cistronic mRNA or the informosomal RNA that have been transcribed during maturation and then translated during germination is uncertain at present. Since the content of this sRNA is very high in mature soybean cotyledons (116 mg/100 gm tissue), it should be possible to discern what enzyme(s) it encodes and to obtain some clue of its function in germination processes. It is also possible that it is the precursor of ribosomal RNA.

A kinetic view of water uptake, polysome quantity, and *in vitro* protein synthesizing ability of isolated ribosomes from wheat embryo is shown in Fig. 2.3 (Marcus *et al.*, 1966). It is clear that ribosome activation follows by polysome formation during imbibition. Since ATP is required for polysome formation (Marcus, 1969), it is possible that the ATP content may be the primary limiting factor for the commencing of germination.

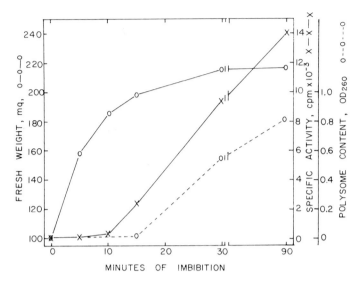

Fig. 2.3. Time course of fresh weight increase, polysome content, and *in vitro* protein-synthesizing ability of ribosomes (× — ×) isolated from 100 mg dry wheat embryo (Marcus *et al.,* 1966).

c. Transfer RNA's. They probably preexist in dry seeds, since the supernatant fraction is sufficient to support protein synthesis in wheat embryo (Marcus and Feeley, 1965; Vold and Sypherd, 1968b; Chen *et al.,* 1968a).

d. Mitochondria. These organelles were observed in dry seeds as early as 1947 (Engel and Bretschneider, 1947). Little information could be found in the literature regarding the drying of these organelles, but aging of these organelles in aqueous suspension even at low temperature is well-known. The preservation of mitochondria in seeds through de-hydration and extended periods of dry storage seems to be a sheer wonder of nature. Nevertheless, instantaneous oxygen uptake could be observed upon wetting seeds (Mayer and Poljakoff-Mayber, 1963). Malate dehydrogenase (EC 1.1.1.37), succinate cytochrome c reductase (EC 1.3.99.1), cytochrome oxidase (EC 1.9.3.1), NADH-cytochrome reductase (EC 1.6.99.3), $\alpha$-ketoglutarate dehydrogenase and most of the TCA cycle enzymes are present in dry seeds (Cherry, 1963; Mayer and Poljakoff-Mayber, 1963; Marré, 1967). Since ATP increased within 15 minutes of imbibition in lettuce seeds (Pradet *et al.,* 1968), enzymes for ATP synthesis in mitochondria must be conserved. A correlation of water uptake, respiration, content of adenosine phosphates, and germination is summarized in Fig. 2.4. Increase of total adenosine phosphates was extremely rapid for the first 30 minutes of imbibition at a rate of 2 nmoles/minute/gm dry seed, the major portion of which was seed coat and reserves. The enzymes responsible for synthesizing the nucleotides must be preexisting or synthesized in the time periods of 30 minutes. This dynamic aspect of energy supply for the awakening step of seed germination had not been explored prior to the investigation of Pradet et al. (1968).

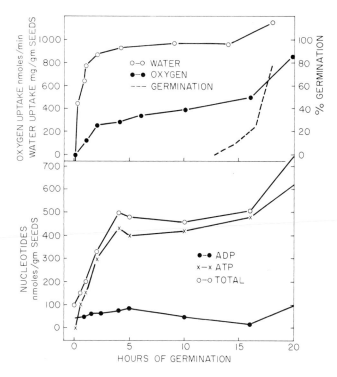

FIG. 2.4. Time course of increases in water uptake, oxygen consumption, seed germination, total adenine nucleotides, adenosine diphosphate, and adenosine triphosphate in lettuce seed (Pradet et al., 1968).

e. ENZYMES. Many enzymes have been reported as preexisting in dry seeds of various species as summarized in Table 2.1. Most of these enzymes appear to be easily reactivated in seeds by hydration. However, some of them, such as the protease in wheat, rye, and barley, require a specific activation agent. The enzyme is located mainly in the aleurone layer and requires activation by reducing agents (Engel and Heins, 1947).

3. THE SEQUENCE OF AWAKENING

The sequence and magnitude of the activity of these preexisting enzymes and organelles become foremost important factors in the

**TABLE 2.I**
Enzymes Found in Dry and Hydrated Seeds

| Enzyme | Seed | Reference |
|---|---|---|
| Aldolase | Carrot, spinach, radish | Takeo (1969) |
| ADPG-Pi transferase | Wheat germ | Dankert et al. (1964) |
| β-Amylase | Wheat, rye, barley | Engel (1947a) |
| | Pea cotyledon | Juliano and Varner (1969) |
| Isoamylase | Pea | Mayer and Shain (1968) |
| Fructose 1,6-diphosphatase | Castor bean endosperm | Scala et al. (1968) Marré (1967) |
| | Wheat embryo | Scala et al. (1968) |
| α-Galactosidase | Watermelon | Ahmed and Cook (1964) |
| | Bean | Mayer and Polizakoff-Mayber (1963) |
| α,β-Glucosyl transferase | Alfalfa | Hutson (1964) |
| β-Galactosidase | Watermelon | Ahmed and Cook (1964) |
| β-glucosidase | Barley | F. B. Anderson et al. (1964) |
| Glucose-6-phosphate dehydrogenase | Castor bean endosperm | Marré (1967) |
| | Peanut cotyledon | Cherry (1963) |
| Invertase | Watermelon | Ahmed and Cook (1964) |
| Maltase | Rice | Nomura et al. (1969) |
| 6-Phosphogluconate dehydrogenase | Castor bean endosperm | Marré (1967) |
| Phosphorylase | Pea cotyledon | Juliano and Varner (1969) |
| UDP-glucuronic acid decarboxylase | Wheat germ | Castanera and Hassid (1965) |
| UDPG transferase | Bean | Leloir et al. (1961) |
| Aminoacylase (aminoacyl-tRNA synthetase, amino acid activating enzymes) | Wheat germ | Moustafa (1963) |
| | | Mounter and Mounter (1962) |
| | | Marcus and Feeley (1965) |
| | Forage seeds | Umebayashi (1968) |
| | Pea | J. W. Anderson and Fowden (1969) |
| Glutamic decarboxylase | Cereal | Linko (1961) |
| | Bean | James (1968) |
| | Corn | Grabe (1964) |
| Homocystine methyl-transferase | Jack bean | Abrahamson and Shapiro (1965) |
| | | Hanabusa (1961) |
| Peptidase | Maize | Scandalios (1969) |
| Protease | Pea | Young et al. (1960) |
| | Squash | Penner and Ashton (1967) |
| | Barley | Engel and Heins (1947) |
| | Peanut | Irving and Fontaine (1945) |
| | Soybean | Ofelt et al. (1955) |
| | Cotton | Rossi-Fanelli et al. (1965) |
| Urease | Jack bean | Abrahamson and Shapiro (1965) |
| | | Hanabusa (1961) |

| Enzyme | Seed | Reference |
|---|---|---|
| Isocitrate lyase | Italian stone pine | Firenzuoli *et al.* (1968) |
| | Peanut cotyledon | Cherry (1963) |
| | Ponderosa pine | Ching (1970) |
| Malate synthase | Italian stone pine | Firenzuoli *et al.* (1968) |
| Lipases | Cottonseed | Ory *et al.* (1968) |
| | Douglas fir | Ching (1968) |
| Lipoxidase | Fatty seeds | Mayer and Poljakoff-Mayber (1963) |
| Phospholipase D | Cotton | Tookey and Balls (1956) |
| Aconitase | Peanut cotyledon | Marcus and Velasco (1960) |
| Citrate synthetase | Peanut cotyledon | Marcus and Velasco (1960) |
| Cytochrome oxidase | Peanut cotyledon | Cherry (1963) |
| DPNH oxidase | Peanut cotyledon | Cherry (1963) |
| | Lettuce | Mayer and Poljakoff-Mayber (1963) |
| Dihydroascorbic acid reductase | Wheat | Carter and Pace (1964) |
| α-Ketoglutarate dehydrogenase | *Vigna* sp. | Mayer and Poljakoff-Mayber (1963) |
| Malate dehydrogenase | *Vigna* sp. | Mayer and Poljakoff-Mayber (1963) |
| NADP-isocitrate dehydrogenase | Peanut cotyledon | Marcus and Velasco (1960) |
| Succinic dehydrogenase | Peanut cotyledon | Cherry (1963) |
| Alcohol dehydrogenase | Wheat | Bhatia and Nilson (1969) |
| Allantoinase | Soybean | K. W. Lee and Roush (1964) |
| Catalase | Maize | Scandalios (1969) |
| | Almond | Mihalyfi (1968) |
| Esterase | Wheat, rye, barley | Engel (1947b) |
| | Corn | Scandalios (1969) |
| | Wheat germ | Moustafa (1963) |
| | | Mounter and Mounter (1962) |
| | Wheat | Bhatia and Nilson (1969) |
| Peroxidase | Maize | Scandalios (1969) |
| | Wheat | Bhatia and Nilson (1969) |
| Phytase | Lettuce | Mayer and Poljakoff-Mayber (1963) |
| Phosphatase | Pea | Young *et al.* (1960) |
| Ribonuclease | Wheat embryo | Vold and Sypherd (1968a) |
| | Peanut cotyledon | Cherry (1963) |
| | Corn | Ingle and Hageman (1965) |
| Adenosine kinase | Wheat embryo | Price and Murry (1969) |
| Inosine phosphorylase | Wheat embryo | Price and Murry (1969) |
| 5-Phosphoribosyl pyrophosphate synthetase | Wheat embryo | Price and Murry (1969) |

awakening stage of seed germination. With tritiated water as the aqueous medium for germination, seeds of mustard *(Sinapis alba)* were used to trace the sequence of metabolic events in the very early stage of germination (Spedding and Wilson, 1968). The first sample was taken at 5 minutes after the beginning of imbibition at 20°C. At that time only γ-aminobutyric and aspartic acids contained radioactivity, with the former having more $^3H$ than the latter. After imbibition for 10 minutes, two more amino acids were radioactive. Activity was in the following order: γ-aminobutyric acid > glutamic acid > aspartic acid > alanine. This order of activity was maintained for 30 minutes after wetting. In addition to the four amino acids, radioactive malic and citric acids were observed after imbibition for 3 hours. Lipid intermediates in this fatty seed were not labeled until later. The authors concluded that amino acid metabolism is important in early stages of germination. They further proposed that seeds store unstable α-oxo acids (α-keto acids) initially required for respiration as the corresponding amino acids. Subsequently, these amino acids are deaminated and through transamination produce various kinds of α-oxo acids needed for the Krebs cycle and respiration which lags just behind the amino acid metabolism.

Providing unstable intermediates, which are probably lacking in the dry seed, for the Krebs cycle to operate and synthesize ATP for anabolic activity is indeed an important step for subsequent physiological processes. Since only catalytic amounts of these α-oxo acids are necessary in operating the Krebs cycle and they have been found in dry seeds (Satoh, 1968), other significant metabolic events might be involved in these seeds, particularly during the first 30 minutes of imbibition. Activity of glutamic decarboxylase, as a well-known enzyme present in dry seeds of cereals, could be detected in wheat germ containing only 18% water (Linko and Milner, 1959). Later the activity of this enzyme was found to be related to seed viability (Linko and Sogn, 1960; Linko, 1961; Bautista and Linko, 1962; Bautista *et al.,* 1964), and such activity was used to measure seedling vigor of cereals (Grabe, 1964). This enzyme is also very active in beans (James, 1968) and soybeans (Grabe, 1970). Apparently glutamic decarboxylase possesses a unique feature in determining subsequent metabolic patterns of seed germination since it is active at very low water content, progressively increases in activity with increasing water content of seeds, and its activity correlates with seed viability and subsequent growth rate. In the mustard seeds, the first and the most intensively labeled soluble metabolite is γ-aminobutyric acid, which might be the product of glutamic decarboxylase activity as glutamic acid is commonly present in dry seeds (Spedding and Wilson, 1968). Carbon dioxide, the other product of the enzyme reaction, may be dissipated as a gas, may dissolve in the cell sap and change the intercellular pH, may be incorporated into Krebs cycle intermediates (Haber and Tolbert, 1959), or may directly contribute to synthesis of the purine skeleton of AMP (Mahler and Cordes, 1968). Since AMP is limiting in dry seeds (Pradet *et al.,* 1968), the synthesis of AMP and further phosphorylation to ADP are necessary prior to formation of ATP via respiration. The fate of the γ-aminobutyric acid has not been traced, and the endeavor might be rewarding in reference to subsequent metabolic events. Again it is entirely possible that glutamic acid decarboxylase reflects the degree of denaturation of the entire population of preexisting enzymes at the earliest stages of germination. The early labeled amino acids also may act as inducers for synthesis of specific enzymes as glutamine induces lipase activity in wheat endosperm (Tavener and Laidman, 1969).

An ingenious design of suspended water uptake by equilibrating seeds at low water potentials makes possible the analysis of events occurring at different low water content (Wilson and Harris, 1968; Wilson, 1970). Wilson and Harris coated crested wheatgrass with $^{32}$Pi, then equilibrated the seeds between paper strips moistened with various saturated salt solutions of known water potential. Five days later, the seeds were analyzed for phosphorylated compounds by column chromatography. The first identifiable compound to become phosphorylated was the hexose phosphate which was found in seeds containing 16.2% water. In seeds containing about 23% water (Fig. 2.5, upper), nicotinamide adenine diphosphate (NAD), uridine diphosphate hexose (UDP-hexose), ATP, and inositol hexa- and tetra-esters also were phosphorylated. At 29.8% water content (Fig. 2.5, lower), seed enzymes phosphorylated AMP, other inositol esters, and many unknowns. Two facts are of significance in relation to the awakening process of seeds; the first is the early phosphorylation of hexose and UDP-hexose which probably indicates the glycolysis of preexisting hexose and sucrose to provide substrate for respiration; the second is the phosphorylation of NAD and ATP. NAD is a common preexisting coenzyme in dry cereal seeds, but it appears to be deficient for normal metabolic demand and is quickly synthesized during the first 24 hours of germination (Mukherji *et al.,* 1968, Fig. 16 in III.D). ATP is the product of oxidative phosphorylation of conserved ADP (Pradet *et al.,* 1968, III.A.) in active mitochondria. Therefore, one may conclude from these data that the first event during resumption of growth in seeds is activation of glycolytic enzymes, phytase and the respiratory apparatus, followed by the synthesis of enzymes for NAD, UDP, and ADP formation.

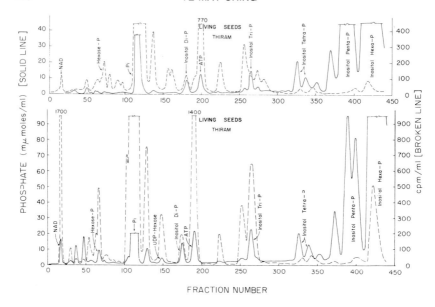

FIG. 2.5. Elution pattern of phosphate esters from crested wheatgrass seeds treated with $^{32}$P-labeled $NaH_2PO_4$ and a microbicide (thiram, tetramethythiuram disulfide), and equilibrated 5 days at 100 atm. of water potential (upper) and at $-40$ atm. (lower) (Wilson, 1970).

If seeds were coated with radioactive individual sugars or amino acids the awakening sequence probably could be traced by the experimental approach developed by Wilson and Harris (1968).

### B. Protein-Synthesizing Machinery

Protein-synthesizing machinery provides basic equipment for cells to express their specific mode of activity by synthesizing required enzymic, structural, and regulatory proteins, even though sometimes enzymes formed at one stage may not be functionally active until the optimal or necessary conditions occur at the same or another stage of development. During imbibition, and the lag period of water uptake, the increase of respiration, ATP content, ribosomal amino acid incorporating ability, polysome content, and *in vivo* protein-synthesizing activity are clearly demonstrated in Figs. 2.3–2.4. The kinetics of these changes indicate an activation of preexisting systems.

### 1. SYNTHESIS DURING THE LAG PERIOD OF WATER UPTAKE

The embryo of decoated cottonseed (Waters and Dure, 1966) synthesizes proteins and ribosomes during the first 16 hours of imbibition as shown by the incorporation of amino acids-[14]C and phosphate-[32]P (Fig. 2.6a). Upon incubation for the first 12 hours with cycloheximide (an inhibitor for protein synthesis at the translation level, 1 mg/ml), the capacity for amino acid-[14]C incorporation is eliminated in the following four hours, but [32]Pi is incorporated into a ribosome precursor as indicated in Fig. 2.6b. This treatment also stops further development of the embryo. If the embryo is incubated with actinomycin D (an inhibitor of RNA synthesis at transcription level, 20 $\mu$g/ml), for the first 12 hours, incorporation of [32]Pi into RNA is impaired (Fig. 2.6c); but the treatment neither decreases the total ribosomal population nor affects protein synthesis and further growth. Thus either actinomycin D does not interfere with normal development during this stage of germination, or new mRNA and rRNA are not essential in producing proteins for the metabolic activities of this early stage. The authors conclude that the protein synthesized during this period is directed by stable messenger RNA and ribosomes which had been preformed in the mature seed. Chakravorty (1969a) observed ribosome synthesis during the first 16–24 hours of germination of black-eye peas *(Vigna unguiculata)*. The synthesis was partially inhibited by either cycloheximide or actinomycin D. Apparently not only the long-lived mRNA, but also new mRNA's are functioning in this seed at this particular stage. An antimetabolite of rRNA synthesis, 5-fluorouracil, and another inhibitor of protein synthesis, puromycin, were ineffective in preventing ribosome synthesis at this stage. The author speculated that the permeability of these two compounds may cause such results.

Using decotyledonized embryos of radish *(Raphanus sativus)* seeds, Fujisawa (1966) also was able to block RNA synthesis with treatment of actinomycin D and thiouracil, but not fresh weight increase during this stage. When puromycin and chloramphenicol, an inhibitor of protein synthesis by organellar ribosomes, were incubated with the embryo axes, RNA synthesis was not affected, but increase in fresh weight and incorporation of leucine-[14]C were inhibited. Treatment by actinomycin D and thiouracil at this stage also inhibited the incorporation of leucine-[14]C, a result not obtained with cotton whole embryo (Waters and Dure, 1966). Whether this difference is specific to tissue or to species is unknown. One interesting finding in this study is that none of these inhibitors suppressed respiration increase, which might indicate a temporal independency of mitochondrial biogenesis and activity.

Another report indicates that a group of new mRNA's synthesized during the sixth hour of imbibition in mung beans *(Phaseolus aureus)* is chemically different from RNA synthesized at 24 and 48 hours germination (Biswas, 1969). The author incubated mung beans that had imbibed water for 5 hours in uridine-[3]H for one hour and extracted total RNA

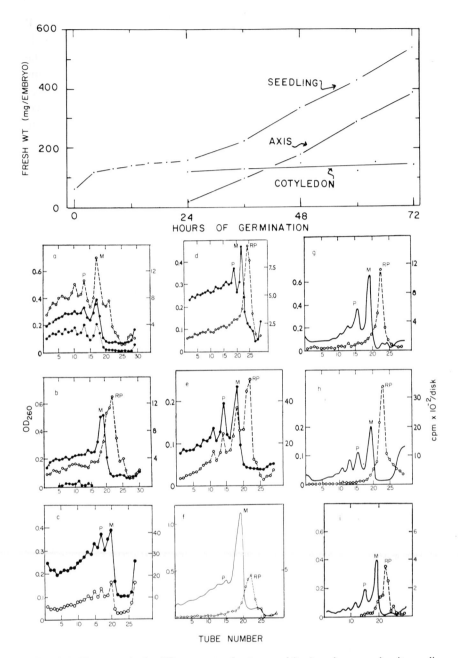

Fig. 2.6. Water uptake by different parts of cotton seed (top), and sucrose density gradient analysis of ribosome population (O—O), newly synthesized cytoplasmic nucleic acids -$P^{32}$ (O--O), and protein-$^{14}C$ (●----●) from one seed. (P, polysomes; M, monosomes; RP, ribosomal precursor.) Whole embryo imbibed in water for 12 hours then exposed 3 hours to tracers (a). Whole embryo imbibed in cycloheximide solution for 12 hours then 3 hours in tracers (b). Whole embryo imbibed in actinomycin D solution for 12 hours then in tracer

from the beans. The extracted RNA was separated by sucrose density gradient centrifugation. The zone containing the highest specific radioactivity was isolated and used for hybridization with mung bean leaf DNA and for competition with RNA isolated from older material. The newly synthesized RNA was different from the total RNA isolated at 24 hours and 48 hours germination, since they competed with the labeled RNA only to the extent of one-third and one-sixth, respectively, of the complementary sites on DNA. This highly labeled RNA was called "mRNA" by the author.

Polysomes were found *in situ* in electron micrographs of imbibed pea seeds (Chapman and Rieber, 1967), and polysomes of 110 S, 140 S, and 170 S were separated by ultracentrifugation of cotyledon extracts from peas that had imbibed water for 17 hours (Barker and Rieber, 1967). About one-half of the monosomes became polysomes in wheat embryos after 16 hours imbibition (Marcus and Feeley, 1965).

In wheat embryos, total proteins, RNA's and rRNA remained constant during imbibition, but sRNA (soluble RNA includes mainly tRNA and free mRNA if present) decreased steadily during this stage because of a preexisting, heat sensitive ribonuclease which showed reduction in activity with time of germination (Vold and Sypherd, 1968b). These investigators observed a slight increase in DNA during this stage (Fig. 2.9b, p. 135) indicating DNA synthesis prior to radicle elongation.

Walton and Soofi (1969) also concluded from a study with excised embryo axes of beans that the capacity for synthesis of nucleic acids and proteins increased in axes prior to the initiation of axis elongation. In intact barley *(Hordeum vulgare)* seeds, a linear increase of protein synthesis was found during 1 to 11 hours of germination. The protein-synthesizing capacity of 5-hour germinants was reduced by puromycin ($10^{-3}$ M) to 65% and by actinomycin D (10 $\mu$g/ml) to 88% of the control (Abdul-Baki, 1969). Apparently long-lived mRNA was present in the seed and most of its translation system was insensitive to puromycin.

During the lag period of water uptake and respiration, it is apparent from the results mentioned above that more polysomes are formed from the long-lived mRNA and preexisting ribosomes. The formation of polysome is probably at the expense of ATP produced by steady state respiration. Active synthesis of RNA polymerases for various RNA's and proteins needed for new ribosomes also commences at this stage.

for 3 hours (*c*). Cotyledon (*d*) and seedling axis (*e*) of seed germinated for 42 hours then exposed 3 hours to tracer. Cotyledon of seed germinated for 42 hours, exposed to cycloheximide for 10 hours then to tracer for 3 hours (*f*). Cotyledon (*g*) and seedling axis (*h*) of seed germinated for 72 hours then exposed 3 hours to tracer. Seedling axis of seed germinated for 42 hours then exposed to actinomycin D for 10 hours and 3 hours to tracer (*i*) (Waters and Dure, 1966).

## 2.   RNA Synthesis after Radicle Emergence

a. Monosomes, Polysomes, and Membrane-Bound Ribosomes.
Monosomes and polysomes increased in cotton seeds during the second
day as shown by the total content ($OD_{260}$) and $^{32}Pi$ incorporation in
cotyledons and axes (Fig. 2.6d,e; Waters and Dure, 1966). Subsequently,
no appreciable increase was found in 72-hour germinants (Fig. 2.6g,h).
The newly synthesized and highly labeled particles were lighter than
monosomes, and they were verified by sequential tracing technique to be
ribosomal and possible mRNA precursors (RP in Fig. 2.6). Most interesting of all is the persisting synthesis of the precursor RNA in the
presence of cycloheximide and actinomycin D (Fig. 2.6f,i).

A progressive increase of ribosomes with germination stage was ob
served in the endosperm of castor beans (Fig. 2.7). Preferential formation

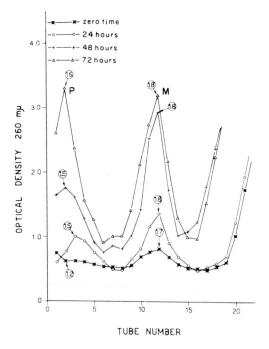

FIG. 2.7.  Change in ribosomal content and composition (figure in circles $= OD_{260}/OD_{280}$)
with germination time in endosperm of castor beans (Marré *et al.*, 1965b). (P, polysomes,
M, monosomes.)

of polysomes at the expense of monosomes in peanut cotyledon during
germination has been demonstrated (Marcus and Feeley, 1965). Active
synthesis of RNA also occurred in wheat embryos at this stage. This is

seen from the observations that exposure of embryos to $^{32}$P; during the first 24 hours of germination resulted in a specific activity as low as 40–60 cpm/mg RNA, whereas, a 24-hour incorporation of $^{32}$Pi by embryos of 1-day-old germinating seeds gave a specific activity of 30,000 cpm/mg RNA (D. Chen *et al.,* 1968a).

The biosynthetic sequence of ribosomes was explored by Chakravorty (1969b) in blackeye peas at this stage of seed germination. In common with animal material, the synthesis of both heavier and lighter ribosomal subunits, including 28 S and 18 S rRNA and their respective proteins, and the association of 5 S rRNA with the heavier subunit occurred in nuclei. The two subunits of ribosome entered cytoplasm as separate entities, and they joined together to form monomeric ribosome only after association with mRNA. The increase of ribosomal content ceased on the second and third day of germination in blackeye peas; only ribosomal precursor was produced during that period. The precursor slowly matured to ribosome after 24 hours' incubation. The slowness of ribosomal maturation also was observed in cotton seedlings (Waters and Dure, 1966).

Without detergent, such as deoxycholate, in the homogenizing medium for isolation of microsomes together with a lower centrifugal force, a fraction of membrane-bound ribosomes in addition to free monosomes and polysomes could be separated. Payne and Boulter (1969) reported that total RNA content in cotyledons of broad beans *(Vicia faba)* was constant during the first 4 days of germination, but the membrane-bound ribosomes were synthesized while free ones were degraded. The authors believed that the two populations of ribosomes might synthesize different groups of proteins and play a role in cellular differentiation.

b. NATURE OF RNA's.    D. Chen *et al.* (1968a) attempted to discern how much of the various RNA's in relation to total genetic information stored in DNA are synthesized at each stage of germination, and how different they are in reference to their base sequences. They extracted total nucleic acids from germinating wheat embryos of different age and separated the extracts by a methylated albumin kieselguhr (MAK) column. Fig. 2.8 illustrates the elution pattern of different nucleic acids extracted from peanut cotyledons and wheat embryos separated by the column. Nucleotide composition of various nucleic acids in germinating seeds is shown in Table 2.II. Using different species of RNA eluted from the MAK column to hybridize homologous DNA, D. Chen *et al.* (1968a) found that the RNA synthesized on the second day was composed of 79% mRNA, 19% rRNA and 2% tRNA. They observed that the total quantity of RNA transcribed on both the second and third days was 2.90% DNA bases, and the distribution of different kinds of RNA synthesized on the

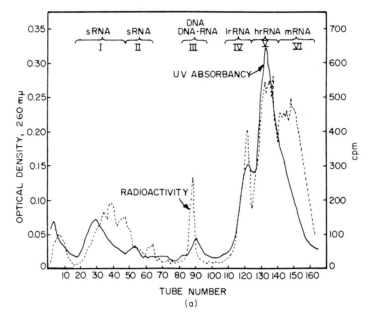

FIG. 2.8a. Profile of methylated albumin kieselguhr column fractionated total nucleic acids from peanut cotyledon of 6-day-old germinant incubated with $^{32}$Pi (Cherry, 1967).

TABLE 2.II

NUCLEOTIDE COMPOSITION (MOLAR PERCENTAGE)
OF VARIOUS NUCLEIC ACIDS IN GERMINATING SEEDS

| Fraction | CMP | AMP | GMP | UMP |
|---|---|---|---|---|
| Peanut cotyledons[a] | | | | |
| s₁RNA | 27.5 | 18.0 | 34.1 | 20.4 |
| s₂RNA | 30.0 | 19.8 | 31.4 | 18.8 |
| DNA-RNA | 26.3 | 21.9 | 31.0 | 20.9 |
| lrRNA | 25.1 | 22.3 | 32.1 | 20.5 |
| hrRNA | 20.0 | 25.4 | 32.7 | 21.7 |
| mRNA | 18.5 | 28.2 | 29.6 | 23.7 |
| Cotton Seedlings[b] | | | | |
| sRNA | 27.7 | 17.2 | 31.0 | 24.3 |
| lrRNA | 21.8 | 22.9 | 30.1 | 25.1 |
| hrRNA | 23.4 | 21.2 | 32.4 | 23.0 |
| Blackeye Pea Seedlings[c] | | | | |
| 5 S RNA | 25.0 | 21.2 | 31.8 | 22.0 |
| 18 S RNA | 20.3 | 24.0 | 30.8 | 24.9 |
| 28 S RNA | 22.2 | 23.8 | 32.3 | 21.4 |

[a]Chroboczek and Cherry (1966). Total nucleic acids, MAK column fractions.
[b]Waters and Dure (1966). Ribosomal fractions.
[c]Chakravorty (1969b). Fractions from ribosomal pellet.

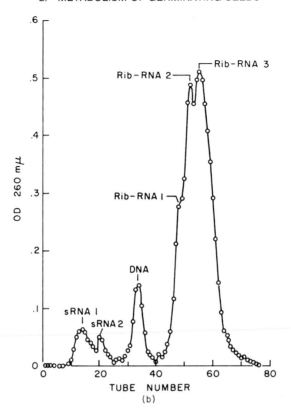

FIG. 2.8b. Profile of methylated albumin kieselguhr column fractionated total nucleic acids from wheat embryo (Vold and Sypherd, 1968a).

second day was similar to that of the third day. The information transcribed, however, was different, since the second-day RNA's competed
to the extent of 50% DNA complementary sites of the third-day RNA's.
However, the RNA's synthesized on the second day competed well with
that of 1-day-old embryos as well as with the conserved RNA's in dry
embryos. This triple dose of similar information delivered from nuclei in
the embryo during seed maturation, on the first and second day of germination might be a device to insure survival of the species. The qualitative
aspect of the information transmitted, however, must be different as
growth patterns are distinctly varied on the first and second day of wheat
germination. Alternatively, the information transmitted may be in the
form of long-lived informosomes which direct the germination process as
a whole and the informosomes produced on the third day are for differentiation, i.e., production of new leaves, roots, etc.

In contrast to the steady information flow in wheat embryos found by D. Chen *et al.* (1968a), a precocious burst of messages in early stages of germination may be operative in tissues with a determined developmental pattern, such as the case of mung bean cotyledons (Biswas, 1969). The total RNA isolated from cotyledons in seeds soaked for 6 hours competed to the extent of 88% of the mRNA extracted from the same material, while the total RNA isolated from seeds after 24 and 48 hours of germination only competed with the 6-hour mRNA to 64% and 15%, respectively. The author concluded from these competition data in hybridization studies that a burst in production of many species of RNA occurs in early stages of seed germination, followed by phasing out of RNA species with germination time.

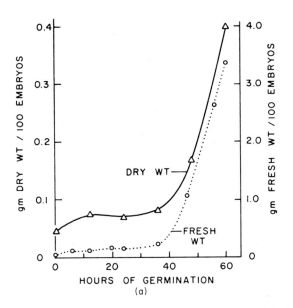

FIG. 2.9a. Changes in weights in germinating wheat embryo (Vold and Sypherd, 1968a).

Based on the direct quantitative analysis of RNA's by MAK column fractionation, Vold and Sypherd (1968b) observed a slow increase of RNA during the first 24 hours of germination, followed by an accelerated pace in wheat embryos (Fig. 2.9b). Protein increase followed the trend of total RNA and rRNA, while soluble RNA (tRNA and free mRNA) decreased during the first 16 hours then increased rapidly up to 24 hours, and was followed by a slower increase (Fig. 2.9b).

The changes in various nucleic acids in germinating seeds of red pine *(Pinus resinosa)* were analyzed by MAK column chromatography of

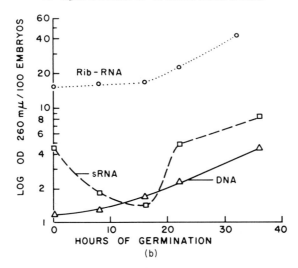

Fig. 2.9b. Changes in various nucleic acids in germinating wheat embryo (Vold and Sypherd, 1968a).

phenol-extracted total nucleic acids (Sasaki and Brown, 1969). Little change in the absolute quantity of sRNA, rRNA, and DNA in the megagametophyte was observed in contrast to a continuous increase in the embryo (Fig. 2.9c). Rapid turnover of megagametophytic RNA's, however, was noted by the results of $^{32}$P incorporation.

These observations probably indicate that the total RNA population is too gross to resolve the delicate question of the kind of information in the form of mRNA transmitted at each stage. Therefore, direct studies of mRNA are imperative. The isolation of undegraded mRNA is difficult because of the common occurrence of ribonuclease in tissue. Phenol extraction of nucleic acid often results in aggregation and partial degradation of RNA (Fedorcsak et al., 1969). Bentonite and polyvinyl sulfate are not always effective in removing ribonuclease, and often result in loss of polysomes (Weeks and Marcus, 1969). Using diethyl pyrocarbonate (DEP), a general protein denaturing compound, as an inhibitor of ribonuclease, Fedorcsak et al. (1969) isolated a fraction of RNA from barley embryos. This RNA fraction had a better template activity than the RNA isolated by other methods or the tobacco mosaic virus RNA extracted by phenol. With the same inhibitor, Weeks and Marcus (1969) found a protective effect on polysome structure of wheat embryos and corn root tips. However, they observed that DEP is deleterious to polysomal ability of amino acid incorporation in vitro. Perhaps combining the procedure of Weeks and Marcus in isolating polysomes and the method of

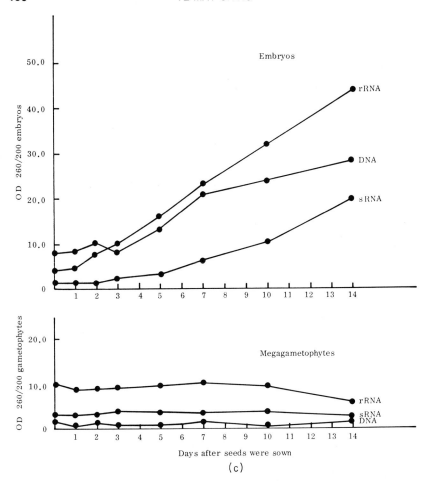

FIG. 2.9c. Changes in various nucleic acids in germinating seed of red pine (Sasaki and Brown, 1969).

Fedorcsak *et al.* in extracting RNA from the isolated polysomes may result in undegraded mRNA. Nevertheless, the task of mRNA isolation from plant materials awaits future research.

    c. CHARACTERISTICS OF RIBOSOMES. The affinity between ribosomal subunits may vary with the stage of germination. Biswas (1969) observed resistance to dissociation treatment of dialyzing isolated 80 S ribosomes to 60 S and 40 S subunits in mung bean seeds soaked for 6 hours, whereas those of cotyledons soaked for 12, 24, or 48 hours were readily dissociated. Ribosomes isolated from different tissue often exhibit compositional variation as shown in Table 2.III.

**TABLE 2.III.**

PROTEIN AND NUCLEIC ACID CONTENT OF PURIFIED RIBOSOMES
ISOLATED FROM GERMINATING SOYBEANS[a] AND WHEAT EMBRYOS[b]

| Ribosomes | Protein (%) | RNA (%) | $OD_{260}/OD_{280}$ |
|---|---|---|---|
| Soybean cotyledons | | | |
| soaked 16 hours | 53.7 | 46.3 | 1.3 |
| germinated 5 days | 56.3 | 43.7 | 1.3 |
| Soybean hypocotyl | | | |
| germinated 5 days | 44.4 | 55.6 | 1.5 |
| Wheat embryos, dry | 45 | 55 | – |

[a]From Matsushita *et al.* (1966).
[b]From Wolfe and Kay (1967).

The 25 S and 18 S RNA of the two subunits of ribosomes in young pea seedlings are not homologous, whereas the RNA of the respective subunits from root and shoot are homologous, as shown by experiments of competitive hybridization with pea DNA. The degree of homology among different species often indicates their taxonomic relationship (Table 2.IV, Trewavas and Gibson, 1968).

At least nineteen to twenty bands of ribosomal proteins were separated

**TABLE 2.IV.**

COMPETITIVE HYBRIDIZATION BETWEEN LABELED ALASKA PEA RIBOSOMAL
RNA AND ALASKA PEA DNA AND RIBOSOMAL RNA FROM OTHER PLANTS[a]

| Species of DNA | Species of [32]P-RNA | Unlabeled competitor | Competitor RNA [32]P-RNA | % Competition (degree of homology) |
|---|---|---|---|---|
| Alaska pea (7 $\mu$g) | Alaska pea (25 S, 3 $\mu$g) | 25 S (Alaska pea) | 200 | 100 |
| | | 25 S (Meteor pea) | 180 | 81 |
| | | 25 S (Runner bean) | 150 | 36 |
| | | 25 S (Broad bean) | 190 | 34 |
| | | 25 S (Barley) | 180 | 9 |
| | | 25 S (Maize) | 180 | 7 |
| | | 18 S (Alaska pea) | 200 | 1 |
| | Alaska pea (18 S, 3 $\mu$g) | 18 S (Alaska pea) | 150 | 100 |
| | | 18 S (Meteor pea) | 160 | 74 |
| | | 18 S (Runner bean) | 200 | 25 |
| | | 18 S (Broad bean) | 170 | 28 |
| | | 18 S (Barley) | 180 | 6 |
| | | 18 S (Maize) | 200 | 4 |
| | | 25 S (Alaska pea) | 200 | 1 |

[a]From Trewavas and Gibson (1968).

by acrylamide gel electrophoresis of mung bean cotyledons; two of the bands were reduced in quantity with germination time (Biswas, 1969). Therefore, qualitative changes may occur in ribosomal constituents during germination.

As shown in Table 2.V, enzymes associated with isolated ribosomes may vary in their specific activity because of their origin. Whether these enzymes were endogenous, absorbed, or newly synthesized is not known. Nevertheless, differences with respect to growth stage and kind of tissue are obvious.

TABLE 2.V.

SPECIFIC ACTIVITY OF ENZYMES FOUND IN PURIFIED RIBOSOMES ISOLATED FROM GERMINATING SOYBEANS[a]

| | Activity units $\times$ $10^3$/mg protein | | |
|---|---|---|---|
| | Cotyledons | | Hypocotyl |
| Enzymes | 16 hr soaked | 5-day germinated | 5-day germinated |
| Ribonuclease | 3.3 | 10 | 109 |
| Phosphodiesterase | 4.8 | 5.0 | 47 |
| Acid phosphatase | 35 | 93 | 243 |
| 5′-Nucleotidase | | | |
| AMP | 9.1 | 44 | 186 |
| GMP | 6.7 | 42 | 193 |
| CMP | 7.4 | 38 | 173 |
| UMP | 5.1 | 36 | 158 |
| dGMP | 7.0 | 50 | 145 |
| dCMP | 4.7 | 47 | 144 |
| Nucleoside triphosphatase | | | |
| ATP | 20 | 35 | 149 |
| CTP | 16 | 33 | 126 |
| Peroxidase | 43 | 8.2 | 67 |
| $\beta$-Glucosidase | 5.5 | 2.6 | 3.3 |

[a]From Matsushita et al. (1966).

Information regarding the molecular architecture of ribosomes appears to be necessary in order to discern their specific activities at different stages of development. Physical and chemical characteristics of the cytosol surrounding the ribosomes also should be analyzed so that the influence of the microenvironment can be revealed.

d. CHANGES IN tRNA'S AND AMINOACYL-tRNA SYNTHETASES. The rate of readout of a particular mRNA to produce a specified protein can be controlled at the translation level. The control can be accomplished not only by the functional ribosomes but also by the quantity and kinds of tRNA's and/or synthetases. Genes might regulate the kinds of enzymes or proteins produced in a particular organ at a designated stage by modulating specific tRNA's and/or synthetases. Such regulation often is reflected in the composition and characteristics of tRNA and synthetases.

Vold and Sypherd (1968a) found the ratio of individual tRNA's in dry and germinating wheat embryos to be different. Particularly lysyl-tRNA, prolyl-tRNA and seryl-tRNA increased during the first 40 hours of germination. The change in ratio indicates a differential production and/or preferential destruction of tRNA's. In another study, Vold and Sypherd (1968b) found one ribonuclease in dry wheat embryos that degraded pre-existing RNA's during early stage of germination, and another ribonuclease synthesized after the lag period of water uptake. The second ribonuclease distributed 60% of its activity in soluble fraction and 40% on ribosomes. According to Hsiao (1968) the differential distribution might be an artifact resulting from the rupturing of cells.

In beans (Fig. 2.10) the total activity of aminoacyl-tRNA synthetases decreased to 50% of that in the cotyledons of dry beans after 7 days'

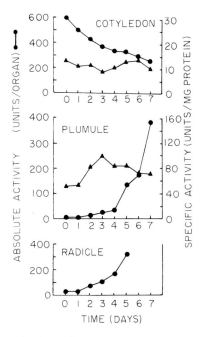

FIG. 2.10. Change in total and specific activity of aminoacyl-tRNA synthetase in various tissue of germinating beans (J. W. Anderson and Fowden, 1969).

germination, while the specific activity of the enzyme system was not changed. In the plumule and radicle, increases of approximately 40- and 20-fold of the total activity, respectively, were observed in the same period of germination. The specific activity of this enzyme system increased in both the plumule and radicle about twofold during the first 2–3 days of germination and then gradually decreased. Only asparaginyl-, valyl-, and histidinyl-tRNA synthetase in the plumule did not conform with this general pattern.

Amino acid-activating enzymes in germinating peas appeared to be synthesized *de novo* since chloramphenicol and *p*-fluorophenylalanine completely inhibited the activity (Henshall and Goodwin, 1964).

Further differences in the kinds of leucyl-tRNA's and synthetases were noted in various organs of germinating soybeans (M. B. Anderson and Cherry, 1969). Six leucine-accepting tRNA's were separated by Freon column chromatography from the soluble RNA extracted from cotyledons and hypocotyls of 5-day-old germinants, and subsequently each of these was charged with radioactive leucine using a soluble fraction as the source of synthetase. The enzyme isolated from hypocotyl tissue could not synthesize the complex with the tRNA number 5 and 6, whereas the enzyme isolated from cotyledons did. The quantity of tRNA number 5 and 6 was also extremely low in hypocotyls. An organ specificity of the kinds of tRNA and the activity of synthetase is well demonstrated by this study.

### C. *Respiration and Mitochondria*

#### 1. GENERAL PATTERN OF RESPIRATION IN GERMINATING SEEDS

Respiration involves oxidative breakdown of certain organic seed constituents — mainly sugars, starches, fatty acids, and triglycerides — to provide biological energy in the form of ATP which facilitates many of the energy-requiring anabolic activities accompanying germination and growth. Generally the higher the rate of oxygen uptake shown by an organism, organ, cell, or mitochondrion under normal circumstances, the more and faster are the various metabolic activities. Thus the rate of oxygen absorption is often used as an index of seed vigor (Woodstock, 1965; Woodstock and Grabe, 1967). Figure 2.11 presents the normal course of respiration during germination under optimal conditions. The pattern of oxygen uptake appears to be similar in three kinds of seeds with different reserves indicating that the metabolic events are taking place similarly. The first phase of germination is the imbibition period when major activity is confined to reactivation of preexisting macromolecules and organelles which results in (*1*) the formation of $\alpha$-oxo acids for operation of the Krebs cycle and oxidative phosphorylation of pre-

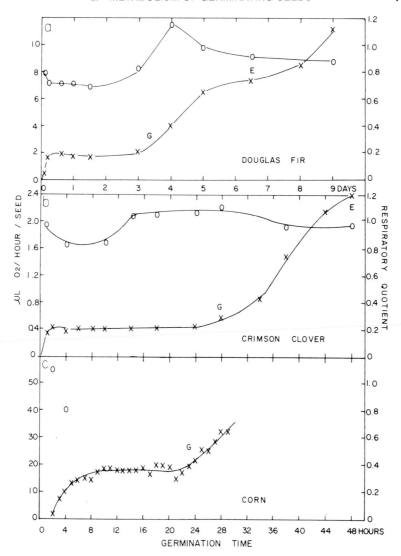

Fig. 2.11. Respiration rate ($\times-\times$) and respiratory quotient ($O-O$) of germinating seeds in Douglas fir (a) (Ching, 1959); crimson clover (b) (Ching, 1970a), and corn (c) (Woodstock and Grabe, 1967). (G, germination; E, epicotyl emergence.)

existing ADP with Pi, (2) the phosphorylation of conserved sugars with Pi, resulted from phytase activity, to provide substrates for respiration and for formation of more ATP, (3) the synthesis of AMP from conserved ribose, glutamine, glycine, $CO_2$, aspartic acid, formic acid, with the aid of ATP, GTP, and tetrahydrofolate, (4) the increase of ADP by adenylate kinase, ATP, and AMP, (5) the synthesis of new proteins using pre-existing amino acids, tRNA's, aminoacyl-tRNA synthetases, activated mRNA's, ribosomes, and newly synthesized ATP. The second phase is a steady state of water uptake and respiration when all the preexisting systems are working at full capacity for the synthesis of substrates toward the biogenesis of new ribosomes, sRNA's, enzymes, mitochondria, glyoxysomes, plastids, nuclei and membranes. The third phase is a non-synchronized cell division and growth stage, characterized by a continuous increase in fresh weight and respiration. After seedlings turn green, gaseous exchange involves photosynthesis as well as respiration, and the rate of exchange becomes meaningless without separate measurements of each process. Respiratory quotient (RQ) usually indicates the nature of the substrate, whether aerobic respiration or fermentation is occurring (e.g., the early stage of respiration in peas).

## 2. DEVELOPMENTAL STUDY OF MITOCHONDRIA DURING GERMINATION

Because cells in storage tissue of seed differentiate but do not divide, many studies are devoted to organellar biosynthesis and senescence in order to characterize mechanisms of cellular differentiation. Among these the most complete picture is presented by the developmental study of mitochondria in endosperm of germinating castor beans (Akazawa and Beevers, 1957a,b). Figure 2.12 summarizes the findings. The amount of mitochondria as measured by milligram of protein N/tissue increased with germination to the fifth day and declined thereafter. Oxygen uptake as milliliter per milligram mitochondrial protein N per hour ($QO_2$) followed the exact trend of mitochondrial quantity. The efficiency of oxidative phosphorylation (P/O ratio) decreased slowly from the second day to the fifth day, then it decreased rapidly to the eighth day. The composition of isolated mitochondria also changed with germination time, with a reduction of acid soluble phosphates per milligram protein N, a slight increase of phospholipids per milligram of protein N, and a constant ratio of RNA to protein in the organelle. Results from mitochondria isolated by sucrose density gradient from endosperm of germinating caster bean verified the above data (Lado and Schwendimann, 1967; Gerhardt and Beevers, 1970).

Mitochondria isolated from the haploid storage tissue of pine seeds exhibited similar changes with germination (Fig. 2.22, p. 179) as the castor bean endosperm. The kinetic aspect of these changes indicates a lack of turnover in these organelles. In other words, they are synthesized in early stages of germination after the lag period of respiration and then become senescent with age. However, the oxygen uptake of mitochondria from

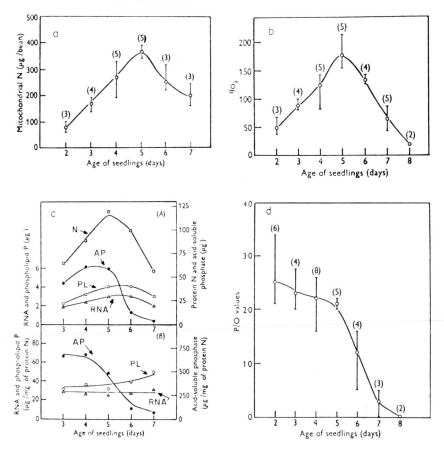

FIG. 2.12. Changes in total content (a), respiration rate (b), composition (c), and oxidative phosphorylation (d), of mitochondria isolated from endosperm of germinating castor beans (Akazawa and Beevers, 1957b).

sugar pine *(Pinus lambertiana)* endosperm followed a sigmoid curve when citrate and α-ketoglutarate were used as substrates. A plateau and a decrease were observed when succinate was used for the substrate during the 10 days of germination (Stanley, 1957). Whether this inconsistency is real or is due to technical difficulties is not clear.

A pattern of rise and fall of oxygen uptake also was observed in mitochondria isolated from soybean cotyledons (Howell, 1961), peanut *(Arachis hypogaea)* cotyledons (Cherry, 1963), and waxbean cotyledons (Stinson and Spencer, 1970). The P/O ratio increased first in the soybean mitochondria, then decreased (Howell, 1961). Mitochondria isolated from cucumber *(Cucumis sativus)* cotyledons increased both in oxygen uptake and P/O ratio during the first 5 days of germination (Hanson *et al.,* 1959). In mitochondria of peanut cotyledons, total activity of succinate cytochrome reductase and reduced diphosphopyridine nucleotide (DPNH) cytochrome reductase increased up to the ninth day of germination then declined. DPNH oxidase activity remained constant throughout the course of germination, whereas cytochrome oxidase activity was high in dry seeds, decreased in the first 9 days, then increased. In cotyledons of germinating peanut, mitochondrial DNA and the activity of succinoxidase and succinic dehydrogenase increased continuously for 8 days (Breidenbach *et al.,* 1967), while the activity of aconitase and condensing enzyme (citrate synthase) increased for 5 days and then declined (Marcus and Velasco, 1960). In wax bean cotyledons, the isolated mitochondria had two kinds of physiological activity and stability during the course of 10-day germination period. From day 1 to day 3, the oxidative phosphorylation efficiency (ADP/0), respiratory control value, and contraction–swelling property of the mitochondria decreased with time of germination. From day 3 to day 5, mitochondria were incapable of phosphorylation and contraction-swelling, had highest ATP hydrolyzing enzyme (ATPase) activity, most leaky membrane, and lowest respiratory control value. From day 5 to day 8, all the attributes increased to a peak on day 7 then declined (Stinson and Spencer, 1970; Malhotra and Spencer, 1970). All these might indicate that the mitochondria isolated from first 3 days of germination were preexisting kind from the maturation period of the seed and those isolated from day 5 to day 8 were synthesized during germination. This type of absolutely synchronized degradation and synthesis of energy-producing apparatus during an active physiological process is obviously detrimental to the well-being of the organism, and probably this condition cannot exist *in vivo.* Mixing of metabolic compartments during tissue grinding may bring about the stimulation of ATPase, the dissociation of membrane constituents, and the inhibition of component enzymes. Nevertheless, these data are difficult to interpret without a complete picture of the *in vivo* metabolic compartmentation.

Mitochondria from the embryo of ungerminated pine seeds showed a higher oxidative activity with various Krebs cycle substrates than the particles from 5-day-old germinated seedlings or from embryos of seeds that had been stratified for 60 or 120 days (Stanley and Conn, 1957). An increase of oxygen uptake for the first 3 days, followed by a decline was found in mitochondria isolated from scutella of germinating corn (*Zea mays*) seeds, while the P/O ratio continuously decreased in the same material (Table 2.VI). There was little change in oxygen uptake by mitochondria of corn embryos, but a definite decrease of P/O ratio was observed (Table 2.VI). The continuous decline of oxygen uptake and P/O

**TABLE 2.VI.**
Oxygen Uptake and Phosphorylation by Mitochondria
Isolated from Various Tissue During Germination

| Material and age | Substrate | QO$_2$(N) | P/O |
|---|---|---|---|
| Corn scutella[a] | α-Ketoglutarate | | |
| 1 day | | 402 | 2.45 |
| 3 days | | 491 | 2.21 |
| 5 days | | 179 | 0.59 |
| Corn scutella[b] | Succinate and pyruvate | | |
| 1.5 days | | 705 | 1.00 |
| 2.0 days | | 843 | 0.87 |
| 2.5 days | | 877 | 0.85 |
| 3.0 days | | 1190 | 0.59 |
| Corn embryo axes[b] | Succinate and pyruvate | | |
| 1.5 days | | 958 | 1.13 |
| 2.0 days | | 920 | 1.11 |
| 2.5 days | | 880 | 0.75 |
| 3.0 days | | 924 | 0.85 |
| Cucumber cotyledons[a] | α-Ketoglutarate | | |
| 1 day | | 131 | 2.30 |
| 3 days | | 771 | 3.18 |
| 5 days | | 627 | 2.84 |
| Soybean cotyledons[c] | αKetoglutarate | | |
| 1 day | | 110 | 0.87 |
| 3 days | | 340 | 1.79 |
| 5 days | | 300 | 1.26 |
| 7 days | | 80 | 0.58 |
| 9 days | | 50 | 0.64 |

[a]From Hanson et al. (1959).
[b]From Cherry et al. (1961).
[c]From Howell (1961).

ratio with germination time in mitochondria isolated from embryo tissue is difficult to explain with respect to development and growth. Five-day-old pine seedlings and 3-day-old corn seedlings had multiplied cell numbers over the embryonic stage and synthesized new mitochondria, and an increase of oxygen uptake and P/O would be expected from the newly formed organelles and increased population. A developmental shift of the energy production site to chloroplasts in green parts of the seedlings may offer partial explanation to the change. A reinvestigation of this aspect using different parts of the seedling and especially using improved isolation techniques would be of value.

The results from an analysis of respiratory substrates coupled with a study of the utilization of acetate-$^{14}$C and glucose-$^{14}$C in buckwheat

*(Fagopyrum esculentum)* seedlings suggested that both the pentose phosphate pathway and the usual glycolytic sequence and the Krebs cycle were participating in total respiratory metabolism of this starchy seed (Effer and Ranson, 1967a). The complexity is indeed intriguing, but it is difficult to determine the extent of each pathway.

### D. Phosphorus Compounds and Supply of Energy and Reducing Equivalents

1. METABOLISM OF PHOSPHORUS COMPOUNDS

To insure adequate supplies of phosphorus, potassium, magnesium, and calcium for the synthesis of metabolites and of functional and structural constituents (particularly ATP, coenzymes, and nucleic acids) for the early growth of seedlings prior to efficient root absorption, seeds usually assimilate phytic acid in the form of magnesium, potassium, and calcium salts during their maturation. These salts are hexaphosphates of inositol (phytin) which may amount to 53% of the total phosphorus in oats to 80% in rice (Preece *et al.,* 1960; Hall and Hodges, 1966; Asada *et al.,* 1969). Phytin has been found in sunflower *(Halianthus annuus),* maize, vetch *(Vicia sativa),* barley, crested wheatgrass *(Agropyron desertorum),* wheat, beans, and cotton (Matheson and Strother, 1969; Mayer and Poljakoff-Mayber, 1963). A detailed account of compositional changes in different phosphorus compounds during germination of oats *(Avena sativa)* in the dark is summarized in Fig. 2.13. It is clearly shown that the increase of various phosphorus compounds in seedlings is at the expense of endosperm reserves.

The utilization of phytate was explored in germinating wheat (Matheson and Strother, 1969). They found an *in vivo* increase of phytase activity during germination from 680 units/gm wheat seeds to 930 at 1 day, 1400 at 2 days, and 2120 at 3 days with a simultaneous decrease of 38 nmoles of phytate per gram of seeds during the first 5 days. The increase in hydrolytic products, inositol, and inorganic phosphate, was only 16 and 24%, respectively, of the theoretical value, indicating rapid utilization of these products. Further incorporation study of myoinositol-U-[14]C by seedlings showed that pectin, hemicellulose, soluble intermediates, and to a small extent, lipids were labeled. One interesting observation was that the inorganic phosphate content (mg/100 gm fresh weight) remained constant in seedlings during the 14 days of germination, which agreed with the results of Hall and Hodges (1966). Matheson and Strother (1969) thought the concentration of inorganic phosphate in tissue might be the controlling mechanism of phytase activity.

Phytase in different seed tissues may not be identical with respect to

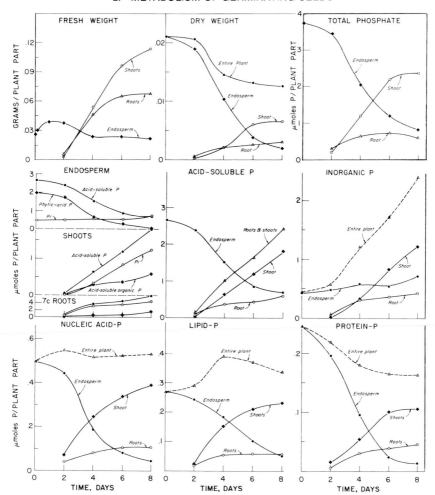

FIG. 2.13. Changes in weights and phosphorous compounds in various parts of germinating oats (Hall and Hodges, 1966).

origin, induction, and inhibition. The following observations support this statement. The increase of phytase in the scutellum of wheat during germination was completely abolished by inorganic phosphate ($2 \times 10^{-2} M$), puromycin ($3 \times 10^{-4} M$), or actinomycin D (80 $\mu$g/ml). Only puromycin repressed the synthesis of wheat embryo phytase, while actinomycin D delayed, and inorganic phosphate had no effect on the embryo phytase (Bianchetti and Sartirana, 1967). Eastwood *et al.* (1969) found that the increase in phytase activity during germination in the aleurone layer of wheat seeds resulted from activation of a preexisting, inactive form.

Hydroxylamine, glutamine, a combination of either one with indoleacetic acid, or purine and pyrimidine nucleotides induced large increases in phytase activity and the induction was not inhibited by actinomycin D, chloramphenicol, puromycin, or cycloheximide. The increase of total phytase during germination, however, was influenced by gibberellin (Srivastava, 1964).

Phosphorylation of soluble metabolites and coenzymes during germination of crested wheatgrass was demonstrated by Wilson and Harris (1966). A total of some thirty acid-soluble phosphorylated compounds was separated by ion exchange resin column chromatography of acid extracts of 2-day germinants; hexose phosphate, UDP-hexose, ADP, ATP, UTP, and degradation products of phytic acid, inositol phosphates were observed after 1 day. A quantitative study of this type would be of value for detailed metabolic investigation.

Approximately 60% of the thiamine was in the cotyledons of dry broad beans of which 30% was in the form of phosphate and the other 30% in the free form. The free form became phosphorylated after 24 hours germination, while no conversion was observed if the cotyledons were detached from the embryo axis at the beginning of germination (Kikuchi and Hayashi, 1969). This indicates that the activation or synthesis of thiamine kinase (EC 2.7.6.2) is dependent on the presence of the embryo axis. The embryo axis of dry broad bean contained 40% (20 $\mu g/gm$ dry weight) of the total thiamine, all of which was in free form. The thiamine was esterified with phosphate after 5 hours soaking of the detached axes, and all the thiamine was phosphorylated to the triphosphate in 17 hours. The fact that the usual coenzyme, thiamine pyrophosphate, was found at later stages of germination led Kikuchi and Hayashi to consider the thiamine triphosphate as a "sparking" substance for seed germination. The significance of this compound awaits further experimentation.

A good correlation of the activity of thymidine kinase (ATP-thymidine 5'-phosphotransferase, EC 2.7.1.21) and uridine kinase (ATP-uridine 5'-phosphotransferase, EC 2.7.1.48) with the accumulation of RNA and DNA was shown in corn seedlings (Wanka and Walboomers, 1966). The uridine kinase at the early stage of germination was different from that at the latter stage in electrophoretic mobility and ammonium sulfate precipitation characteristics. Hence, some of the kinase may also participate in a catabolic role. Two components of the thymidine kinase were separated by electrophoresis, and a complex consisting of one each of these two components gave maximum activity. Component T was low in dry seeds but increased rapidly after 36 hours germination. Component P was higher in activity than component T in dry seeds but maintained its activity up to 48 hours germination, then increased rapidly. These findings indicate that the two components are synthesized by different mechanisms, and their ratio controls the total activity of the enzymes. Uridine kinase behaved similarly. More work apparently will be needed to resolve whether these components are tissue specific, coenzyme form specific, etc., before their significance can be evaluated.

The fact that $MgATP^{2-}$ was the substrate of nucleotide monophosphokinases (ATP-ribonucleoside monophosphate phosphotransferase, EC 2.7.4.6) isolated from cotyledons of germinating cucumber seed, whereas free $ATP^{4-}$ was an inhibitor, shows the *in vivo* importance of cofactor concentration in addition to the availability of substrate (May and Symons, 1968). Localized organellar concentrations of cofactor and substrate would also control the activity of the enzymes, since May and Symons found that 30–80% of the nucleotide kinase activity was contained in the chloroplast fraction, 20–75% in cytoplasm, and 3–8% in mitochondria of the cotyledon cells. Furthermore, many peaks were separated from the enzyme extract, indicating that organellar-specific enzyme may be present.

The rapid synthesis of coenzymes and nucleotides during germination is shown by the sixfold increase of 5-methyltetrahydrofolic acid in 48 hours in 1-day-old pea seedlings (Roos *et al.,* 1968).

## 2. ENERGY SUPPLY

Seed germination requires a tremendous amount of biological energy (ATP) not only for biogenesis of new cellular constituents in seedlings, but also for the formation of protein-synthesizing machinery in producing enzymes for degradation and conversion of storage compounds. The availability of ATP is a definite controlling factor in germination and seedling growth. Unfortunately, only a few reports could be found in the literature regarding the energy level in seeds during germination. In lettuce *(Lactuca sativa)* seeds, Pradet *et al.* (1968) elegantly determined water and oxygen uptake, germination percent, and content of adenine nucleotides for the first 20 hours of germination at 20° C. The results are summarized in Fig. 2.4 (Section III,A,2). A net increase of total adenine nucleotides in the first 4 hours and after radicle emergence is shown in the figure. A high ratio of ATP to ADP from 4–10 was maintained in the 20-hour experimental period. Apparently there is no limit in energy supply for the myriad of synthetic activities of seed germination.

A high ATP content was maintained in pea cotyledons during germination. ADP content remained low, and AMP was intermediate in the early stage of germination. The ratio of ATP/ADP was as high as 6–12 in 1- to 3-day germinants (Brown and Wray, 1968, Fig. 2.14). The high AMP content might result from anaerobic conditions in the interior of the large cotyledons during the early stage of germination and from rapid synthesis.

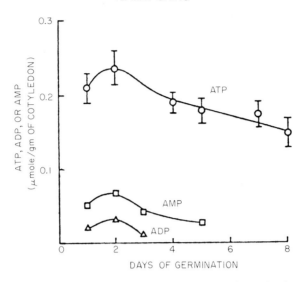

Fig. 2.14. Content of adenine nucleotides in cotyledons of germinating peas (Brown and Wray, 1968).

In Norway spruce *(Picea abies)* and Scotch pine *(Pinus silvestris)* each gametophyte contained a negligible quantity of ATP in ungerminated seeds with embryos of 1 mg in weight. In both seeds germinated at 20°–30° C, ATP increased from zero to 2 to 3 nmoles per gametophyte at the time of radicle emergence, then gradually decreased to 1–2 nmoles when seedling fresh weight was 18 mg. There was a gradual increase in ATP from about zero in ungerminated embryos to a peak of 2 nmoles when seedling fresh weight was 6 mg and they had a water content of 80%. ATP then decreased somewhat when seedling weight reached 18 mg. The highest ATP/ADT ratio in both species when seeds were germinated at low temperature (15°C), was 3.0–3.5 in the gametophyte and 2.3–2.7 in the embryo, while the average was about 2 for different seed sources. The high ATP content was not related to optimum temperature for germination, however (Bartels, 1968).

In germinating barley seeds, the ratio of ATP/ADP was found to be 0.83, 1.85, and 3.9 for 1-, 3- and 5-day germination, respectively (Latzko and Kotze, 1965). Scutellum slices of 4-day-old corn germinants contained 450, 190, and 60 nmoles/gm fresh weight of ATP, ADP, and AMP, respectively. An ATP/ADP ratio of 2.4 is shown in this material (Garrard and Humphreys, 1969).

In almost all known cases, ATP supply does not appear to be limiting during germination in suitable temperature range and with normal oxygen supply. For survival reasons, seeds are provided with apparatus (mitochondria) for ATP production and systems for ADP synthesis and production of inorganic phosphate to meet the energy requirement of growth activities. If the environmental conditions are changed to adverse ones, such as very low temperatures and anaerobic condition, ATP would be limiting and germination arrested. A good example is the rapid induction of adenylate kinase by a nitrogen atmosphere in imbibed lettuce seeds to change ADP to AMP and ATP so that the energy demand could be temporarily relieved until all the energy charge is exhausted (Pradet *et al.,* 1968; Pradet and Bomsel, 1969, Fig. 2.15).

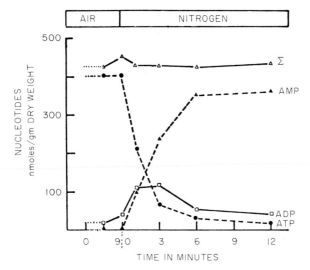

FIG. 2.15. Change in content of adenosine phosphates in response to anaerobiosis in imbibed lettuce seeds (Pradet *et al.,* 1968).

Atkinson defined "energy charge" as half of the number of anhydride-bound phosphate groups per adenine moiety (Atkinson, 1969). ATP contains two anhydride-bound phosphates, thus it is fully charged or has an energy charge of one. ADP accordingly contains 0.5 energy charge. In a biological system, the relative energy charge is calculated by the sum of the mole fraction of ATP plus half the mole fraction of ADP or in terms of individual concentrations.

$$\frac{[ATP]+\frac{1}{2}[ADP]}{[ATP]+[ADP]+[AMP]} = \text{Energy charge}$$

In animal material, Atkinson observed that enzymes involved in ATP-regenerating and ATP-utilizing sequences respond to the energy charge of the tissue (Atkinson, 1968). When the energy charge is less than 0.5, ATP-regenerating systems are dominantly active; when the energy charge is greater than 0.5, ATP-utilizing systems become active until the energy charge approaches one where ATP-utilizing systems are predominant. Thus, energy charge is an index regulating the activity of various sequences related with energy utilization and regeneration of the tissue.

Atkinson's hypothesis is confirmed by the data obtained with lettuce seeds which show (Fig. 2.15) that anaerobiosis prevents oxidative phosphorylation of ADP but induces adenylate kinase for regeneration of ATP (Pradet *et al.,* 1968; Pradet and Bomsel, 1969). The metabolism of seed germination apparently can be modified by environmental conditions. The regulation by energy charge is one of the examples in suppressing some sequences and inducing others. For better understanding of growth pattern and developmental control, energy charge of the tissue should be explored.

3. REDUCING EQUIVALENTS

Reducing equivalents stored in the form of reduced nicotinamide adenine dinucleotide (NADH) or reduced nicotinamide adenine dinucleotide phosphate (NADPH) and as easily oxidized substrates coupled to NAD or NADP linked dehydrogenases are another biological currency needed for synthesis of many cellular components involving one or more reductive steps (Ragland and Hackett, 1965). In embryonic tissue they are particularly in demand for growth of new cells and tissues. NADPH is required for the synthesis of deoxynucleosides from ribonucleosides, fatty acids from acetyl units, glutamic acid from $\alpha$-ketoglutarate, and reduction of nitrate, glutathione, and sulfate. NADH assists in hexose formation from precursors containing three carbons, in steroid biosynthesis, etc. While hexoses and glutamic acid might be ample in seeds, deoxynucleosides, fatty acids, ammonia, reduced glutathione, sulfide, and steroids are needed for biogenesis of new cellular constituents, particularly in meristematic tissues.

Several key intermediates in the *de novo* synthetic pathway and ensymes required for NAD synthesis have been found in plants (Hadwiger *et al.,* 1963), and the conversion of nicotinamide or nicotinic acid to NAD has also been noted (Waller *et al.,* 1966; Mukherji *et al.,* 1968). Further phosphorylation of NAD to NADP was catalyzed by NAD kinase with ATP and $Mg^{2+}$, and the enzyme activity was enhanced by red light and inhibited by far-red light in cotyledons of germinating seeds of *Pharbitis nil* (Tezuka and Yamamoto, 1969). A phytochrome preparation from the first internode of Alaska pea seedlings contained NAD kinase, the specificity of which increased as the phytochrome preparation was further purified. The $K_m$ of the enzyme decreased from $1.84 \times 10^{-3}$ $M$ to $0.9 \times 10^{-3}$ $M$ by irradiating the preparation with red light (Yamamoto and Tezuka, 1969). Many of NAD or NADP dependent dehydrogenases were observed in germinating seeds (Yamamoto, 1967), and the conversion of NADH to NADPH by transhydrogenase (EC 1.6.1.1) has been demonstrated in pea mitochondria (Davies, 1956). All these reports indicate that germinating seeds are capable of synthesizing these coenzymes and their contents in tissue are probably controlled by the concentrations and activities of synthetic enzymes, ATP, $Mg^{2+}$, phytochrome, and substrates.

During seed germination, each organ seems to exhibit a specific pattern of nicotinamide coenzymes. Little resemblance can be seen between the cotyledons of germinating Japanese beans *(Vigna sesquipedalis)* and peas as shown in Fig. 2.16a and d. In bean hypocotyls, NADP was the main coenzyme (Fig. 2.16b), while NAD was the major one in the storage organ (Fig. 2.16a). When whole rice seeds *(Oryza sativa)* were analyzed a combined pattern of NAD dominance at the early stage and then NADP at the latter stages of germination was observed (Fig. 16c). A parallel, sigmoid increase of NADPH and NAD, however, was found in germinating corn seeds from 1 to 14 days, and the highest total content of $2.5 \times 10^{-3}$ $M$ of these coenzymes was reached in 2-day-old germinants (Fritz *et al.,* 1963). In 3-day-old Japanese bean seedlings, a total content of $1.3 \times 10^{-3}$ $M$, $1.2 \times 10^{-4}$ $M$, and $2.9 \times 10^{-5}$ $M$ of nicotinamide nucleotides was observed respectively in the tip, middle, and base of roots. These amounts decreased with seedling age to about 30% of that on the sixth day (Yamamoto, 1963). Based on the data reported by Brown and Wray (1968), a gradual decrease of total nicotinamide coenzymes from the highest of $2.7 \times 10^{-4}$ $M$ on the second day to $1.1 \times 10^{-4}$ $M$ on the tenth day was shown in germinating pea cotyledons. These concentrations fall within the range of these coenzymes generally found in active cells (Sund, 1968). Table 2.VII summarizes more data on the contents of nicotinamide coenzymes in various organs of germinating seeds.

The data in Table 2.VII, particularly the fifth column, afforded evidence which led Yamamoto (1963) to conclude that NAD and NADH are related with catabolic activities in storage tissue of seeds, and NADP and NADPH correlate with anabolic metabolism of embryo tissues. From enzyme studies, Yamamoto speculated that both coenzymes could be rate limiting for specific enzyme reactions *in vivo*. The content of these coenzymes was rapidly influenced by oxygen concentration, light conditions, low temperature, hormones, nutrition, and developmental stages (Effer and Ranson, 1967b; Yamamoto, 1969). These coenzymes

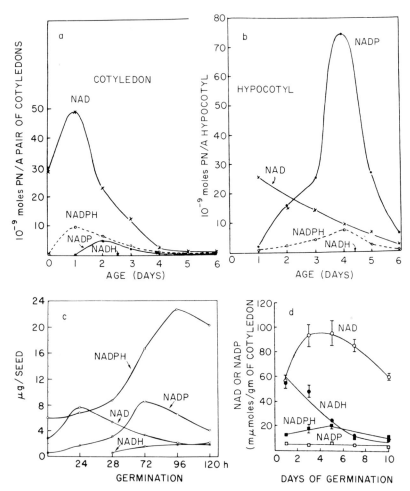

Fig. 2.16. Content of nicotinamide adenine dinucleotide (NAD) and nicotinamide adenine dinucleotide phosphate (NADP) in: cotyledon (a), and hypocotyl of germinating beans (b) (Yamamato, 1963); germinating rice (c) (Mukherji *et al.*, 1968); and cotyledon of germinating peas (d) (Brown and Wray, 1968).

must play an important role in metabolic control of sequences related with seed germination. The activation of NADP-dependent enzymes by NADP has been indicated (Yamamoto, 1969) but what controls the total concentration in various tissue is not precisely known. Genetic programming by regulating enzyme concentration for synthesis, metabolic modulation by the availability of ATP and $Mg^{2+}$, and environmental regulation via the effect of light quality on phytochrome may all play a role in the dynamic changes in nicotinamide coenzymes in cells. The bio-

**TABLE 2.VII.**

CONTENT OF OXIDIZED AND REDUCED NICOTINAMIDE COENZYMES IN VARIOUS PARTS OF 5-DAY-OLD GERMINATING SEEDS[a]

| Kind | Organ | NAD | NADH | NADP | NADPH | $\dfrac{\text{NADP}+\text{NADPH}}{\text{NAD}+\text{NADH}}$ | $\dfrac{\text{NADH}+\text{NADPH}}{\text{NAD}+\text{NADP}}$ |
|---|---|---|---|---|---|---|---|
| | | | nmoles/part | | | | |
| Vigna sesquipedalis | Cotyledon | 3.86 | 0.00 | 0.29 | 0.27 | 0.15 (1)[b] | 0.06 |
| | Hypocotyl | 7.20 | 0.00 | 26.80 | 2.12 | 4.02 (27) | 0.06 |
| Phaseolus angularis | Cotyledon | 34.70 | 6.56 | 4.99 | 1.89 | 0.17 (1) | 0.21 |
| | Epicotyl | 3.00 | 2.81 | 4.66 | 5.77 | 1.80 (11) | 1.12 |
| Soybean | Cotyledon | 36.90 | 21.20 | 1.80 | 4.03 | 0.10 (1) | 0.65 |
| | Hypocotyl | 0.58 | 0.29 | 1.02 | 1.86 | 3.31 (33) | 1.34 |
| Castor bean | Endosperm | 25.30 | 14.75 | 8.57 | 4.73 | 0.33 (1) | 0.58 |
| | Hypocotyl | 2.24 | 0.01 | 0.05 | 8.24 | 3.69 (11) | 3.71 |
| Sunflower | Cotyledon | 11.10 | 0.79 | 0.70 | 2.58 | 0.28 (1) | 0.29 |
| | Hypocotyl | 0.38 | 0.01 | 2.63 | 9.50 | 31.10 (111) | 3.15 |
| Corn | Endosperm | 5.72 | 1.20 | 1.05 | 0.64 | 0.24 (1) | 0.27 |
| | Coleoptile | 0.62 | 0.77 | 4.62 | 1.89 | 4.69 (20) | 0.51 |
| Watermelon | Cotyledon | 7.60 | 0.02 | 0.01 | 1.30 | 0.17 (1) | 0.17 |
| | Hypocotyl | 0.16 | 0.29 | 2.09 | 3.83 | 13.16 (77) | 1.83 |
| Rice | Endosperm | 1.48 | 0.16 | 0.15 | 0.00 | 0.09 (1) | 0.10 |
| | Coleoptile | 0.04 | 0.02 | 0.13 | 0.28 | 6.83 (76) | 1.76 |
| Wheat | Endosperm | 1.88 | 0.04 | 0.44 | 0.06 | 0.26 (1) | 0.04 |
| | Coleoptile | 0.10 | 0.05 | 0.74 | 0.91 | 11.01 (42) | 1.14 |

[a] From Yamamoto (1963).
[b] relative ratio of growing part to storage tissue.

synthesis of NAD was recently demonstrated in the first leaf of barley seedling (Ryrie and Scott, 1969). The procedure used in that study may be applied to germinating seeds, and the factors affecting NAD biogenesis may be discerned.

## IV. Specific Metabolic Processes

### A. Metabolism of Fatty Seeds

The fatty seeds are discussed first for two reasons: (a) 75% of the spermatophytes contain fats as reserve in seeds, and (b) more recent information is available in the literature.

### 1. COMPOSITIONAL CHANGES

As a background for further detailed metabolism of fatty seeds, the compositional changes during germination of Douglas fir (*Pseudotsuga menziesii* Franco) seeds are shown in Figs. 2.17–2.19 (Ching, 1966). The seed is composed of an outer thin woody seed coat, a middle thick haploid megagametophyte, and a small embryo in the center.

Germination is indeed a growth process of the embryo as shown by the 35-fold increase in fresh weight and six-fold increase in dry weight of the seedling during the first 2 weeks of growth. In contrast, little change in fresh weight and a 70% reduction in dry weight were observed in gametophytic tissue (Fig. 2.17, upper and middle left). Apparently the embryo grew at the expense of the gametophyte with a loss of 8% of the total weight, probably for energy supply (Fig. 2.17, middle left). The seed-coat seemed to assist in water uptake during the early stages of germination (Fig. 2.17, lower left).

Little change of DNA content was observed in the gametophyte during early stages of germination, but a 60% reduction was found at the end of germination. The embryo showed a doubling of DNA content at the radicle emerging stage, 4th day in germination, and a 4.5-fold increase toward the end of germination. The DNA content of the gametophyte was very low and constituted only 0.13% of its dry weight, whereas that of the embryo was 0.25% (Fig. 2.17, upper right).

In dry seeds, 5 $\mu$g and 19 $\mu$g of RNA were found in each embryo and gametophyte, respectively (Fig. 2.17, middle right). These quantities constituted 0.50 and 0.21% of their dry weight, respectively. A slight increase in RNA in both embryo and gametophyte was found in stratified seed. A rapid rate of RNA synthesis was observed in seedlings and a moderate increase followed by a rapid decrease were observed in the gametophyte. At about the time of plumule emergence, seedlings showed a 12-fold increase in RNA while the gametophyte contained only two-

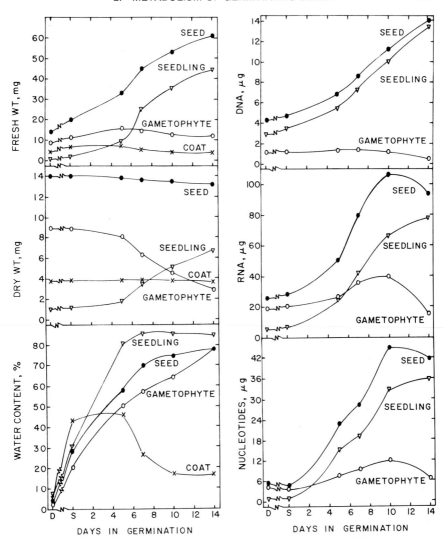

FIG. 2.17. Changes in fresh weight, dry weight, water content, DNA, RNA, and nucleo-tides in individual embryo, female gametophyte, seed coat, and whole seed of Douglas fir seed during germination. (D = air-dried seed, S = stratified seed.) (Ching, 1966.)

thirds of its original RNA content. The sigmoid curve of RNA increase in seedlings interestingly paralleled the increase in fresh and dry weight and DNA. This perhaps indicates accumulation mainly of structural RNA (ribosomal) and to a lesser extent, messenger RNA and soluble RNA. In gametophytic tissue, the moderate increase of RNA coincided with the stages of rapid transfer of dry matter from gametophyte to seedling. This increase may reflect an active synthesis of enzymes for rapid degradation of proteins and lipids, and conversion of acetyl CoA to sugars. At the end of germination, the gametophyte apparently nourished the seedling by hydrolyzing RNA and DNA to nucleotides and presumably transferring them to the seedling. Hydrolysis of RNA and DNA at this stage of development in gametophytic tissue was further verified by the marked increase of soluble phosphorus compounds and inorganic phosphate (Fig. 2.19, lower right, p. 160).

The change in soluble nucleotides was similar to that in RNA except for small reduction shown by the gametophyte after stratification (Fig. 2.17, lower right). This reduction perhaps provides substrate for the small increase in DNA in the embryo and in RNA in both the embryo and gametophyte during stratification. A 24-fold increase of nucleotides in the seedling at the end of germination was observed, while only a twofold increase was found in the gametophyte at the cotyledon emerging stage. These changes in DNA, RNA and soluble nucleotides indicated *de novo* synthesis of nucleic acids and are in general agreement with the findings on corn, peanuts, peas, and castor beans (Section IV,B,1 and IV,C,1).

Lipids were the major food reserves in both the gametophyte and embryo in dry seeds (Fig. 2.18, upper left). Lipids made up 48 and 55% of the dry weight of the gametophyte and embryo, respectively. During stratification, gametophytic lipids were reduced by 0.3 mg but little change occurred in the embryo. During germination, lipids in the gametophyte continued to decrease from 4.1 to 0.5 mg. After an initial drop to 0.3 mg, the lipid content of the seedling gradually increased to 0.9 mg.

Changes in metabolites soluble in aqueous methanol and ethanol during germination are summarized in Fig. 2.18, middle left. Soluble compounds made up 13% of the dry weight of the embryo. They increased slightly during stratification, and toward the end of germination, increased rapidly to 45% of seedling dry weight. Soluble compounds in the gametophyte constituted 14–24% of the dry weight. The data for the residue insoluble in chloroform-methanol, aqueous methanol, and ethanol are plotted in Fig. 2.18, lower left. Rapid accumulation of insoluble residue in the seedling coincided with a marked reduction in the gametophyto after radicle emergence. Sugars, sugar nucleotides, other organic phosphates, inorganic phosphate, amino acids, amides, peptides, soluble proteins, organic acids, etc., were detected in the soluble fraction by trichloroacetic acid precipitation, ion exchange resin column chromatography, and paper chromatography.

A slow accumulation of sugars and starch during stratification, a plateau prior to radicle emergence, then a rapid increase is shown in Fig. 2.18, right column. During early stages of germination, the increase was in

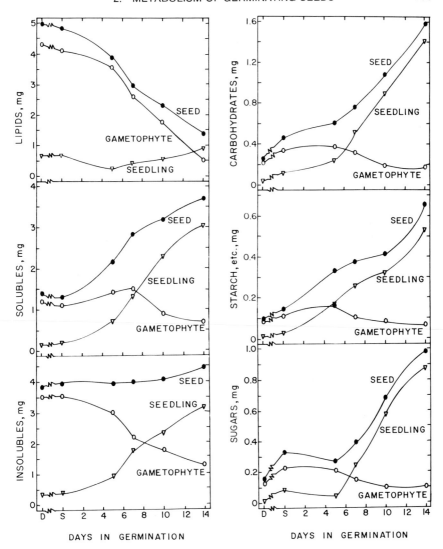

FIG. 2.18. Changes in lipids, total aqueous alcohol-soluble compounds, insoluble residues, total carbohydrates, starch and sugars in individual embryo, female gametophyte, and seed without coat of Douglas fir during germination. (D = air-dried seed, S = stratified seed.) (Ching, 1966.)

both the embryo and gametophyte, while at later stages, a rapid increase was observed in the seedling. These sequential changes agree with the findings in angiosperms and verify the means of food transfer by sugars during seed germination. A pronounced increase of sugars in both the embryo and gametophyte after stratification was of interest and may be considered a beneficial effect of stratification.

Utilization of nitrogenous reserves is shown in Fig. 2.19, left column. A slight reduction (9%) in total N was observed, probably as a result of root diffusional losses. Rapid translocation of soluble compounds from

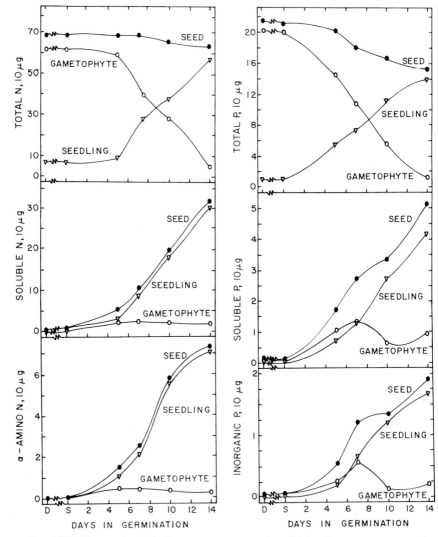

FIG. 2.19. Changes in total nitrogenous compounds, soluble nitrogenous compounds, and α-amino N-compounds, total, soluble, and inorganic phosphorous compounds in individual embryo, female gametophyte, and seed without coat of Douglas fir during germination. (D = air-dried seed, S = stratified seed.) (Ching, 1966.)

the gametophyte to seedling was accomplished with little accumulation in storage sites. Solvation of insoluble protein, activation of preexisting enzymes, degradation of storage protein, active transport and synthesis to new protein, and other nitrogenous compounds occurred as a chain of events mainly in the gametophyte during the early stages, prior to radicle emergence, and mostly synthesis in the seedling during the later stages. It is also interesting to note that the soluble nitrogenous compounds in the seedling comprised about 50% of the total nitrogenous compounds of which only 10 to 15% were free amino acids and amides. These changes agree with those reported for corn and beans and represent a general trend for seed germination (Sections IV,B,1) and IV,C,1).

Phosphorus compounds are important in the dynamic process of synthesis and energy supply. The changes during germination are summarized in Fig. 2.19, right column. A reduction of 29% in total phosphorus in the seed at the end of germination is puzzling; perhaps this may be attributed to root diffusion. Soluble phosphorus compounds in the seed increased 40-fold and constituted 30% of the total P by the end of germination, with 80% found in the seedling and 20% in the gametophyte. Inorganic phosphate increased similarly to soluble phosphorus compounds, and constituted about 30 to 50% of the soluble P compounds. The reserve form of phosphorus probably is phytin since it is commonly found in angiosperm seeds. Change in phosphorus compounds during germination were in general agreement with those in cotton seed and oats (Section III,D,1; Hall and Hodges, 1966).

It is clear from Figs. 2.17–2.19 that the haploid gametophyte enveloping the embryo is the storage tissue where lipids, proteins, and phosphorus reserve are hydrolyzed, converted, and translocated to the embryo for synthesis of various cellular constituents. The well-studied castor beans are similar in structure to Douglas fir seeds except that the storage organ of the former is the triploid endosperm. The changes of dry weight, nitrogenous compound, carbohydrates, and lipids in castor beans during germination are comparable to those in Douglas fir seeds (Yamada, 1955a,b; Canvin and Beevers, 1961; Kriedmann and Beevers, 1967a,b; Stewart and Beevers, 1967; Marré, 1967).

## 2. General Lipid Metabolism

Most seed lipids are comprised of triglycerides. The seeds of quaking grass *(Briza spricata),* however, contain 20% lipids of which 49% are digalactosylglycerides and 29% monogalactosylglycerides with little triglycerides (Smith and Wolff, 1966).

The quantitative changes of major lipid classes during germination as exemplified by Douglas fir seeds are given in Table 2.VIII (Ching, 1963).

**TABLE 2.VIII.**
WEIGHT OF SEED, TOTAL LIPID (TL), TRIGLYCERIDES (TG), FREE FATTY ACIDS (FFA),
PHOSPHOLIPIDS (PL), AND UNSAPONIFIABLES (US) DURING
GERMINATION OF A DOUGLAS FIR SEED[a]

| Germination days | Dry wt. (mg) | TL (mg) | TG (mg) | FFA ($\mu$g) | PL ($\mu$g) | US ($\mu$g) |
|---|---|---|---|---|---|---|
| 0 | 13.1 | 4.2 | 3.7 | 126 | 210 | 45 |
| 5 | 12.6 | 4.0 | 3.4 | 80 | 340 | 60 |
| 7 | 12.5 | 3.4 | 2.8 | 170 | 360 | 92 |
| 10 | 12.4 | 2.7 | 2.1 | 108 | 370 | 87 |
| 14 | 12.1 | 1.4 | 0.8 | 56 | 378 | 96 |

[a]From Ching (1963).

Similar changes were observed in seeds of flax *(Linum usitatissimum)* (Zimmerman and Klosterman, 1965).

These data indicate that (*1*) triglycerides are utilized, (*2*) phospholipids are synthesized presumably for cellular and organellar membranes in the seedling, (*3*) a small fluctuation of free fatty acids occurs, and (*4*) a small increase in the unsaponifiable fraction (e.g., waxes, pigments, and sterols) is observed in Douglas fir seeds probably for the cuticle layer, chloroplasts, and cellular and organellar membranes.

Recent techniques of lipid chemistry (McKillicam, 1966; Allen *et al.,* 1966) combined with modern macromolecular and organellar fractionation of material collected at different stages of germination would be of value in elucidating biosynthetic pathways and explaining the relationship of structure and function in different organelles. At present, no complete data for germinating seeds are available.

Varietal differences of composition of triglycerides, fatty acids, and unsaponifiables in various economically important seed crops have been shown with the aid of gas–liquid chromatography and thin layer chromatography (Herb, 1968; Earle *et al.,* 1968; Fedeli *et al.,* 1968). While this information is useful in economic and taxonomic studies at present, these varieties may provide material for quantitative genetic and metabolic studies in the future.

a. STEROL COMPOUNDS.   Kemp *et al.* (1967) found no change in total sterol compounds in germinating corn, an increase of esterified sterols in the scutellum, and rapid accumulation of free sterols in shoots and roots. Table 2.IX illustrates quantitative changes of different sterols in various tissues during germination. An interesting trend shown in the table is the reduction of both free and esterified $\beta$-sitosterol and an increase of stigmasterol in root tissue. Lack of compositional changes in

**TABLE 2.IX.**
COMPOSITION OF FREE AND ESTERIFIED STEROLS OF GERMINATING MAIZE[a]

| Germination days | Sterol[b] | Total mg/100 seedling | % of Sterol weight[c] | | |
|---|---|---|---|---|---|
| | | | β-Sitosteral | Stigmasterol | Campesterol |
| | | Shoot | | | |
| 4 | F | 1.17 | 51 | 29 | 20 |
| | E | 1.47 | 65 | 13 | 22 |
| 9 | F | 4.10 | 50 | 29 | 21 |
| | E | 1.27 | 64 | 17 | 19 |
| 13 | F | 8.92 | 50 | 30 | 20 |
| | E | 1.45 | 69 | 13 | 18 |
| | | Root | | | |
| 4 | F | 1.08 | 37 | 37 | 26 |
| | E | 0.43 | 67 | 17 | 16 |
| 9 | F | 3.97 | 21 | 49 | 30 |
| | E | 0.60 | 57 | 26 | 17 |
| 13 | F | 6.23 | 20 | 51 | 29 |
| | E | 0.74 | 49 | 32 | 19 |
| | | Scutellum | | | |
| 4 | F | 4.36 | 57 | 17 | 26 |
| | E | 2.30 | 70 | 10 | 20 |
| 9 | F | 3.36 | 59 | 14 | 27 |
| | E | 3.35 | 69 | 9 | 22 |
| 13 | F | 3.30 | 57 | 15 | 28 |
| | E | 5.95 | 69 | 10 | 21 |
| | | Endosperm | | | |
| 4 | F | 9.57 | 69 | 10 | 21 |
| | E | 3.56 | 76 | Tr | 24 |
| 9 | F | 8.35 | 69 | 11 | 20 |
| | E | 5.00 | 77 | Tr | 23 |
| 13 | F | 8.65 | 71 | 10 | 19 |
| | E | 3.00 | 77 | Tr | 23 |

[a]From Kemp *et al.* (1967).
[b]F = free sterols, E = esterified sterols.
[c]Tr = trace

other compounds probably means uniform synthesis or they are at a steady state of metabolism.

In light or dark, steroids and free sterols increased in the radicle, hypocotyl, and epicotyl of germinating beans. In the cotyledons, free sterols decreased while steroids increased considerably (Duperon and Duperon, 1969).

In a study involving incubation of shoots of germinating bean seedlings with mevalonic-2-$^{14}$C for various lengths of time, and subsequently isolation of sterols from cellular fractions of chloroplasts, mitochondria, microsomes and supernatant, Knapp *et al.* (1969) concluded that the site of sterol biosynthesis is microsomal and supernatant since the total and specific radioactivities were highest in those two fractions and the activities increased with the time of incubation.

b. FATTY ACIDS. Seed lipid reserve, triglyceride, is first hydrolyzed to glycerol and fatty acids by the action of lipases. Recent information on the intracellular distribution of lipase will be discussed in the next section. With the advent of elegant analytic techniques in recent years, thirty-five new fatty acids have been discovered (Wolff, 1966), some of which are of economic importance in addition to their general nutritional value. Some of them are oxygenated ($-CH-CH-$), acetylenic ($-C{\equiv}C-$),

$$-CH-CH- \quad \overset{\diagdown\diagup}{O}$$

and allenic ($-CH{=}C{=}CH-$). These structural variations of fatty acids at present could only be ascribed to genetic variations; in the future, some functional roles might be found to be associated with the different molecular structures of fatty acids.

The composition of fatty acids in different classes of lipids has been analyzed by gas–liquid chromatography. Table 2.X reveals differences between seeds and seedlings of flaxseed. Similar results were observed in germinating Douglas fir seed (Ching, 1963).

Several points of metabolic interest are indicated in this type of study: (*1*) fatty acids with longer chains are synthesized during germination, particularly in phospholipids, (*2*) preferential utilization of the major reserve component is indicated (e.g., linoleic acid for Douglas fir seeds and linolenic acid for flaxseeds), and (*3*) $\alpha$ oxidation of fatty acids in addition to $\beta$ oxidation is operative in flaxseeds.

The enzymes for $\alpha$ oxidation of fatty acids probably are located in the mitochondria and supernatant (Stumpf, 1969) and the $\beta$ oxidation definitely takes place in glyoxysomes that will be discussed in the next section.

Some hydrolyzed fatty acids are reutilized directly for synthesis of phospholipids and glycolipids that are needed as organellar constituents, but the majority are partially oxidized and converted to sugars which are then transported to the seedling axis for growth. The c backbone of fatty acids could be used also to synthesize amino acids in sunflower, pumpkin *(Cucurbita pepo)*, flaxseed, and watermelon seeds *(Citrullus vulgaris)* (Sinha and Cossins, 1965).

The synthesis of fatty acids appear to be associated with particles. Macey and Stumpf (1968) found that two fractions, a $100,000 \times g$ sedimented pellet (microsomal) and another $10,000 \times g$ precipitated pellet

**TABLE 2.X.**

PERCENTAGE DISTRIBUTION OF FATTY ACIDS IN FREE FORM (FFA), PHOSPHOLIPIDS
(PL), AND TRIGLYCERIDES (TG) IN SEEDS (S) AND SEEDLINGS (L) OF FLAX[a]

| | % Distribution[b] | | | | | |
| | FFA | | PL | | TG | |
| Fatty Acid | S | L | S | L | S | L |
|---|---|---|---|---|---|---|
| 14:0[c] | 0.8 | 0.2 | 2.5 | 0.7 | Tr | Tr |
| 15:0 | Tr | Tr | Tr | – | – | – |
| 15:1 | – | – | – | – | – | – |
| 16:0 | 9.5 | 12.3 | 26.0 | 18.5 | 7.3 | 6.6 |
| 16:1 | 1.2 | 0.8 | 5.5 | 1.9 | – | 0.2 |
| 17:0 | – | 0.1 | Tr | 0.5 | – | Tr |
| 17:1 | – | Tr | 1.0 | 0.4 | – | Tr |
| 18:0 | 1.9 | 3.3 | 6.6 | 3.4 | 1.9 | 3.2 |
| 18:1 | 37.3 | 28.5 | 22.5 | 12.8 | 35.2 | 36.6 |
| 18:2 | 23.2 | 14.6 | 18.4 | 21.9 | 10.9 | 15.2 |
| 18:3 | 26.1 | 19.9 | 17.5 | 37.2 | 44.7 | 38.0 |
| 20:0 | Tr | 0.7 | Tr | 0.4 | – | 0.1 |
| 22:0 | – | 1.7 | Tr | 1.2 | – | Tr |
| 24:0 | – | 13.8 | – | 0.6 | – | – |
| 26:0 | – | 3.9 | – | – | – | – |

[a]From Zimmerman and Klosterman (1965).
[b]Tr = trace, less than 0.1%.
[c]Carbon chain length and number of double bonds.

(mitochondrial) from isotonic homogenate of pea seedlings, were responsible for synthesis of long-chain fatty acids of different lengths. The $100,000 \times g$ pellet synthesized $C_{16}$ to $C_{24}$ whereas the $10,000 \times g$ pellet was limited to $C_{16}$ and $C_{18}$. All systems required acyl carrier protein (ACP), NADH, and NADPH if malonyl CoA was the substrate. If acetyl CoA was the substrate, additional $Mg^{2+}$, $CO_2$, and ATP were required. A major fatty acid-synthesizing system was, however, isolated by $1000 \times g$ centrifugation of developing and germinating safflower *(Carthamus tinctorius)* seed homogenate (McMahon and Stumpf, 1966). The requirements for substrates, coenzymes, and cofactors are similar to the pea system. The site of oleic acid biosynthesis in developing castor bean endosperm was found in glyoxysome-like particles (Zilkey, 1969).

In view of the partial autonomy of mitochondria and chloroplasts discovered in recent years, it is logical to expect localized synthesis of structural constituents in the organelles. With improved analytic techniques, a more clear-cut concept of the division of labor will evolve in the near future.

c. Conversion of Fatty Acids to Sugars. The conversion of fatty reserve to sugars in germinating seeds has been known for 15 years (Yamada, 1955b) and the pathways were established about 10 years ago (Bradbeer and Stumpf, 1959; Canvin and Beevers, 1961; Fig. 2.20). Since

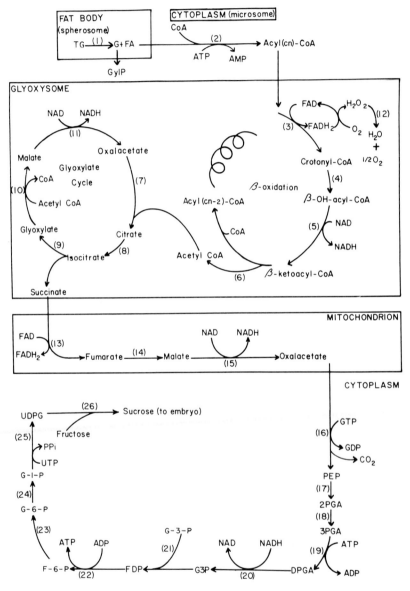

Fig. 2.20. Compartmental pathways of lipid utilization in fatty seeds. (1) Lipase (Glycerol ester hydrolase, EC 3.1.1.3); (2) Fatty acid thiokinase [Acyl CoA synthase, acid:

this is a unique feature and is the predominant metabolic activity in storage tissue of fatty seeds, the key enzymes responsible for the conversion, i.e., isocitrate lyase and malate synthase, have been studied extensively. Most investigations were based on total soluble enzymes extracted from seeds or seedlings. A compartmental distribution of these enzymes in glyoxysome (cytosome, microbody, peroxisome) was discovered by Breidenbach and Beevers in 1967. Details of compartmentation of lipid metabolism will be discussed in the next section. This section will review some of the pertinent information on isocitrate lyase and malate synthase.

*i. Change of enzyme activity during germination.* Carpenter and Beevers (1959) found that isocitrate lyase activity was very low in ungerminated seeds of cotton, castor bean, and pumpkin, but it increased quickly to a peak when fat breakdown was at its highest and then declined again. They also observed that isocitrate lyase was very active in cotyledons of germinating seeds of sunflower, soybean, peanut, and watermelon. Purified isocitrate lyase in the endosperm of castor bean had an optimum pH of 7.5 and an absolute requirement for $Mg^{2+}$ and sulfhydryl

---

CoA ligase (AMP), EC 6.2.1.3]; (3) Acyl-CoA dehydrogenase (Acyl CoA:NAD oxidoreductase, EC 1.3.99.3); (4) Crotonase (L-3-hydroxyacyl-CoA hydro-lyase, EC 4.2.1.17); (5) β-Hydroxyacyl-CoA dehydrogenase (L-3-hydroxyacyl-CoA:NAD oxidoreductase, EC 1.1.1.35); (6) β-Ketoacyl thiolase (Acyl-CoA c-acyl transferase EC 2.3.1.16); (7) Citrate synthase [citrate oxalacetate lyase (CoA-acetylating) EC 4.1.3.7]; (8) Aconitase [citrate (isocitrate) hydro-lyase, EC 4.2.1.3]; (9) Isocitrate lyase (*threo*-Ds-isocitrate glyoxylate-lyase, EC 4.1.3.1); (10) Malate synthase (L-malate glyoxylate-lyase, EC 4.1.3.2); (11) Malate dehydrogenase (L-malate:NAD oxidoreductase, EC 1.1.1.37); (12) Catalase (hydrogen peroxide:hydrogen peroxide oxidoreductase, EC 1.11.1.6); (13) Succinate dehydrogenase (succinate:FAD oxidoreductase, EC 1.3.99.1); (14) Furmarase (fumarate hydratase, EC 4.2.1.2); (15) Malate dehydrogenase (L-malate:NAD oxidoreductase, EC 1.1.1.37); (16) Phosphoenolpyruvate carboxykinase [GTP, oxalacetate carboxy-lyase (transphosphorylating), EC 4.1.1.32]; (17) Enolase (phosphopyruvate hydratase, EC 4.2.1.11); (18) Phosphoglyceromutase (2,3-diphospho-D-glycerate:2-phospho-D-glycerate phosphotransferase, EC 2.7.5.3); (19) Phosphoglycerate kinase (ATP:3-phospho-D-glycerate 1-phosphotransferase, EC 2.7.2.3); (20) Glyceraldehyde-3-phosphate dehydrogenase [D-glyceraldehyde-3-phosphate:NAD oxidoreductase (phosphorylating), EC 1.2.1.12]; (21) Aldolase (fructose-1,6-diphosphate D-glyceraldehyde-3-phosphate-lyase, EC 4.1.2.13); (22) Phosphofructokinase (ATP:D-fructose-6-phosphate 1-phosphotransferase, EC 2.7.1.11); (23) Phosphohexoisomerase (D-glucose-6-phosphate ketol-isomerase, EC 5.3.1.9); (24) Phosphoglucomutase (α-D-glucose-1,6-phosphate: α-D-glucose-1-phosphate phosphotransferase, EC 2.7.5.1); (25) UDPG pyrophosphorylase (UTP: α-D-glucose-1-phosphate uridyly-transferase, EC 2.7.7.9); (26) Sucrose synthetase (UDP glucose-fructose glucosyltransferase, EC 2.4.1.13). (TG, triglyceride; G, glycerol; FA, fatty acids; FAD, flavine adenine dinucleotide; PEP, phosphoenol pyruvate; 2 PGA, 2-phosphoglyceric acid; 3 PGA, 3-phosphoglyceric acid; DPGA, 1,3-diphosphoglyceric acid; G-3-P, glyceraldehyde 3-phosphate; FDP, Fructose 1,6-diphosphate; F-6-P, Fructose 6-phosphate; G-6-P, glucose 6-phosphate; G-1-P, glucose 1-phosphate; UDPG, Uridine diphosphate glucose; GY1P, glycerol 1-phosphate; GTP, guanosine triphosphate; GDP, guanosine diphosphate.

compounds, and the enzyme was inhibited by malonate. The rise and fall of malate synthase activity also was noted in castor bean endosperm (Yamamoto and Beevers, 1960). A similar activity curve of isocitrate lyase and malate synthase was found in germinating peanut cotyledons (Marcus and Velasco, 1960; Marcus and Feeley, 1964b), mustard seeds (Karow and Mohr, 1967), lettuce seeds (Mayer *et al.*, 1968), watermelon seeds (Hock and Beevers, 1966), and the megagametophyte of Italian stone pine *(Pinus pinea)* (Firenzuoli *et al.*, 1968) and ponderosa pine *(Pinus ponderosa)* (Ching, 1970b).

*ii. Genesis of isocitrate lyase and malate synthase.* Hock and Beevers (1966) treated dry watermelon seeds for 1 day with actinomycin D (10 $\mu$g/ml). They found that normal development of the two enzymes and growth of the seedling were impaired. If the treatment was applied after 1.5 days of the germination, little inhibition was observed in enzyme activities. When cycloheximide (10 $\mu$g/ml) was applied at any time during the germination course, a consistent inhibition of these two enzymes and growth of the seedling was observed. Thus they concluded that normal increases in isocitrate lyase and malate synthase might be due to a re-utilization of relatively stable RNA produced only during the first day of germination, and the decline phase of enzyme activity was a consequence of the limited half-life (2–3 days) of the enzymes. The results suggest that the synthesis of both isocitrate lyase and malate synthase is not directed by long-lived mRNA in watermelon seeds. A differential inhibition of isocitrate lyase by actinomycin D in endosperm and in cotyledons of imbibed castor bean was observed by Lado *et al.* (1968). In endosperm slices the chemical at a concentration of 100 $\mu$g/ml prevented 30–40% of the normal increase of the enzyme activity in 24 hours, whereas in detached cotyledons, the same concentration inhibited 86% of the enzyme activity. The reason for this difference is unknown. This type of incomplete inhibition of isocitrate lyase and RNA synthesis by actinomycin D was further demonstrated in intact peanut seeds (Gientka-Rychter and Cherry, 1968). Soaking the seeds in actinomycin D at 50 $\mu$g/ml for 18 hours reduced both the enzyme activity and the incorporation of uridine-2-$^{14}$C in RNA by 55% in cotyledons of 2-day-old germinants. Cycloheximide (100 $\mu$g/ml) also decreased enzyme activity by 50% of peanut seedlings. These results indicated that the synthesis of RNA and protein is essential to increase the activity of both enzymes in germinating seeds. Whether the newly synthesized RNA contains specific mRNA for isocitrate lyase is unknown. The observed reduction of enzyme activity and RNA synthesis could result from a reduced population of ribosomes and inhibited protein synthesis for RNA polymerase, structural proteins, and other enzymes responsible for the synthesis of various substrates.

Lado *et al.* (1968) had indicated, however, the specific mRNA of iso-citrate lyase was indeed synthesized, since other respiratory and glyolytic enzymes were not inhibited as much as isocitrate lyase after treatment with actinomycin D.

Direct evidence for *de novo* synthesis of isocitrate lyase and malate synthase was provided by density labeling of the enzymes in excised peanut cotyledons (Longo, 1968). Malate synthase had a buoyant density of 1.270 and isocitrate lyase of 1.290 when the excised cotyle-dons were incubated in water. Their densities changed to 1.315 and 1.365, respectively, when the cotyledons were incubated in 100% $D_2O$. When the soluble fractions of water and $D_2O$-cultured cotyledons were centri-fuged, two peaks having densities 1.270 and 1.315 were observed for malate synthetase. Because of the single-peaked profile of each enzyme in the cesium chloride gradient, Longo concluded that "They do not pre-exist in some inactive form in dry cotyledons but are completely syn-thesized after the onset of germination." Judging from the percent increase in density, 3.6% for malate synthase and 5.5% for isocitrate lyase, 70–90% of the stable H in the enzyme molecules had probably been re-placed by D. If the amino acids in the enzymes were coming from hy-drolysis of reserve proteins, replacement would be only at peptide-N sites which would constitute a very small increase of the mass. Hence, Longo (1968) speculated that the substrate amino acids of the two en-zymes were from a pool which did not derive directly from the hydrolysis of storage proteins but rather, for instance, from glucose.

*iii. Metabolic control.* When glucose (0.1 *M*) was used as a germination medium for castor beans, growth, isocitrate dehydrogenase, and glucose-6-phosphate dehydrogenase were not affected, while isocitrate lyase in endosperm was reduced to 60% of the water control in 24 hours. In cotyledons of squash *(Cucurbita maxima)* seeds, glucose at the same con-centration inhibited isocitrate lyase activity by 50% in 24 hours (Lado *et al.,* 1968). Sucrose (0.1 *M*) was found to increase the RQ from 0.65 to 1, to stop the degradation of reserve fatty acids in the axis, and to en-hance the synthesis of fatty acids from acetate-2-[14]C in embryo axes isolated from 2-day-old germinants of squash seeds (Cocucci and Caldog-no, 1967). Alberghina (1967) found that 0.1 *M* glucose, fructose, and sucrose all changed the RQ of castor bean cotyledons to 1 and increased lipid synthesis. Incubation of the cotyledons in 0.1 *M* sucrose increased the level of glucose 6-phosphate, soluble sugars, and starches. Glycerol did not alter the RQ. Nagamachi *et al.* (1967) observed inhibition of the purified isocitrate lyase from castor bean endosperm by glucose 6-phosphate and the inhibition appeared to be noncompetitive for sub-strate. These results indicate clearly that fatty acid oxidation and the glyoxylate cycle could be controlled by glycolytic intermediates (e.g., G-6-P) and by the end products, e.g., glucose, fructose, sucrose, by a feedback mechanism at the transcription level (Marré, 1967). This type of control probably occurs *in vivo;* when the sucrose concentration in cotyledons of 3-day germinants of castor beans was as high as 0.3 $M$ and that of glucose 6-phosphate was 0.1 $M,$ a concurrent change of RQ from 0.7 to 1 was observed (Marré, 1967; Yamada, 1955b).

Another fatty seed, hazel *(Corylus avellana)* was used to study the effect of GA on soluble metabolites in cotyledons during germination (Pinfield, 1968). The seed required GA for breaking dormancy, but the incorporation rate of acetate-2-[14]C in GA-treated and water-incubated seeds was similar prior to radicle emergence. GA$_3$ ($3 \times 10^{-4} M$) treatment doubled the incorporation of acetate-2-[14]C into sucrose, and significantly increased the radioactivity in glutamate. Furthermore, GA promoted isocitrate lyase activity in hazel cotyledon slices. This is an interesting finding and probably more detailed work will be forthcoming.

Gibberellic acid also stimulated lipase activity to 2.5-fold in cotton seed, while actinomycin D and aflatoxin prevented the stimulation by GA treatment (Black and Altschul, 1965).

### 3. COMPARTMENTATION OF LIPID UTILIZATION

a. DEVELOPMENTAL COURSE OF THE CONCEPT. The association of the glyoxylate cycle with the mitochondrial fraction was first demonstrated in germinating endosperm of castor beans (Marcus and Velasco, 1960). Subsequently H. J. Lee *et al.* (1964) found the same association in endosperm and cotyledons of sesame *(Sesamum orientale)* seeds. Lee and his associates obtained radioactive intermediates of the glyoxylate cycle by incubating the organellar fraction with acetate-U-[14]C. Furthermore the incorporation in various intermediates indicated a perferential utilization of acetate in the glyoxylate cycle over the TCA cycle. Radioactive glutamate was also observed, indicating amination of oxalacetate and the presence of transaminase in the isolated particles. A developmental trend of the particulate material with germination of sesame seeds is shown in Table 2.XI. It demonstrates that not only the quantity of the particles but also the specific activity of the key enzymes of the glyoxylate cycle increased to a peak and then decreased. Tanner and Beevers (1965) have also shown that significant proportions of glyoxylate cycle enzymes sedimented with mitochondria.

$\beta$-Oxidation of fatty acids has been known to occur in mitochondria for many years. In 1964 a soluble system of $\beta$ oxidation in addition to mitochondria was reported and subsequently confirmed (Rebeiz and Castelfranco, 1964; Rebeiz *et al.,* 1965a,b; Yamada and Stumpf, 1965). This soluble system probably originated from ruptured glyoxysomes due to

TABLE 2.XI.

DEVELOPMENTAL CHANGES IN THE SPECIFIC ACTIVITY OF GLYOXYLATE CYCLE
ENZYMES ASSOCIATED WITH 16,500 × $g$ PELLETS OF
SESAME SEEDS DURING GERMINATION[a]

| Days at 25°C | Radicle length (mm) | Soluble protein (mg/ml) | Isocitrate lyase[b] | Malate synthase[b] |
|---|---|---|---|---|
| 0 | – | 4.8 | – | 0.027 |
| 2 | 1 | 5.5 | 0.28 | 0.097 |
| 4 | 3 | 13.0 | 0.87 | 0.295 |
| 6 | 12 | 25.2 | 1.84 | 0.455 |
| 8 | 21 | 31.8 | 2.71 | 0.489 |
| 10 | 38 | 33.7 | 2.74 | 0.360 |
| 12 | 42 | 34.3 | 2.81 | 0.223 |
| 14 | 45 | 30.5 | 2.48 | 0.191 |

[a] From H. J. Lee *et al.* (1964).
[b] $\mu$Moles glyoxylate or malate produced/min/mg protein.

abrasives, high speed shearing, and hypotonic media used in homogenizing the plant material. Nevertheless, the possibility of divorcing β-oxidation enzymes from mitochondria provided the clue that they might be cosedimented and the particle-containing β-oxidation enzymes must be easily ruptured. Meanwhile, sufficient knowledge concerning lysosomes in animal tissues had accumulated, whereby the organelle could be confidently separated from mitochondria by differential centrifugation, sucrose density gradient centrifugation, or zonal centrifugation. Application of the sucrose density gradient centrifugation on a crude mitochondrial pellet obtained from the endosperm of germinating castor bean led to the discovery of glyoxysomes (Breidenbach and Beevers, 1967).

b. COMPARTMENTAL PATHWAYS. In recent years, much information has been accumulated on the compartmentation of lipid utilization in fatty seeds. The pathways are summarized in Fig. 2.20. The reserve triglycerides stored in fat bodies (spherosomes) are first hydrolyzed *in situ* to glycerol and free fatty acids by lipase (Ching, 1968; Ory *et al.,* 1968; Ory, 1969). The glycerol is phosphorylated in the cytoplasm and enters the later part of glycolysis and the TCA cycle for complete oxidation or conversion to cellular constituents. The free fatty acid is esterified with coenzyme A (CoASH) in postmitochondrial supernatant (Jacks *et al.,* 1967), in microsome (Takeuchi *et al.,* 1967), or directly in the glyoxysome (Cooper and Beevers, 1970). The acyl CoA enters the glyoxysomes and is oxidized in a β-oxidation sequence to acetyl CoA, two of which condense via glyoxylate cycle intermediates to form one succinate (Hutton and Stumpf, 1969; Cooper and Beevers, 1969b). Succinate leaves the glyoxysomes to join the TCA cycle in the mitochondria and is oxidized to oxalacetate, which then shunts out of the TCA cycle and is converted to sugars by a reversed sequence of glycolysis in the cytoplasm (Cooper and Beevers, 1969a; Breidenbach *et al.,* 1968).

With this introduction, each organelle will be discussed in more detail in the following paragraphs.

c. FAT BODIES AND LIPASES. As revealed by electron micrographs, the storage cells of fatty seeds consist mainly of fat bodies, protein bodies, and nuclei. Mitochondria, endoplasmic reticula, ribosomes, proplastids, and glyoxysomes are present but are very sparse. During germination, the fat bodies decrease in size and number, and acquire a vesicular, proteinaceous outer membrane indicating *in situ* lipolysis. Concurrently, protein bodies become granulated, fragmented, and finally solubilized, indicating *in situ* hydrolysis. Other organelles become conspicuous and some increase in number (mitochondria and glyoxysomes) or in development (chloroplast and amyloplast) as the storage particles diminish (Bagley *et al.,* 1963; Ching, 1965, 1970b; Yatsu, 1965; Horner and Arnott, 1966). Further separating megagametophytes of dry and germinating Douglas fir seeds into protein bodies, mitochondria, microsomes, light and heavy fat bodies, and soluble fractions and testing their lipolytic activity at pH 5.2 and pH 7.1, Ching (1968) found that lipase activity was present in dry seeds and that the fat bodies, particularly the heavy kind, were the site of lipase activity. Ory *et al.* (1968) also found high acid lipase activity in spherosomes (fat bodies) of dry castor beans. The pH optimum of the acid lipase was 4.2. The enzyme was heat stable and comprised three components: the apoenzyme, a lipid cofactor (a cyclic tetramer of ricinoleic acid), and a protein activator (a small heat stable glycoprotein) (Ory, 1969).

Induction of lipase activity in starch endosperm of germinating wheat can be accomplished by incubation of the embryo axis or scutellum with hydroxylamine, glutamate, and to a lesser degree by nucleotides, nitrate, nitrite, ammonium salts, and a mixture of amino acids. Actinomycin D, puromycin, cycloheximide, and azaserine all strongly inhibited the induction of lipase by hydroxylamine. GA, indoleacetic acid (IAA), or kinetin, or any combination of them did not induce the lipase activity (Tavener and Laidman, 1969). It appears, therefore, that lipase in nonfatty seeds is not preexisting and requires induction and the synthesis of RNA and protein.

Even though lipases are found in dry fatty seeds, their activity probably needs to be induced *in vivo* by some soluble metabolites or hormones that are quickly produced upon hydration of preexisting cellular constituents.

In cotton seeds, the lipase activity can be stimulated by the application of gibberellin and the stimulation requires the synthesis of RNA and protein (Black and Altschul, 1965). Even though both total and specific lipase activity increased several-fold during germination (Ching, 1968), *de novo* synthesis of additional lipase(s) during germination for increased lipolysis has not been rigorously proven as has been done for isocitrate lyase and malate synthase.

The decline of lipase activity with the reduction of lipid content in germinating seeds is well established (Yamada, 1957; Ching, 1968), although the mechanism of the decline is unknown.

d. GLYOXYSOME (CYTOSOME, PEROXISOME, MICROBODY). Using discontinuous sucrose density gradient centrifugation on unwashed mitochondrial pellet isolated from the endosperm of germinating castor beans, Breidenbach and Beevers (1967) discovered the association of glyoxylate cycle enzymes with particles heavier than mitochondria. The particles thus were named glyoxysomes (Breidenbach *et al.,* 1968). Subsequent elegant work from Beevers' laboratory and from Stumpf's laboratory established the association of $\beta$-oxidation of fatty acids with glyoxysomes in endosperm of germinating castor beans. The exclusive association of $\beta$-oxidation enzymes with glyoxysomes appears to be a unique feature of plant materials since fatty acids are known to be oxidized in mitochondria of animal tissue. The peroxisomes (glyoxysomes) of protozoa *(Tetrahymena pyriformis)* contain isocitrate lyase and malate synthase, but have not yet been examined for $\beta$ oxidation enzymes (Muller *et al.,* 1968).

*i. Morphology and characteristics of glyoxysomes.* Glyoxysomes are present in dry seeds of Douglas fir, ponderosa pine, and soybeans (Ching, 1970a, 1970b). Glyoxysomes have been observed in germinating yucca *(Yucca schidigera)* seed (Horner and Arnott, 1966) and squash cotyledons (Breidenbach *et al.,* 1968). In germinating seeds, the number of glyoxysomes increases many times (Ching, 1970b). Their *in situ* morphology is shown in Fig. 2.21. They usually are 0.5–2 $\mu$ in diameter, surrounded by a single unit membrane, and contain dense granular substances. Occasionally, an electron opaque amorphous center core is present (Fig. 2.21d). They generally are larger than animal peroxisomes (0.5 $\mu$ in diameter, de Duve and Baudhuin, 1966), and are in the size range of leaf peroxisomes (cytosomes 0.5 $\mu$ in diameter, Tolbert *et al.,* 1968; Frederick and Newcomb, 1968). No crystalline structures like those found in leaf peroxisomes have been observed to date in glyoxysomes. In contrast to animal peroxisomes, glyoxysomes are easily ruptured in a hypotonic medium.

Glyoxysomes of different origin have a density of 1.25, and they can be easily isolated from a crude mitochondrial pellet in a discontinuous

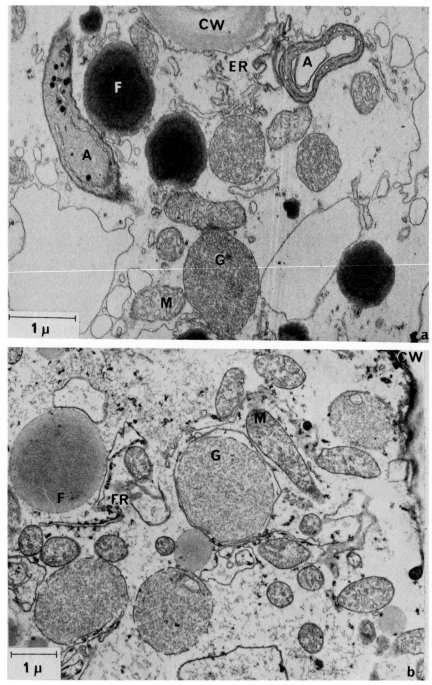

Fig. 2.21. Electron micrographs of glyoxysome in megagametophyte of Douglas fir (a), and ponderosa pine (b), in cotyledon of soybean (c), and in megagametophyte of Jeffrey

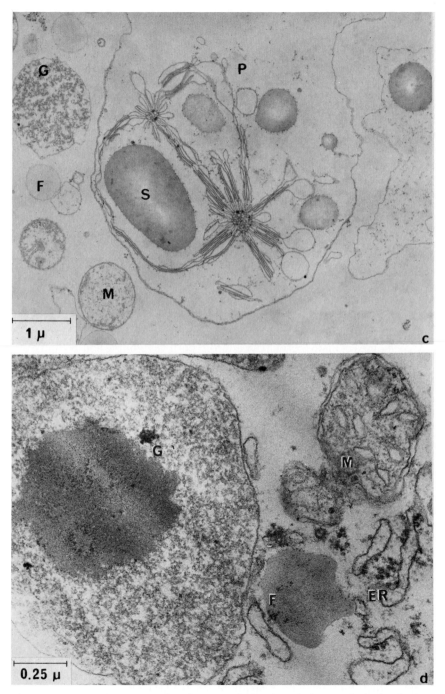

pine (d). (A, amyloplast; CW, cell wall; ER, endoplasmic reticulum; F, fat body; G, glyoxysome; M, mitochondrion; P, proplastid; S, starch grain.

sucrose density gradient of 35–60% (w/w) for 4–6 hours at 90,000–105,000 × $g$. The occurrence of glyoxysomes was demonstrated also by the presence of glyoxysomal enzymes in corn scutellum, and in watermelon and peanut cotyledons. No indication of glyoxysomal enzyme activities was found in hypocotyls of castor beans, peanut, and watermelon, corn roots, or avocado and zucchini squash fruit (Cooper and Beevers, 1969a).

*ii. Function of glyoxysomes.* Pathways of $\beta$-oxidation and glyoxylate cycle in glyoxysomes were established by Cooper and Beevers (1969a,b) with five independent procedures: (*1*) reduction of NAD upon addition of palmitoyl CoA to glyoxysomes; (*2*) thiolase activity by measuring NAD reduced upon adding CoA and acetoacetyl CoA into glyoxysomes which were enriched with exogenous malate and citrate synthase; (*3*) malate-$^{14}$C formation from glyoxylate-$^{14}$C and acetyl CoA resulted from $\beta$-oxidation of exogenous palmitoyl CoA in glyoxysomes; (*4*) direct assay of enzyme activities; (*5*) conversion of isocitrate-5,6-$^{14}$C to succinate-$^{14}$C in isolated glyoxysomes. Table 2.XII shows the distribution and function of various enzyme activities in endosperm of germinating castor bean.

Concurrently Hutton and Stumpf (1969) found $\beta$-oxidation activity in glyoxysomes of endosperm isolated from both developing and germinating castor beans. They used two methods to illustrate $\beta$-oxidation: (*1*) production of acetyl CoA by incubating free fatty acids with glyoxysomes, and (*2*) direct assay of the component enzymes of $\beta$-oxidation (i.e., crotonase, $\beta$-hydroxyacyl dehydrogenase and $\beta$-ketothiolase). The mitochondrial fraction of their preparation also had $\beta$-oxidation activity. However, both total and specific activity of the overall $\beta$-oxidations and three individual enzymes showed that the participation of mitochondria in $\beta$-oxidation of fatty acids was minor and possibly due to contaminations.

The major function of glyoxysomes is therefore $\beta$-oxidation of fatty acids to acetyl CoA and condensation of the product to succinate (Fig. 2.20). The organelle is not self-sufficient with respect to the regeneration of reduced NAD, while FADH$_2$ could be oxidized by oxygen in the particle to hydrogen peroxide, which is broken down by catalase present in the particle. Cooper and Beevers (1969b) reached the above conclusion on the basis of stoichiometric ratio of 0.5 : 1 : 1 of oxygen uptake, NADH accumulation, and acetyl CoA production upon additon of palmitoyl CoA to glyoxysomes of castor bean endosperm. Glyoxysomes do not have an electron transport system like those of mitochondria (Breidenbach and Beevers, 1967), thus NADH accumulates in isolated glyoxysomes (Cooper and Beevers, 1969a,b). Therefore, Cooper and Beevers concluded that the reducing equivalent of NADH must be used in other cellular compartments.

**TABLE 2.XII.**

DISTRIBUTION OF VARIOUS ENZYME ACTIVITY ($\mu$MOLES/GM FRESH TISSUE/HR) IN
DIFFERENT FRACTIONS ISOLATED FROM ENDOSPERM OF 5-DAY GERMINATING
CASTOR BEANS[a]

| Enzyme | Supernatant[b] | Crude particulate[c] | Mitochondria[d] | Proplastids[d] | Glyoxysomes[d] |
|---|---|---|---|---|---|
| **TCA and GA cycle Enzymes:** | | | | | |
| Citrate synthase | 2 | 155 | 218(73) | 2(7) | 74(25) |
| Aconitase | 840 | 62 | 18(41) | 2(5) | 4(10) |
| Malate dehydrogenase | 22,150 | 16,700 | 10,700(62) | 311(2) | 4,475(26) |
| GA cycle enzymes: | | | | | |
| Isocitrate lyase | 66 | 342 | 8(3) | 9(3) | 207(85) |
| Malate synthase | 44 | 450 | 46(5) | 27(3) | 790(89) |
| TCA cycle enzymes: | | | | | |
| Succinate dehydrogenase | 0 | 87 | 65(100) | 0 | 0 |
| Fumarase | 20 | 334 | 394(92) | 4 | 2 |
| $\alpha$-Ketoglutarate oxidase | 0 | 25.2 | 220(100) | 0 | 0 |
| NAD isocitrate dehydrogenase | 0 | 36 | 25(100) | 0 | 0 |
| NADP isocitrate dehydrogenase | 28 | 17 | 9(73) | 0.4(3) | 1.2(10) |
| $\beta$-Oxidation enzymes: | | | | | |
| $\beta$-Oxidation[e] | 6.8 | 25.8 | 0 | 2.8(19) | 11.4(81) |
| Thiolase | 35.8 | 115 | 0 | 6.3(15) | 22.9(57) |
| Accessory enzymes: | | | | | |
| Glutamate oxalacetate transaminase | 244 | 1850 | 862(30) | 36(1) | 1380(60) |
| Catalase | 3860 | 2990 | 258(5) | 1039(20) | 3360(66) |
| P-Enolpyruvate carboxykinase | 183 | 1 | 0.1 | 0 | 0 |
| P-Enolpyruvate carboxylase | 11 | 0 | 0 | 0 | 0 |
| Glutamic dehydrogenase | 0 | 3.2 | 2.8(100) | 0 | 0 |
| Glycolic oxidase | 0 | 3.7 | 0 | 0.8(22) | 2.9(78) |

[a]From Copper and Beevers (1969a,b).

[b]Supernatant fraction of tissue homogenate centrifuged at 9500 $\times$ g for 10 minutes.

[c]Pellet fraction from above. This was used for the density gradient centrifugation to separate mitochondria (density 1.19), proplastids (density 1.23), and glyoxysomes (density 1.25).

[d]The figures in parenthesis show the percent of enzyme activity recovered from the gradient after centrifugation.

[e]Production of NADH by adding palmitoyl CoA.

The association of RNA (Gerhardt and Beevers 1969; Ching, 1970b), DNA (Ching, 1970b), and protein synthesizing ability (Ching, 1969, 1970b) with the glyoxysomal fraction indicates some degree of autonomy in this organelle. More research is in progress.

*iii. Developmental pattern of glyoxysomes.* A change of β-oxidation activity with germination time of castor bean was first noted by Hutton and Stumpf (1969). The data are shown in Table 2.XIII.

**TABLE 2.XIII.**
DEVELOPMENTAL CHANGES IN SPECIFIC ACTIVITIES OF β OXIDATION AND COMPONENT ENZYMES IN GLYOXYSOMES OF CASTOR BEAN ENDOSPERM DURING GERMINATION[a]

| Days of germination | β-Oxidation[b] | Crotonase[c] | β-OH-dehydrogenase[c] | β-Ketothiolase[c] |
|---|---|---|---|---|
| 2 | 18.8 | 349.0 | 773.0 | 22.4 |
| 4 | 187.5 | 3210.0 | 3350.0 | 58.5 |
| 6 | 0 | 526.0 | 1125.0 | 3.2 |

[a] From Hutton and Stumpf (1969).
[b] $\mu$Moles acetyl CoA produced/hr/ml glyoxysomal fraction.
[c] $\mu$Moles product produced/hr/ml glyoxysomal fraction.

A similar pattern is shown in both mitochondria and glyoxysomes of the megagametophyte of germinating ponderosa pine seeds (Ching, 1970b). The profiles of protein content and marker enzyme activities in sucrose-density gradients of $10,000 \times g$ pellets isolated from 250 gametophytes on different days of germination are shown in Fig. 2.22. A quantitative account of this developmental pattern is summarized in Table 2.XIV. A similar pattern was observed in the endosperm of castor beans (Gerhardt and Beevers, 1970).

The rise and fall of the quantity and the marker enzyme activity of both mitochondria and glyoxysomes parallel the rate of lipid disappearance in pine seed (Fig. 2.23). A simple explanation of substrate stimulation and end product inhibition for this pattern could account for the biosynthesis but not for the disappearance of organelles, since end products are not accumulated. A genetically programmed developmental pattern must also be involved. However, execution of the program could be triggered and altered by exogenous and endogenous signals. The induction of lipase activity by gibberellins (Black and Altschul, 1965) might stimulate the biogenesis of glyoxysomes, and the inhibition of isocitrate lyase by sucrose, glucose, fructose, and glucose 6-phosphate (Lado *et al.,* 1968; Nagamachi *et al.,* 1967) might prevent the formation of glyoxysomes. It would be interesting to test these speculations individually and in com-

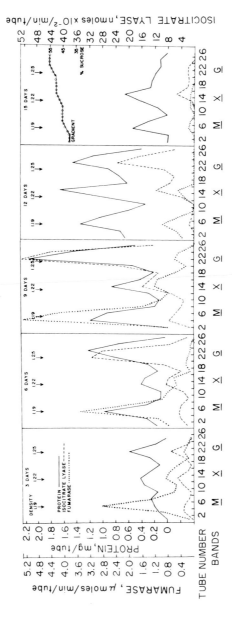

FIG. 2.22. Developmental pattern of mitochondria (M), glyoxysomes (G), and mixture of mitochondria, glyoxysomes, and proplastides (X) in 250 megagametophytes of germinating ponderosa pine seeds (Ching, 1970b).

**TABLE 2.XIV.**

Total Organellar Protein Content and Specific Activity
of Enzyme in Mitochondria and Glyoxysomes Isolated from
100 Megagametophytes of Germinating Ponderosa Pine Seeds[a]

| Days of germination | Protein content | | Specific activity[b] | |
|---|---|---|---|---|
| | Mitochondria ($\mu$g) | Glyoxysome ($\mu$g) | Fumarase | Isocitrate lyase |
| 3 | $153 \pm 11$ | $690 \pm 93$ | $2.97 \pm 0.40$ | $0.12 \pm 0.01$ |
| 6 | $373 \pm 82$ | $1258 \pm 176$ | $3.75 \pm 0.13$ | $0.21 \pm 0.06$ |
| 9 | $645 \pm 193$ | $1503 \pm 105$ | $4.97 \pm 0.51$ | $0.35 \pm 0.06$ |
| 12 | $1087 \pm 93$ | $1411 \pm 132$ | $0.85 \pm 0.21$ | $0.17 \pm 0.04$ |
| 15 | $479 \pm 105$ | $339 \pm 29$ | $0.27 \pm 0.04$ | $0.06 \pm 0.01$ |

[a] The data are the mean and standard deviation of four replications (Ching, 1970b).

[b] Specific activity was estimated in the peak fraction of respective bands as $\mu$mole fumarate or glyoxylate produced per mg organellar protein per minute.

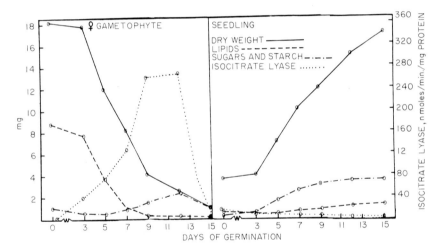

Fig. 2.23. Changes in dry weight, lipids, sugars and starch, and isocitrate lyase activity in megagametophyte and seedling of germinating ponderosa pine seeds (Ching, 1970b).

bination to delineate the interaction of genetic information and metabolic modulation.

### B. Metabolism of Reserve Protein and Starch in Seeds

Almost all seeds contain protein reserves for the nitrogenous supplies required by the young seedlings before they are able to absorb nitrogen by roots. The lowest protein contents usually are found in starchy seeds (about 10% of the seed dry weight), and highest protein content (35–45%

of seed dry weight) are often associated with fatty seeds. So-called proteinaceous seeds such as peas and beans actually contain 20–40% of their dry weight as protein, and 40–60% as carbohydrate. This section will, therefore, discuss metabolism of both nitrogenous compounds and carbohydrate reserves. Another reason for reviewing carbohydrate metabolism in peas and beans is that some degree of metabolic control in enzyme synthesis could be observed whereas the synthesis of catabolic enzymes in cereal seeds appears to be a passive consequence of hormonal mediation from the embryo.

### 1. OVERALL COMPOSITIONAL CHANGES

Overall compositional changes during germination of pea seeds are summarized in Figs. 2.24, 2.25, and 2.28. The transfer of dry weight from

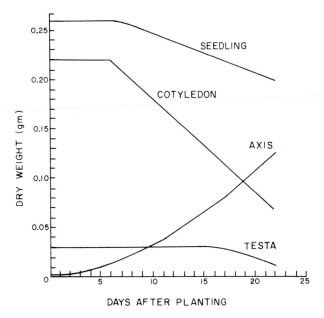

FIG. 2.24. Changes in dry weight of whole seedling, axis, pair of cotyledons, and testa in germinating peas (Bain and Mercer, 1966).

cotyledons to the seedling axis at the expense of stored starch and protein in cotyledons is clearly shown. Sugars, soluble nucleotides, and amino acids appear to be the metabolites transported to the seedling axis where dry weight, protein, soluble nitrogenous compounds, nucleotides and nucleic acids accumulate. The general trend of compositional changes follows that of fatty and starchy seeds (Sections IV,A,1 and IV,C,1).

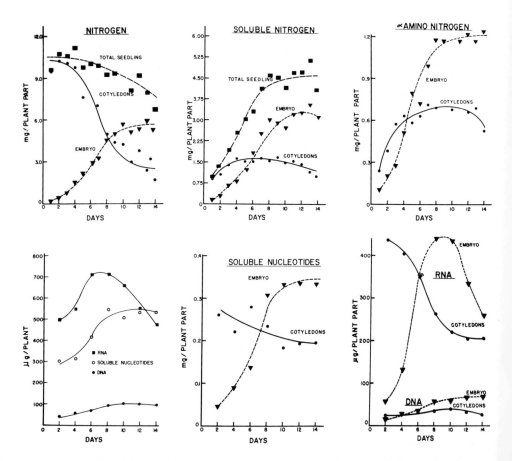

FIG. 2.25. Changes in various nitrogenous compounds in the cotyledons and embryo axis of peas during germination (Beevers and Guernsey, 1966).

## 2. DEGRADATION OF RESERVE PROTEIN

a. PROTEINASES AND METABOLITES. The hydrolysis of reserve protein is catalyzed by (*1*) proteinases (proteases) to amino acid units and peptides, depending upon their specificity, and (*2*) peptidases which hydrolyze endogenous and hydrolytically produced peptides to amino acid units. To study these enzyme activities, usually a cell-free extract is prepared from germinating seeds and incubated with commercially available proteins or peptides under different pH conditions. The reaction products, amino acids, are measured by optical absorption at 280 m$\mu$ or by the quantity of ninhydrin products as estimated by optical absorption at 570 m$\mu$. Because of variations in species and age of materials, as well as in extraction procedures and the assay methods used by different investigators, a unified picture is difficult to draw. Nevertheless some generalizations can be made. The proteolytic activity increases with *in situ* protein hydrolysis and then decreases when reserve proteins are exhausted. In the meantime, other proteolytic enzymes are synthesized for the turnover of existing enzymes and structural protein of organelles. Finally, senescence of the storage tissue is effected by some nonspecific lysosomal type proteinases.

Fig. 2.26 presents the changes in total and specific activities of protease and peptidase in peas during germination. Two substrates, $\alpha$-benzoyl-DL-arginine-$p$-nitroanilide (BAPA) and L-leucine-$p$-nitroanilide (LPA), were used for the assay of peptidase. The time used was for maximum activity under the assay conditions. These figures show that in cotyledon extracts of germinating peas, both total and specific activity of protease increased during germination whereas the total activity of peptidase was reduced. Because of the opposite trends exhibited by these two enzymes in relation to age, and the parallel reduction in total cotyledon N and peptidase activity, Beevers and Splittstoessar (1968) speculated that the peptidase activity might be associated with utilization of reserve proteins, whereas the protease activity might be related to protein degradation as well as to turnover-type hydrolysis that generally occurs with growth and development.

A linear increase of neutral peptidase activity from zero to 30 units for 70-hour-old germinating barley was reported. About 80% of this activity was associated with the seedling portion, and the remainder occurred in the endosperm. Only one-third of the low acidic peptidase activity, and about one-twentieth of the neutral activity, was found in endosperm of the germinating barley (Prentice *et al.,* 1969). These findings indicate that peptidases apparently are related to growth and development, since the protein storage site of barley is the endosperm where peptidases are not at all active.

In squash cotyledons of germinating intact seeds, proteinase activity increased to a maximal level at 2–3 days, then declined to about the initial value after 7 days (Wiley and Ashton, 1967).

A small portion of some of the total proteinases preexists in dry seeds (Section III,A,2,e). The majority however, appears to be synthesized during early stages of germination, since puromycin, cycloheximide and prolonged treatment with actinomycin D inhibited the increase of proteinase activity (Penner and Ashton, 1966; Ihle and Dure, 1969; Fig. 2.27).

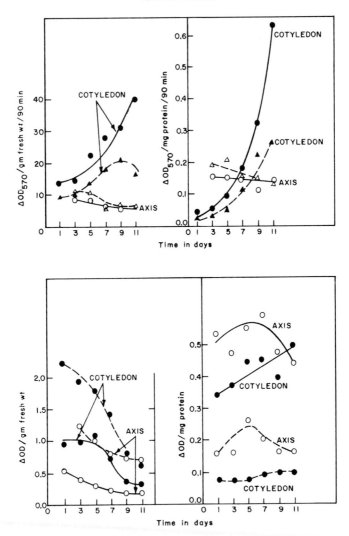

FIG. 2.26. Activity of protease (upper) and peptidase (lower) of germinating pea seedlings (Beevers, 1968; Beevers and Splittstoesser, 1968). Protease pH 5.5 (—) pH 7.0 (----). Peptidase, LPA 5 minutes (—); BAPA, 15 minutes (----).

A restraining effect of the embryo axis on the activity of proteinases in cotton cotyledons is shown in Fig. 2.27. Curve a shows the cotyledon proteinase activity of intact seeds which exhibits only two-thirds of the total activity reached in cotyledons excised after 1 or 2 days of germination and incubated 2 more days (curves c and d, respectively). Intact seeds treated with cycloheximide at 0, 1, 2, and 3 days of germination

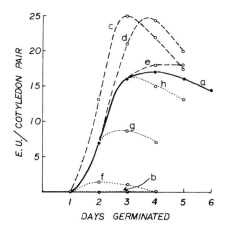

FIG. 2.27. Effect of excision and cycloheximide on the appearance and quantity of proteinase in cotton seeds during germination (Ihle and Dure, 1969). Control (a); excised cotyledon after 1, 2, and 3 days germination in intact seedling (c, d, and e, respectively); cycloheximide treated cotyledon excised after 0, 1, 2, and 3 days germination in intact seedling (b, f, g, and h, respectively).

(curves b, f, g, and h, respectively) produced proportionally less protease than the control. These results further prove that proteinase is synthesized cumulatively during germination and an inhibitor of protein synthesis stops the accumulation (Ihle and Dure, 1969).

In contrast to cotton seeds, the induction of proteinase activity in squash cotyledons apparently originates from the embryo axis. Furthermore, benzyladenine, kinetin, 6-(benzylamino)-9-(tetrahydropyranyl)-9-purine, or phenyladenine at a concentration of 100 $\mu M$ could replace the embryo axis in inducing proteinase activity in excised cotyledons (Penner and Ashton, 1967). It is possible that the cotton proteinase from excised cotyledons consists of two kinds, one of which is responsible for degradation of reserve proteins and the other is induced by excision, resulting in hydrolysis of a senescence-type.

In order to assign functional roles to different enzymes, it is imperative to determine how many proteinases are present *in vivo* in different tissues, what their specificities and activities are at different stages of germination, and how they are controlled. It is also important to know if any of them are bound on particles such as protein bodies (organelles for reserve protein in seeds, see next section). More sophisticated work apparently will be needed to answer these questions.

b. ULTRASTRUCTURAL CHANGES OF PROTEIN BODIES. Protein bodies, (protein granules, aluerone grains) are single-membrane spheres which may become angular shaped upon packing during dehydration of ripening seeds. Their diameter varies with the species, ranging from 1–22 $\mu$. Protein bodies appear to be granular or crystalline and sometimes contain electron opaque inclusions of phytin and calcium oxalate. On seed germination, the insoluble protein in the body becomes fragmented, granulated, and finally solubilized with a concomitant enlargement of the whole body. Often several protein bodies converge to form the cell vacuole (Setterfield *et al.*, 1959; Bagley *et al.*, 1963; Paleg and Hyde, 1964; Perner, 1965; Horner and Arnott, 1966; Bain and Mercer, 1966; Opik, 1966).

These *in vivo* changes indicate that reserve insoluble proteins are degraded in a localized organelle of the cell. Unfortunately, the single-membraned protein bodies are easily ruptured by ordinary homogenizing procedures. Therefore, isolation of intact protein bodies from different stages of germination is difficult.

c. CHEMICAL COMPOSITION OF PROTEIN BODIES. Physical and chemical properties of storage proteins of seeds vary in different species of plants. A detailed review of isolation procedures and properties of various seed proteins is given by Altschul *et al.* (1966).

Isolation of protein bodies has been conducted by several procedures all of which were successful to some degree in retaining organellar functions or contents. Table 2.XV summarizes the composition of isolated protein bodies from various sources.

TABLE 2.XV.

COMPOSITION IN PERCENT OF DRY WEIGHT OF ISOLATED PROTEIN BODIES FROM VARIOUS SOURCES

| | Maturing wheat endosperm[a] | Dry peanut[b] | Dry soybean meal[c] | Dry cottonseeds[d] |
|---|---|---|---|---|
| Proteins | 72 | 72 | 82.5 | 62 |
| RNA | 10.4 | 0.2 | 1.2 | –[e] |
| Phosphorolipids | 0.8 | – | 1.4 | 1.5 |
| Phytic acid | 8.4 | 3.1 | 1.4 | 10.0[f] |
| Total lipids | – | – | 11.3 | – |
| Carbohydrates | – | 6.9 | 3.0 | 10.0 |
| Metals | | | | |
| K | – | – | – | 2.6 |
| Ca | – | – | – | 0.3 |
| Mg | – | – | – | 0.6 |

[a] Data from Morton *et al.* (1964).
[b] Data from Altschul *et al.* (1961).
[c] Data from Tombs (1967).
[d] Data from Lui and Altschul (1967).
[e] –, not determined.
[f] Calculated by percent inositol in globoid × 5 × % globoid in protein body.

The protein bodies isolated from soybean meal had a density of 1.28–1.32 gm/ml, contained RNA, carbohydrates which yielded glucose, fructose, galactose, mannose, arabinose, and ribose upon hydrolysis, and inositol, phospholipids, triglycerides and free fatty acids. The protein in the bodies was glycinin which was separated by electrophoresis to one monomer and one dimer. The well-known soybean trypsin inhibitor was found as a soluble protein which did not associate with protein bodies (Tombs, 1967).

The high content of RNA in protein bodies of maturing wheat endosperm is the result of their protein synthesizing ability (Morton and Raison, 1964; Morton et al., 1964). Apparently the protein synthesizing machinery is no longer present after maturation in dry seeds.

Numerous attempts using $Ca^{2+}$ for hardening membranes, polyamines or albumin for protectant, and various osmoticants had been explored for the isolation of intact protein bodies from gametophytic tissue of germinating Douglas fir seeds, were not completely successful due to the varied size and density of protein bodies, their adsorbing surfaces, and their delicate single membrane (Ching, 1970a). Some characteristics of the isolated protein bodies, and a trend of reduction in protein bodies during germination of Douglas fir seeds were reported (Ching, 1968). Loss of membrane and contamination of other cellular materials also were observed in soybean protein bodies (Tombs, 1967).

d. ENZYME COMPLEMENT AND ACTIVATION IN PROTEIN BODIES. Yatsu and Jacks (1968) isolated intact protein bodies from dehulled dry cotton seeds homogenized in glycerol by differential centrifugation. The isolated protein bodies constituted 75% of the total protein in the homogenate which confirms the results obtained with Douglas fir seeds using aqueous medium containing $CaCl_2$ (Ching, 1968). The cottonseed protein bodies contained 77% of the cellular acid phosphatase (EC 3.1.3.2.) activity (pH 5.0) and 100% of the cellular acid proteinase activity (pH 2.0). The coincidence of protein and phytin (Liu and Altschul, 1967) and acid phosphatase and acid proteinase in isolated protein bodies suggested that the protein bodies are autolytic repository organelles, (Yatsu and Jacks, 1968). Whether these enzymes require activation *in vivo* or change in specific activity during germination, and how the low pH condition is maintained in the organelles are not known at present.

In lettuce seeds, trypsin-like enzyme activity develops during germination, and the increase of activity is abolished by inhibitors of protein synthesis. The inhibitors, however, do not actually inhibit synthesis of proteolytic enzymes but rather prevent the destruction of trypsin inhibitor, which normally inhibits the preexisting proteolytic enzymes (Shain and Mayer, 1968). Further evidence of the activation of a preexisting trypsin-like enzyme is shown by incorporation of $^{35}$S in other enzymes during germination but not in purified trypsin-like enzymes (Shain and Mayer, 1968).

Ghetie (1966), on the other hand, contended that reserve proteins in seeds are stored in protein bodies in the form of phytin-protein, carbo-hydrate-protein, and lipid-protein complexes. In early stages of germina-tion, these reserve protein complexes were activated prior to hydrolysis by proteases. Again, much additional research is needed on the function, characteristics, and degradational pattern of protein bodies in seeds.

### 3. UTILIZATION OF STARCH GRANULES

Peas contain 45–55% and beans may be composed of 50–60% starch (Crocker and Barton, 1957). The starch is deposited in plastids during maturation, and the lamellar structure of the plastids disintegrates as the starch granules grow to full size (Opik, 1968). The size of starch granules varies with species ranging from a diameter of 15 $\mu$ in rice to 50 $\mu$ in beans. During seed germination, starch grains are in direct contact with cytoplasm and appear rugged on the surface, indicating *in situ* hydrolysis (Bain and Mercer, 1966). The major reduction of starch content in ger-minating peas occurs after 8 days of germination as determined by chemi-cal analysis and ultrastructural examination (Bain and Mercer, 1966; Juliano and Varner, 1969). Figure 2.28 illustrates the changes in starch and sugar content in cotyledons of germinating Alaska peas, which con-tain about 35% amylose and 65% amylopectin in the starch granules (Akazawa, 1965). The physical and chemical changes of the starch gran-ules in pea cotyledons during seed germination are shown in Table 2.XVI. The reduction of granular size and viscosity of starch with germi-nation time indicated not only that the aggregates became smaller but also that the component molecules were shorter, with a preferential scission of inner chain $\alpha$-1,4-glucoside linkage by $\alpha$-amylase (Juliano and Varner, 1969).

Starchy or cereal seeds contain 70–80% starch. The pattern of starch utilization during germination of such seeds is illustrated in Fig. 2.29. The hydrolysis of starch in rice endosperm provides energy and synthetic re-quirements of the seedling. Only a very small accumulation of sugars occurs in the endosperm. The sequence of starch breakdown is similar to that in peas. Phosphorylase activity is high in the early stages of germina-tion, increases up to eight days, and then declines. The $\alpha$-amylase activity is negligible at early stages of germination, increases rapidly to 80-fold of the phosphorylase activity in starch breakdown at 10–12 days, and then decreases (Fig. 2.29; Murata *et al.*, 1968). Maltase activity coincides

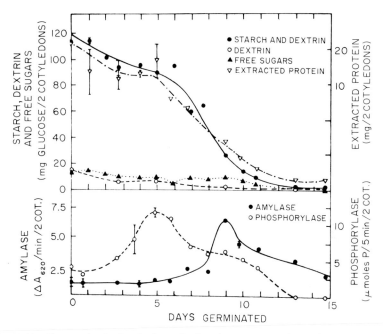

FIG. 2.28. Changes in the level of starch, dextrin, sugars, total protein, amylases, and phosphorylase in the germinating pea cotyledons (Juliano and Varner, 1969).

**TABLE 2.XVI.**

SOME PROPERTIES OF COTYLEDON STARCH GRANULES OF GERMINATING PEA[a]

| | Days germinated | | |
|---|---|---|---|
| Property | 0 | 5 | 11 |
| Starch | | | |
| Mean granule size, $\mu$ | $28.9 \pm 2.2$ | $25.0 \pm 1.2$ | $20.8 \pm 1.4$ |
| Blue value, Abs at 680 m$\mu$ | 0.351 | 0.368 | 0.366 |
| Amylopectin | | | |
| Intrinsic viscosity, ml/gm | 182 | 112 | 95 |
| Mean chain length, anhydro glucose units | $26.6 \pm 0.9$ | $26.5 \pm 0.6$ | $21.8 \pm 0.0$ |
| $\beta$-Amylolysis limit, % | 57 | 53 | 47 |
| Apparent inner chain length, anhydro glucose units | 9 | 10 | 9 |
| Amylose | | | |
| Blue value, Abs at 680 m$\mu$ | 1.02 | 0.92 | 1.05 |
| Intrinsic viscosity, ml/gm | 170 | 163 | 151 |
| $\beta$-Amylolysis limit, % | 84 | 89 | 96 |

[a] Data from Juliano and Varner (1969).

with α-amylase activity, indicating a cooperative role of both enzymes in the tissue (Fig. 2.29; Nomura *et al.*, 1969).

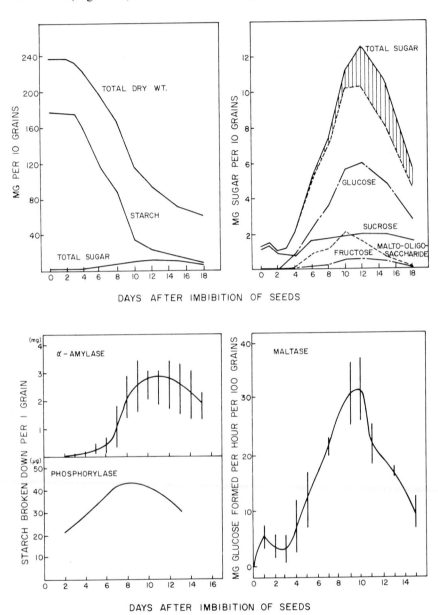

FIG. 2.29. Changes in weight, carbohydrates, and enzyme activities with germination in rice seeds (Murata *et al.*, 1968; Nomura *et al.*, 1969).

a. PATHWAYS. Two catabolic pathways of starch to glucose are known in pea seedlings (Swain and Dekker, 1966a,b), one of which is hydrolytic:

$$\text{Starch} \xrightarrow{\alpha\text{-amylase}} \text{soluble oligosaccharides (dextrins)}$$

$$\text{Soluble oligosaccharides} \xrightarrow{\beta\text{-amylase}} \text{maltose}$$

$$\text{Maltose} \xrightarrow[\text{(EC 3.2.1.20)}]{\alpha\text{-glucosidase}} \text{glucose}$$

and the other is phosphorolytic:

$$\text{Starch} + \text{Pi} \xrightarrow{\text{phosphorylase}} \text{glucose 1-phosphate}$$

$$\text{Glucose 1-phosphate} + \text{UTP} \xrightarrow{\text{UDPG pyrophosphorylase}} \text{UDPG} + \text{PPi}$$

$$\text{UDPG} + \text{fructose 6-phosphate} \xrightarrow{\text{sucrose synthetase}} \text{sucrose 6-phosphate}$$

$$\text{Sucrose 6-phosphate} \xrightarrow{\text{phosphatase}} \text{sucrose} + \text{Pi}$$

$$\text{Sucrose} \xrightarrow[\text{(EC 3.2.1.26)}]{\beta\text{-fructofuranosidase}} \text{glucose} + \text{fructose}$$

All the necessary enzymes are found in both cotyledons and seedling axis with $\alpha$-amylase exclusively localized in cotyledon tissue. The levels of phosphorylase activity present in the cotyledon during early stages of germination can account for the starch reduction, whereas the hydrolytic pathway becomes increasingly important in the later stages of germination and plays the major role in starch degradation (Swain and Dekker, 1966a,b). These findings are further substantiated by data of Fig. 2.28 (Juliano and Varner, 1969) and of Swain and Dekker (1969).

The sequence of phosphorolytic pathway first, followed by hydrolytic pathway in germinating seeds is logical with respect to cell economy. Phosphorylase is a preexisting enzyme in pea cotyledons (Juliano and Varner, 1969), the activation of which provides the substrate for glycolysis and respiration without using ATP, since the energy of the glucosidic bond is retained in this reaction. This degradative route quickly produces greatly needed ATP at the very early stage of germination for the synthesis of enzymes and organelles necessary for further mobilization of reserves. $\alpha$-Amylase is one of the enzymes synthesized in later stages of germination, probably at the expense of amino acids from hydrolyzed protein bodies, with the energy supply from the phosphorylase activity, glycolysis, and respiration.

A debranching enzyme of starch, amylopectin-1,6-glucosidase (isoamylase, (EC 3.2.1.9) is found in zymogen granules (small enzyme-containing particles) of pea seeds. The granules which are isolated by Ficoll density gradient centrifugation, contain an inactive form of the enzyme which can be released and activated by the soluble fraction of seed homogenate or trypsin. The enzyme is preexisting and little additional synthesis occurs during germination (Mayer and Shain, 1968).

Free oligosaccharides such as sucrose, raffinose, and stachyose which often amount to several percent of the dry seed weight, are found in Leguminosae, Cruciferae, and Liliaceae. During germination, some of the raffinose and stachyose is converted to sucrose in pea cotyledons and the rest presumably is transported to the seedling axis for growth (C. Y. Lee and Shallenberger, 1969). The degradative pathway of raffinose was proposed as a two-stage reaction by Pridham *et al.* (1969).

$$\text{Raffinose} \xrightarrow[\text{(EC 3.2.1.22)}]{\text{$\alpha$-glactosidase}} \text{galactose} + \text{sucrose}$$

$$\text{Sucrose} + \text{UDP} \xrightarrow{\text{glucosyltransferase}} \text{UDPG} + \text{fructose}$$

The glucosyltransferase activity increases with germination time in the embryo of broad beans up to the third day and then declines. This enzyme is specially inhibited by D-glucose and D-fructose, illustrating end product inhibition and substrate competition.

b. ENZYMES, INTERMEDIATES, AND CONTROL. It is difficult to discern metabolic regulation in the storage tissue of seeds under normal germination conditions. Probably the determined developmental pattern of metabolism dominates and overshadows the delicate regulatory mechism of individual enzymes in cells, or the direction of mass flow removes the reaction products from reaction sites so quickly that inhibition cannot be exerted. Specifically in what form(s) is the determined metabolic pattern installed in cells? It is the very question of current interest in developmental biology. In seeds, long-lived mRNA's and hormones might be the likely candidates among other possibilities, since conserved mRNA appears to direct the synthesis of protease in cotton embryo (Ihle and Dure, 1969, 1970), and GA stimulates the synthesis of α-amylase and many hydrolytic enzymes in germinating barley (Groat and Briggs, 1969; Pollard, 1969) (Section IV,C,3,b).

In germinating pea cotyledons, the peak of hexokinase activity occurs on the fourth day and precedes that of glucose content which occurs on the ninth day (Brown and Wray, 1968). In contrast, an inducible situation by either glucose or fructose in the cotyledon of castor bean for the activity of hexokinase and fructokinase has been reported (Marré *et al.*, 1965a). The hexokinase in storage tissue such as the pea cotyledons may be a different isoenzyme from the hexokinase found in growing tissue such as the cotyledon of castor bean. The former is synthesized according to the developmental pattern and is insensitive to the control of metabolites due to stabilized conformation, or it resists degradation, whereas the latter turns over rapidly and its activity is regulated by the concentration of substrate, coenzymes, cofactors, and end products. Different tissues containing specific isozymes are known in both animal and plant materials. Aldolase, for instance, in ungerminated radish, spinach *(Spinacia oleracea),* or wild carrot *(Daucus carota)* seeds has a slower electrophoretic mobility than does aldolase extracted from their respective roots and leaves (Takeo, 1969). As revealed by zymograms, the preexisting phosphorylase in pea cotyledons is different from synthesized phosphorylase after 3 days of germination, and the two preexisting $\beta$-amylases have lower mobility than the $\beta$-amylase formed during germination (Juliano and Varner, 1969). Therefore, measurement of overall enzyme activity is not sensitive enough in seed storage tissue to demonstrate any regulatory mechanism, which of course must exist above and beyond the developmental control. The case of repression in isocitrate lyase activity in endosperm of castor bean and squash cotyledons by glucose (Lado *et al.,* 1968) illustrates this point (Section IV,A,3,c).

Another type of enzyme regulation is the activation of preexisting $\beta$-amylase in grains by cysteine (Spradlin *et al.,* 1969). Experimental data suggest that a small conformational change occurs when sulfhydryl groups in $\beta$-amylase are modified, and this causes unfavorable repositioning of the catalytic amino acids. The change is reversible and the *in vivo* enzyme activity could be inhibited by oxidation of sulfhydryl groups of $\beta$-amylase or restored by reduction.

The $\alpha$-amylase in pea cotyledons is synthesized *de novo* during germination, and the microsomal fraction catalyzes the synthesis of the enzyme (Swain and Dekker, 1969). This enzyme-synthesizing ability appears to be related to stage of development just as the protein-synthesizing ability of glyoxysome in pine seeds occurs only when the tissue is ready for mass mobilization of reserves (Ching, 1970b). This type of temporal control in synthetic ability of a specific functional enzyme or organelle at a particular developmental age probably is manifested through a specific mRNA.

The production of $\alpha$-amylase also involved the proper balance of different hormones since application of benzyladenine to dry peas inhibited $\alpha$-amylase activity and growth under germination conditions, whereas, GA treatment stimulated growth (Sprent, 1968).

Better clarity and precision on properties and control mechanisms of individual enzymes and their catalytic activities could be obtained by studying purified enzyme rather than the total soluble enzyme extracts of the tissue. Excellent examples are provided by studies of ADP glucose-pyrophosphorylase (EC 2.7.7.b.) (Dickinson and Preiss, 1969) and various phosphorylases (Tsai, 1969) in starch synthesis of maturing corn seeds. Application or integration of such investigations *in vivo* conditions presents another problem. Phosphorylase I isolated from corn endosperm is competitively inhibited by ATP at 5 m$M$ (Tsai and Nelson, 1968, 1969). The authors thought this inhibition may be a control mechanism of starch degradation during seed germination. Whether such a high concentration of ATP exists in the cytoplasm is questionable since germinating lettuce seeds contain about 0.8 m$M$ (Pradet *et al.*, 1968), pea cotyledons of 1-day-old germinants have 0.33 m$M$ (Brown and Wray, 1968), mega-gametophytes of young seedlings of Norway spruce and Scotch pine have about 0.4 m$M$ (Bartels, 1968), and scutellum tissue of four-day-old corn seedling contains about 0.6 m$M$ ATP (Garrard and Humphreys, 1969). However, enzymes usually operate *in vivo* at an efficiency one or two magnitudes lower than their isolated form (Brown and Wray, 1968), so ATP concentrations could be a factor in phosphorylase activity. It is entirely possible that the phosphorolytic pathway of starch degradation is subject to direct metabolic control, whereas the hydrolytic pathway is influenced by hormone levels (such as GA in barley) and the efficiency of protein-synthesizing machinery in producing hydrolytic enzymes. Perhaps a division of labor is involved in which the phosphorylase provides substrate for respiration with the respiratory product, ATP, inhibiting the enzyme activity, whereas the hydrolytic pathway mobilizes reserves for direct transport to the plant axis. This view is substantiated by observations reported by Latzko and Kotze (1965). They analyzed the concentrations of glycolytic intermediates, ADP, ATP, and hexoses in germinating spring barley (Table 2.XVII) and calculated equilibrium constants of various enzymes in that tissue. Comparing these constants with published data, they noted that phosphorylase activity was very high while hexokinase activity was extremely limiting in barley.

## 4. FATE OF HYDROLYTIC PRODUCTS

Radioactive tracers are commonly used in this type of study. The relative distribution of radioactivity generally indicates the fate of a metabolite as to the extent of its translocation and contribution to the synthesis of various components in the embryo. The distribution of radioactivity in intermediates usually reflects the pool size when a steady state is reached.

Amino acids and glucose, respectively, are major hydrolytic products of protein and starch reserves. Analytic data on free amino acids in cotyledons of germinating peas are given in Table 2.XVIII. The data exhibit quantitative as well as qualitative changes which indicate that the

**TABLE 2.XVII.**

CONTENT OF METABOLIC INTERMEDIATES IN GERMINATING SEEDS

| Intermediates | $\mu$Moles/100 gm fresh weight | | | | | |
|---|---|---|---|---|---|---|
| | Buckwheat[a] | | | Spring Barley[b] | | |
| | 2[c] | 3[c] | 4[c] | 1[c] | 3[c] | 5[c] |
| Glucose 1-phosphate | $-^d$ | – | – | 0.13 | 0.20 | 0.25 |
| Glucose 6-phosphate | 10 | 15 | 14 | 7.76 | 17.50 | 22.20 |
| Fructose 6-phosphate | 1.8 | 1.5 | 0.8 | 3.75 | 8.75 | 12.00 |
| Fructose 1,6-diphosphate | 0.3 | 0.4 | 0.5 | 0.32 | 1.48 | 3.75 |
| 3-Phosphoglyceric acid | 2.4 | 3.5 | 3.6 | – | – | – |
| Glyceraldehyde phosphate | – | – | – | 0.18 | 1.15 | 1.15 |
| Dihydroxyacetone phosphate | 0.3 | 0.4 | 0.4 | 1.35 | 1.72 | 3.75 |
| Phosphoenol pyruvate | 0.9 | 1.0 | 1.5 | – | – | – |
| 6-Phosphogluconate | 0.4 | 0.4 | 0.5 | – | – | – |
| Glucose | – | – | – | 173 | 360 | 1620 |
| Fructose | – | – | – | 362 | 355 | 600 |
| Pi | – | – | – | 1750 | 2625 | 3250 |
| ATP | – | – | – | 3.75 | 3.25 | 7.76 |
| ADP | – | – | – | 4.50 | 1.75 | 2.00 |
| NAD | 5.28 | – | – | – | – | – |
| NADH | 0.16 | – | – | – | – | – |

[a] Data from Effer and Ranson (1967b).
[b] Data from Latzko and Kotze (1965).
[c] Germination days.
[d] –, not determined.

pool is not derived directly from hydrolytic products of reserve protein, and/or that a differential utilization of amino acids is occurring in that tissue.

The fate of amino acids during germination was traced by injecting L-leucine-U-[14]C into cotyledons of germinating peas (Beevers and Splittstoesser, 1968). The tissue was analyzed after 6 hours incubation in the dark. Results are summarized in Table 2.XIX. Transport of amino acids from cotyledons to the seedling axis increased with germination time, and rapid utilization of amino acids for insoluble residue (proteins) and for respiration is shown. On further fractionation of the water soluble fraction, several percent of the radioactivity was found in sugars, 30–40% in organic acids with citrate as the major component, 35–65% in amino acids with leucine, glutamic acid, homoserine, serine, and glycine as the major ones (Beevers and Splittstoesser, 1968). Differential labeling in cotyledons and the seedling axis suggests that different metabolic pathways operate in the two organs. The distribution of [14]C indicates that the carbon skeleton of amino acids could be used as substrates for lipids,

**TABLE 2.XVIII.**

CHANGE IN FREE AMINO ACIDS IN COTYLEDONS OF GERMINATING PEA[a]

|  | μMoles free amino acid/seedling | | |
|---|---|---|---|
|  | 0 days | 2.5 days | 10 days |
| Aspartic and asparagin | 0.08 | 0.47 | 1.33 |
| Threonine | 0.18 | 0.65 | 1.91 |
| *Unknown* | 0.10 | – | 1.46 |
| Serine | 0.04 | 1.59 | 3.04 |
| Homoserine | – | 2.71 | 0.59 |
| Glutamic and glutamine | 0.69 | 5.76 | 1.86 |
| Proline | 0.01 | 0.63 | 0.50 |
| Glycine | 0.12 | 0.28 | 1.29 |
| Alanine | 0.11 | 0.65 | 1.02 |
| Valine | 0.03 | 0.57 | 1.06 |
| Methionine | 0.01 | 0.07 | – |
| Leucine | 0.02 | 0.17 | 0.67 |
| Isoleucine | 0.01 | 0.19 | 0.58 |
| Tyrosine | 0.01 | 0.08 | 0.46 |
| Phenylalanine | – | 0.32 | 0.38 |
| λ-Aminobutyric | 0.01 | 0.40 | 1.07 |
| Lysine | 0.01 | 0.28 | 0.86 |
| Histidine | 0.01 | 0.41 | 0.49 |
| Arginine | 0.01 | 1.94 | 0.05 |
| Ammonia | 0.19 | 3.50 | 2.34 |
| Total amino acids | 1.45 | 17.17 | 18.61 |

[a] From Larson and Beevers (1965).

**TABLE 2.XIX**

DISTRIBUTION OF $^{14}$C FROM LEUCINE-U-$^{14}$C METABOLIZED BY PEA SEEDLINGS[a]

| Days of germination | 1 | 3 | 5 | 7 | 9 | 11 |
|---|---|---|---|---|---|---|
| Fraction | | | cpm $\times 10^{-3}$ | | | |
| $CO_2$ | 2 | 6 | 12 | 11 | 10 | 9 |
| Cotyledons | | | | | | |
|   Lipids | 12 | 9 | 8 | 5 | 3 | 2 |
|   Water-soluble | 658 | 227 | 351 | 466 | 370 | 370 |
|   Nucleic acids | 2 | 3 | 3 | 3 | 4 | 10 |
|   Insoluble residue | 305 | 626 | 461 | 257 | 304 | 392 |
|   Total | 987 | 885 | 823 | 731 | 681 | 774 |
| Axis | | | | | | |
|   Lipids | 0 | 1 | 1 | 1 | 1 | 1 |
|   Water-soluble | 1 | 32 | 39 | 55 | 66 | 94 |
|   Nucleic acids | 0 | 1 | 2 | 2 | 3 | 2 |
|   Insoluble residue | 5 | 45 | 76 | 106 | 137 | 77 |
|   Total | 6 | 79 | 108 | 164 | 207 | 174 |

[a] From Beevers and Splittstoesser (1968).

carbohydrates, protein, and nucleic acids of the cotyledons and axis. The rate of transport and synthesis in the axis increased for 9 days during germination and then declined. Similar results were found by Larson and Beevers (1965) using glutamate-U-$^{14}$C as tracer, by Splittstoesser (1969) using arginine-U-$^{14}$C in pumpkin seeds, and by Sane and Zalik (1968b) using proline-U-$^{14}$C in barley.

Differential utilization of various amino acids for sugars and organic acids was shown in the endosperm of 5-day-old germinating castor beans (Stewart and Beevers, 1967). Sucrose formation from fat reserve occurs in this tissue (Section IV,A,2,c). Thus the facility of converting carbon skeletons to sugars is ample to accommodate the influx of amino acids from hydrolysis of reserve protein. The differential utilization of various amino acids is partly due to the composition of amino acids in the reserve protein; so the higher the content of a particular amino acid the more utilization occurs. The metabolic pathways prevailing in the tissue also regulate the differential utilization. For example, following the mass flow of metabolites in endosperm of germinating castor beans, aspartate is deaminated to oxalacetate which is then converted to sucrose (Section IV,A,2,c) so the major conversion of aspartate-U-$^{14}$C is to organic acids and sugars.

S. S. C. Chen and Varner (1969) noted that 47% of the radioactivity from maltose-U-$^{14}$C absorbed by the endosperm of wild oats (*Avena fatua*) was transported to the embryo in 4 hours. Of this amount 12% was found in alcohol-insoluble proteins, nucleic acids, and carbohydrates, 18% in amino acids and organic acids, 9% in sucrose, 5% in glucose, and 4% remained as maltose.

## C. Metabolism of Cereal Seeds

Physiology and metabolism of germinating cereal seeds have been under investigation for about 100 years, but are still fascinating today because of the continuous progress made in various scientific disciplines, the unique morphological structure, and the genetic makeup of different organs in grains. The degradation of starch granules in cereal seeds has been discussed in previous sections of this chapter. This section is devoted to the division of metabolic activities which are specific to seeds of Gramineae.

### 1. OVERALL COMPOSITIONAL CHANGES

Figure 2.30 presents the chemical changes occurring in germinating corn seeds. The general pattern of starch utilization during seed germination is similar in all species, and the pattern is shown in Fig. 2.29 (Section IV,B,4,a). The distribution and changes in DNA, RNA, and soluble

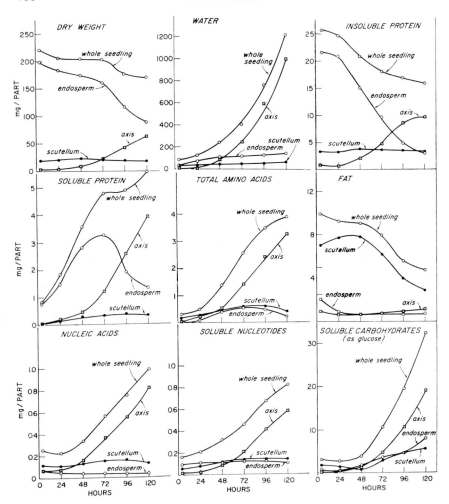

FIG. 2.30. Compositional changes in germinating corn seeds (Ingle *et al.*, 1964).

nucleotides in various parts of germinating corn are summarized in Table 2.XX.

All these data indicate that the endosperm is the storage site for protein and starch, both of which are hydrolyzed with the aid of enzymes to sugars and amino acids which are then transported through the scutellum to the axis where nucleotides, proteins, nucleic acids and other structural components are synthesized *in situ*. The scutellum is the storage site of fat, most of which is utilized during germination (Ingle *et al.*, 1964; Ingle and Hageman, 1965). Similar results are found during germination of

**TABLE 2.XX.**

The Soluble Nucleotide, RNA, and DNA Content of the Axis
Scutellum, Endosperm, and Whole Corn Seedling During a 5-Day
Period of Germination[a]

| | μg/part | | | | | |
|---|---|---|---|---|---|---|
| | 4[b] | 23[b] | 48[b] | 71[b] | 95[b] | 121[b] |
| Axis | | | | | | |
| Nucleotide | 18.2 | 24.3 | 97.1 | 206.0 | 413.0 | 585.0 |
| RNA | 63.3 | 66.6 | 142.5 | 323.0 | 485.0 | 710.0 |
| DNA | 6.3 | 6.4 | 17.2 | 39.6 | 63.5 | 106.5 |
| Scutellum | | | | | | |
| Nucleotide | 53.0 | 74.0 | 114.0 | 138.0 | 141.0 | 138.0 |
| RNA | 105.0 | 95.0 | 117.5 | 136.3 | 146.0 | 114.0 |
| DNA | 12.7 | 11.4 | 15.0 | 15.3 | 15.1 | 15.8 |
| Endosperm | | | | | | |
| Nucleotide | 92.2 | 107.0 | 107.0 | 116.5 | 121.5 | 100.0 |
| RNA | 47.5 | 25.4 | 28.5 | 31.7 | 25.4 | 25.4 |
| DNA | 21.4 | 21.0 | 14.0 | 11.9 | 11.9 | 9.1 |
| Whole seedling (sum of parts) | | | | | | |
| Nucleotide | 163.4 | 205.3 | 318.1 | 460.5 | 675.0 | 823.0 |
| RNA | 215.8 | 187.0 | 288.5 | 491.0 | 656.4 | 849.4 |
| DNA | 40.4 | 38.8 | 46.2 | 66.8 | 90.5 | 131.4 |

[a] From Ingle and Hageman (1965).
[b] Age of seedlings (hours).

barley (Sane and Zalik, 1968a) and a comprehensive review of utilization of cereal seed reserves has been prepared by MacLeod (1969). She summarizes, "The reserve materials of cereal grain can be divided into two categories: those which are present in the embryo and are immediately available for use by the growing seedling and those which are stored in an insoluble form in the endosperm and require to be hydrolyzed and translocated through the scutellum before they can be utilized. Reserves which are used by the seedling during its first 24 hours of growth include sucrose and raffinose (which together account for 20 percent of the dry weight of the embryo), lipids and amino acids."

"The reserve of endosperm includes the hemicelluloses which form the cell walls, starch and protein, and in barley, their hydrolysis is carried out sequentially by β-glucanases and xylanases, amylases and peptidases. Apart from β-amylase, which is present in the endosperm of ungerminated grain, the hydrolytic enzymes originate in the aleurone layer. They are synthesized in response to gibberellins which migrate from the embryo, possibly via the upper half of the scutellum to reach the aleurone normally between 12 and 20 hours after the grain is moistened."

## 2. Function of the Scutellum

The scutellum of cereal seeds has long been known as the organ in which sucrose is first synthesized from glucose resulting from starch degradation in the endosperm and then is transported to the seedling. The necessary enzymes for synthesis of sucrose from glucose, hexokinase, phosphoglucoisomerase, phosphoglucomutase, UDP-glucose pyro-phosphorylase, sucrose synthetase, and UDP-ATP kinase, have been found in the scutellum of germinating wheat and barley seeds (Edelman *et al.*, 1959). Total activity of sucrose synthetase in the scutellum of germinating rice increased for 4 days then declined when UDPG and fructose phosphate were used as substrates, whereas a slow but continuous decline occurred when UDPG and fructose were the substrates (Nomura *et al.*, 1969). S. S. C. Chen and Varner (1969) injected maltose-U-$^{14}$C into the endosperm of 3-day-old maize seedlings and incubated them for 4 hours. About 70% of the total $^{14}$C extracted in alcohol from the scutellum was in sucrose and only 13% and 3% were in glucose and maltose, respectively. In the endosperm of these seedlings, the distribution of $^{14}$C in sucrose, glucose and maltose was 21%, 61%, and 12%, respectively. These findings indicate that maltase is mainly located in the endosperm and sucrose synthetase in the scutellum.

The storage and synthesis of sucrose in the scutellum are both energy-requiring processes (Garrard and Humphreys, 1969). Thus mitochondria are very active and they multiply quickly during germination to supply the requirement of ATP in the scutellum (Hanson *et al.*, 1959). Another function of scutellum is, of course, transferring hormones from embryo to aleurone layer of the endosperm.

## 3. Induction of Enzymes in Endosperm

The degradation of reserves in the endosperm depends mainly on enzymes secreted by the cells in the aleurone layer. Synthesis of these enzymes is stimulated *in vivo* by gibberellins. Synthesis of hydrolytic enzymes can also be evoked by exogenous gibberellins in excised aleurone layers from hydrated and chilled seeds (3°C for 3–7 days) or from soaked embryoless half-seeds (30°C for 1–2 days) (Pollard, 1969).

a. Sequential Events *in Vitro*. When aleurone layers from imbibed embryoless half-seeds were incubated with GA$_3$ (0.5 μg/ml), mobilization of protein bodies (aleurone grains) was observed after 2 hours incubation. Increased endoplasmic reticulum and its associated ribosomes and polysomes was evident after 8–10 hours incubation. α-Amylase started to be released from the layer after 10–12 hours incubation, and RNase after 18–20 hours (Jones, 1969; Jones and Price, 1970).

When aleurone layers from soaked embryoless half-seeds were incubated with $GA_3$ (20 or 100 $\mu g/ml$), more ultrastructural changes were brought about in 18 hours. Similar changes did not occur in water incubated controls until after 24 hours of incubation. Ultrastructural changes accelerated by GA treatment included enlargement, fusion, and vacuolation of aleurone grains (protein bodies), disappearance of globoids within the aleurone grains, disappearance of spherosomes (fat bodies), extensive erosion of cell walls, and very importantly, increase in numbers of mitochondria, Golgi bodies, vesicles, and endoplasmic reticulum (Paleg and Hyde, 1964). Nevertheless, GA accelerated degradation of reserves and promoted synthesis of organelles that were related to energy supply (mitochondria), secretion (Golgi bodies), and synthesis (endoplasmic reticulum, vesicles, and ribosomes).

GA probably does not act at the level of gene transcription because (*1*) no difference was found in RNA synthesis between GA treated aleurone layers and control material in the early stage of incubation (up to 24 hours), (*2*) there was less than 5% increase in total protein synthesis in GA-treated material after 8 hours of incubation even though the amount of radioactivity in protein secreted from GA-treated material was double that of controls after 6 hours of incubation, (*3*) GA enhanced the secretion of sugars, ATPase, GTPase, and phytase, after only 6 hours of incubation and these enzymes did not appear to be newly synthesized, and (*4*) GA stimulated the respiration of aleurone layers, and the stimulation was observed after 4 hours incubation (Pollard and Singh, 1968). Using embryoless half-seeds, Pollard (1969) observed more enzymes secreted in sequence but not in concert. Lack of synchronized secretion of enzymes further nullifies the possibility that GA turns genes on as a primary action.

GA treatment causes sequential increases in secretion of soluble carbohydrates, respiration, secretion of ATPase, GTPase, phytase, phosphomonoesterase, phosphodiesterase, inorganic phosphate, carbohydrate hydrolyzing enzymes other than amylases, peroxidase, and then amylase (Pollard, 1969).

b. *In Vivo* PRODUCTION OF GIBBERELLINS AND $\alpha$-AMYLASE. A quantitative sequential relationship between gibberellins produced and $\alpha$-amylase synthesized was established *in vivo* in malting barley (Groat and Briggs, 1969). Figure 2.31 shows the kinetics of the relationship. A small amount of GA was present in dry seeds, a rapid increase was noted on the second day of malting, followed by a sharp decline on the fourth day. A lag of 20 hours was observed between the peak of GA production and the peak of $\alpha$-amylase synthesis. This lag period encompassed the time of translocation from the site of GA production in the embryo to its site of action in the aleurone layer as well as the time needed for aleurone to re-

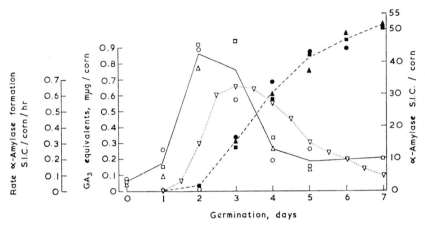

Fig. 2.31. Changes in levels of α-amylase (●-■-▲) and gibberellin-like materials (O-□ -Δ) in barley germinated at 14.4°C (Groat and Briggs, 1969). Calculated rate of α-amylase production, ---Δ---.

act and produce the enzyme. A causal relationship of GA quantity present in the grains and the α-amylase activity ultimately produced was established. Production of enzyme appeared to be temperature dependent as α-amylase was not formed in detectable quantities until after 19–26 hours in grains incubated at 14.4°C or 15–17 hours in grains incubated at 25°C. The endogenous GA behaved more like $GA_3$ than $GA_1$ or a mixture of the two. The quantity of GA in different tissue of malting barley was estimated in the first 2 days of malting (Table 2.XXI).

TABLE 2.XXI.
DISTRIBUTION OF GIBBERELLIN-LIKE MATERIAL IN BARLEY
MALTED FOR DIFFERENT PERIODS[a]

| Days germination after steeping | $GA_3$ equivalents ng/part | | |
|---|---|---|---|
| | Embryo | Endosperm | Whole grain |
| 0 | 0.11 | 0.08 | 0.19 |
| 1 | 0.09 | 0.11 | 0.20 |
| 2 | 0.29 | 0.62 | 0.91 |

[a] From Groat and Briggs (1969).

   Where and how GA exerts its effect on enzyme production is still an enigma. The metabolic events evoked by GA, the quantitative response of α-amylase produced to GA applied, and the timing of GA effects have been narrowed down, and perhaps the precise site of action will soon be brought into focus.

## 4. GENETIC–ENVIRONMENTAL CONTROL OF ISOZYMES

Isozymes refer to multiple molecular forms of an enzyme, with similar or identical catalytic activities, occurring within the same organism (Scandalios, 1969). Genetic variants of one enzyme could be found in varieties or hybrids of one species or among different species (Scandalios, 1969; Takeo, 1969), and these isozymes often afford good markers for studies of breeding behavior and theoretical inheritance. Particular significance of isozymes in metabolic studies lies in the fact that they often are specific to an organelle, organ, or developmental stage. This specificity reflects the result of genetic–metabolic or genetic–environmental interplay. Thus, a study of different isozymes provides clues about their functional requirements and magnitude, from which their role in metabolic activity and development can be discerned.

With improved electrophoretic procedures, the detection of isozymes has been facilitated in recent years. In dry wheat seeds, twelve esterases, two peroxidases, and three alcohol dehydrogenases were found. Upon germination, there was no change in alcohol dehydrogenases, but one new esterase and two more peroxidases were synthesized. The new esterase was localized in the endosperm while the additional peroxidases were found only in the embryo. In older seedlings more tissue-specific patterns were observed (Bhatia and Nilson, 1969; Fig. 2.32). Acid peroxidase increased from one in dry wheat to five isozymes in 2-day-old germinants (Marchesini *et al.*, 1968). The aldolase in carrot, spinach, and radish seeds had different electrophoretic mobility than that in roots or shoots of germinated seedlings (Takeo, 1969). A total of four peptidases occurred in maize seedlings, two of which were common to all tissues. One was found only in the endosperm of ripening seeds, and the other in both the endosperm and embryo of ripening seeds (Scandalios, 1969). One type of catalase was very organ-specific in corn and one isozyme was formed in each tissue of endosperm, silk, ear, and pollen. During seed germination, the quantity of catalase increased with time and new isozyme developed in shoots (Scandalios, 1969). These findings reflect a developmental change in isozyme patterns. Whether the presence of a new isozyme indicates direct gene action and subsequent translation, or direct translation from long-lived mRNA as a result of a metabolic triggering mechanism, or condensation or dissociation of enzyme subunits due to the influence of a specific metabolite, coenzyme, or cofactor is not known.

Organellar specificity of isozymes is indicated in the zymograms of malate dehydrogenase extracted from maize (Scandalios, 1969). A total of twelve isozymes were separated by electrophoresis, seven of which were found in one parental type and eight in the other. Two soluble and two glyoxysomal isozymes were common to both parental types, and all

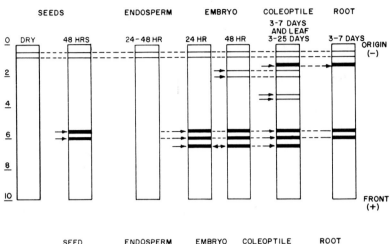

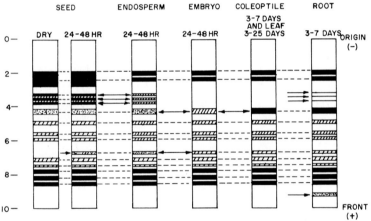

Fɪɢ. 2.32. Electrophoretic patterns of peroxidase (upper) and esterase (lower) in germinating wheat (Bhatia and Nilson, 1969).

of them did not form hybrids in reciprocal heterozygotes. Three mitochondrial isozymes in one parental type, however, hybridized with four in other to form five isozymes in heterozygotes. Although genetic inference is evident in these observations, the technique will be invaluable in establishing developmental patterns of the organelle, and in isolating specific isozyme for further characterization, from which the interaction of genetic and metabolic control may be understood.

## V. Epilogue

It is my wish to summarize original experimental data related with the metabolism of germinating seeds in this chapter so that the reader will have an overall picture of the present state of knowledge, some specific quantitative changes in macromolecules, organelles, coenzymes, metabolites, and enzyme activities, and lastly, some idea about the direction of future research. The picture is certainly not complete and is inevitably painted with conjecture. As concepts in biological sciences change continuously with new discoveries and accumulated knowledge, only constant inventory can enhance progress in any area of biology.

## Acknowledgment

I wish to express my sincere thanks to Drs. R. S. Bandurski, E. Hanson, T. T. Kozlowski, A. Marcus, and E. J. Trione for their critical review of this chapter.

## REFERENCES

Abdul-Baki, A. A. (1969). Metabolism of barley seed during early hours of germination. *Plant Physiol.* **44,** 733.

Abrahamson, L., and Shapiro, S. K. (1965). The biosynthesis of methionine; partial purification and properties of homocysteine methyltransferase of Jack bean meal. *Arch. Biochem. Biophys.* **109,** 376.

Ahmed, M. V., and Cook, F. S. (1964). Separation of an $\alpha$-galactosidase from watermelon seeds. *Can. J. Biochem.* **42,** 605.

Akazawa, T. (1965). Starch, inulin, and other reserve polysaccharides. *In* "Plant Biochemistry" (J. Bonner and J. E. Varner, eds.), 2nd ed., p. 271, Academic Press, New York.

Akazawa, T., and Beevers, H. (1957a). Some chemical components of mitochondria and microsomes from germinating castor beans. *Biochem. J.* **67,** 110.

Akazawa, T., and Beevers, H. (1957b). Mitochondria in the endosperm of the germinating castor bean: A developmental study. *Biochem. J.* **67,** 115.

Alberghina, F. A. M. (1967). Metabolic control of lipid utilization in castor bean cotyledons. *G. Biochim.* **16,** 145.

Allen, C. F., Good, P., Davis, H. F., Chisum, P., and Fowler, S. D. (1966). Methodology for the separation of plant lipids and application to spinach leaf and chloroplast lamellae. *J. Amer. Oil Chem.* **43,** 223.

Allende, J. E., and Bravo, M. (1966). Amino acid incorporation and amino-acyl transfer in a wheat embryo system. *J. Biol. Chem.* **241,** 5813.

Allerup, S. (1958). Effect of temperature on uptake of water in seeds. *Physiol. Plant.* **11,** 99.

Altschul, A. M., Snowden, J. E., Manchon, D. D., and Dechary, J. M. (1961). Intracellular distribution of seed proteins. *Arch. Biochem. Biophys.* **95,** 402.

Altschul, A. M., Yatsu, L. Y., Ory, R. L., and Engleman, E. M. (1966). Seed Proteins. *Annu. Rev. Plant Physiol.* **17,** 113.

Anderson, F. B., Cunningham, W. L., and Manners, D. J. (1964). Studies on carbohydrate-metabolizing enzymes. 10. Barley $\beta$-glucosidase. *Biochem. J.* **90,** 30.

Anderson, J. W., and Fowden, L. (1969). A study of the aminoacyl-sRNA synthatase of *Phaseolus vulgaris* in relation to germination. *Plant Physiol.* **44,** 60.

Anderson, M. B., and Cherry, J. H. (1969). Differences in leucyl-transfer RNA's and synthetase in soybean seedlings. *Proc. Nat. Acad. Sci. U.S.* **62,** 202.

App, A. A. (1969). Involvement of aminoacyl-tRNA transfer factors in polyphenylalanine synthesis by rice embryo ribosomes. *Plant Physiol.* **44**, 1132.

Asada, K., Tanaka, K., and Kasai, Z. (1969). Formation of phytic acid in cereal grains. *Ann. N. Y. Acad. Sci.* **165**, 801.

Atkinson, D. E. (1968). The energy charge of the adenylate pool as a regulatory parameter. Interaction with feedback modifier. *Biochemistry* **7**, 4030.

Atkinson, D. E. (1969). Regulation of enzyme function. *Annu. Rev. Microbiol.* **23**, 47.

Austin, R. B., Longden, P. C., and Hutchinson, J. (1969). Some effects of "hardening" carrot seed. *Ann. Bot. (London)* [N.S.] **33**, 883.

Bagley, B. W., Cherry, J. H., Rollins, M. L., and Altschul, A. M. (1963). A study of protein bodies during germination of peanut seed. *Amer. J. Bot.* **50**, 523.

Bain, J. M., and Mercer, F. V. (1966). Subcellular organization of the cotyledons in germinating seeds and seedlings of *Pisum sativum* L. *Aust. J. Biol. Sci.* **19**, 69.

Barker, G. R., and Rieber, M. (1967). The development of polysomes in the seed of *Pisum arvense*. *Biochem. J.* **105**, 1195.

Bartels, H. (1968). Trockensubstanz, Wasser- und Zwischenstoffgehalte in Keimlingen von *Picea abies* Karst, und *Pinus silvestris* L. unter verschiedenen temperaturbedingungen. *Flora (Jena), Abst. A* **159**, 391.

Barton, L. V. (1961). "Seed Preservation and Longevity." Leonard Hill, London.

Bautista, G. M., and Linko, P. (1962). Glutamic acid decarboxylase activity as a measure of damage in artificially dried and stored corn. *Cereal Chem.* **39**, 455.

Bautista, G. M., Lugay, J. C., Cruz, L. J., and Juliano, B. O. (1964). Glutamic acid decarboxylase activity as a viability index of artificially dried and stored rice. *Cereal Chem.* **41**, 188.

Beevers, L. (1968). Protein degradation and proteolytic activity in the cotyledons of germinating pea seeds (*Pisum sativum*). *Phytochemistry* **7**, 1837.

Beevers, L., and Guernsey, F. S. (1966). Changes in some nitrogenous components during the germination of pea seeds. *Plant Physiol.* **41**, 1455.

Beevers, L., and Splittstoesser, W. E. (1968). Protein and nucleic acid metabolism in germinating peas. *J. Exp. Bot.* **19**, 698.

Beevers, L., Schrader, L. E., Flesher, D., and Hageman, R. H. (1965). The role of light and nitrate in the induction of nitrate reductase in radish cotyledons and maize seedlings. *Plant Physiol.* **40**, 691.

Bekhor, J., Bonner, J., and Dahmus, G. (1969). Hybridization of chromosomal RNA to native DNA. *Proc. Nat. Acad. Sci. U.S.* **62**, 271.

Bhatia, C. R., and Nilson, J. P. (1969). Isozyme changes accompanying germination of wheat seeds. *Biochem. Genet.* **3**, 207.

Bianchetti, R., and Sartirana, M. L. (1967). The mechanism of repression by inorganic phosphate of phytase synthesis in the germinating wheat embryo. *Biochim. Biophys. Acta* **145**, 484.

Biswas, B. B. (1969). Ribosomes in cotyledons of mung bean seeds at different stages of germination. *Arch. Biochem. Biophys.* **132**, 198.

Black, H. S., and Altschul, A. M. (1965). Gibberellic acid induced lipase and α-amylase formation and their inhibition by aflatoxin. *Biochem. Biophys. Res. Commun.* **19**, 661.

Bonner, J. (1965). "The Molecular Biology of Development." Oxford Univ. Press, London and New York.

Bopp, M. (1968). Control of differentiation in Fern-allies and Bryophytes. *Annu. Rev. Plant Physiol.* **19**, 361.

Bradbeer, C., and Stumpf, P. K. (1959). Fat metabolism in higher plants. XI. The conversion of fat into carbohydrate in peanut and sunflower seedlings. *J. Biol. Chem.* **234**, 498.

Brahmachary, R. L. (1968). Information transfer in embryonic development. *Progr. Biophys. Mol. Biol.* **18**, 109.

Breidenbach, R. W., and Beevers, H. (1967). Association of the glyoxylate cycle enzymes in a noval subcellular particle from castor bean endosperm. *Biochem. Biophys. Res. Commun.* **27**, 462.

Breidenbach, R. W., Castelfranco, P., and Criddle, R. S. (1967). Biogenesis of mitochondria in germinating peanut cotyledons. II. Changes in cytochromes and mitochondrial DNA. *Plant Physiol.* **42**, 1035.

Breidenbach, R. W., Kahn, A., and Beevers, H. (1968). Characterization of glyoxysomes from castor bean endosperm. *Plant Physiol.* **43**, 705.

Britten, R. J., and Davidson, E. H. (1969). Gene regulation for higher cells: A theory. *Science* **165**, 349.

Brown, A. P., and Wray, J. L. (1968). Correlated changes of some enzyme activities and cofactor and substrate contents of pea cotyledon tissue during germination. *Biochem. J.* **108**, 437.

Burdon, R. H. (1971). Ribonucleic acid maturation in animal cells. *Progr. Nucl. Acid Res. Mol. Biol.* **11**, 33.

Canvin, D. T., and Beevers, H. (1961). Sucrose synthesis from acetate in germinating castor bean: Kinetics and pathway. *J. Biol. Chem.* **236**, 988.

Carpenter, W. D., and Beevers, H. (1959). Distribution and properties of isocitratase in in plants. *Plant Physiol.* **34**, 403.

Carter, J. E., and Pace, J. (1964). Distribution of dehydroascorbic acid reductase in wheat grain. *Nature (London)* **201**, 503.

Castanera, E. G., and Hassid, W. Z. (1965). Uridine diphosphate D-glucouronic acid decarboxylase from wheat germ. *Arch. Biochem. Biophys.* **110**, 462.

Chakravorty, A. K. (1969a). Ribosomal RNA synthesis in the germinating blackeye pea (*Vigna unguiculata*). I. The effect of cycloheximide on RNA synthesis in the early stages of germination. *Biochim. Biophys. Acta* **179**, 67.

Chakravorty, A. K. (1969b). Ribosomal RNA synthesis in the germinating blackeye pea (*Vigna unguiculata*). II. The synthesis and maturation of ribosomes in the later stages of germination. *Biochim. Biophys. Acta* **179**, 83.

Chapman, J. A., and Rieber, M. (1967). Distribution of ribosomes in dormant and imbibed seeds of *Pisum arvense:* Electron-microscope observations. *Biochem. J.* **105**, 1201.

Chen, D., Sarid, S., and Katchalski, E. (1968a). Studies on the nature of messenger RNA in germinating wheat embryos. *Proc. Nat. Acad. Sci. U.S.* **60**, 902.

Chen, D., Sarid, S., and Katchalski, E. (1968b). The role of water stress in the inactivation of messenger RNA of germinating wheat embryos. *Proc. Nat. Acad. Sci. U.S.* **61**, 1378.

Chen, S. S. C., and Varner, J. E. (1969). Metabolism of [14]C-maltose in *Avena fatua* seeds during germination. *Plant Physiol.* **44**, 770.

Cherry, J. H. (1963). Nucleic acid, mitochondria, and enzyme changes in cotyledons of peanut seeds during germination. *Plant Physiol.* **38**, 440.

Cherry, J. H. (1967). Nucleic acid biosynthesis in seed germination: Influences of auxin and growth-regulating substances. *Ann. N.Y. Acad. Sci.* **144**, 154.

Cherry, J. H., Hageman, R. H., Collins, F. I., and Flesher, D. (1961). Effects of x-irradiation of corn seed. *Plant Physiol.* **36**, 566.

Ching, T. M. (1959). Activation of germination in Douglas fir seed by hydrogen peroxide. *Plant Physiol.* **34**, 557.

Ching, T. M. (1963). Fat utilization in germinating Douglas fir seed. *Plant Physiol.* **38**, 722.

Ching, T. M. (1965). Metabolic and ultrastructural changes in germinating Douglas fir seeds. *Plant Physiol.* **40**, Suppl. VIII.

Ching, T. M. (1966). Compositional changes of Douglas fir seeds during germination. *Plant Physiol.* **41**, 1313.

Ching, T. M. (1968). Intracellular distribution of lipolytic activity in the female gametophyte of germinating Douglas fir seeds. *Lipids* **3**, 482.

Ching, T. M. (1970a). Unpublished data.

Ching, T. M. (1970b). Glyoxysomes in megagametophyte of germinating ponderosa pine seeds. *Plant Physiol.* **45**, 475.

Ching, T. M., and Ching, K. K. (1964). Freeze-drying pine pollen. *Plant Physiol.* **39**, 705.

Chroboczek, H., and Cherry, J. H. (1966). Characterization of nucleic acids in peanut cotyledons. *J. Mol. Biol.* **19**, 28.

Clegg, J. S., and Golub, A. L. (1969). Protein synthesis in *Artemia salina* embryos. II. Resumption of RNA and protein synthesis upon cessation of dormancy in the encysted gastrula. *Develop. Biol.* **19**, 178.

Cocucci, M. C., and Caldogno, F. R. (1967). Inhibition by sucrose of utilization of storage lipids in germinating seeds of *Cucurbita maxima*. *G. Bot. Ital.* **101**, 231.

Cooper, T. G., and Beevers, H. (1969a). Mitochondria and glyoxysomes from castor bean endosperm. *J. Biol. Chem.* **244**, 3507.

Cooper, T. G., and Beevers, H. (1969b). $\beta$-Oxidation in glyoxysomes from castor bean endosperm. *J. Biol. Chem.* **244**, 3520.

Cooper, T. G., and Beevers, H. (1970). Fatty acid activation in glyoxysomes. *Fed. Proc., Fed. Amer. Soc. Exp. Biol.* **29**, 868 (Abstr.).

Criddle, R. S. (1969). Structural proteins of chloroplasts and mitochondria. *Annu. Rev. Plant Physiol.* **20**, 239.

Crocker, W., and Barton, L. V. (1957). "Seed Physiology." Chronica Botanica, Waltham, Massachusetts.

Dankert, M., Ruth, I., Gonclaves, J., and Recondo, E. (1964). Adenosine diphosphate glucose: Orthophosphate adenylyl transferase in wheat germ. *Biochim. Biophys. Acta* **81**, 78.

Davies, D. D. (1956). Soluble enzymes from pea mitochondria. *J. Exptl. Bot.* **7**, 203.

de Duve, C., and Baudhuin, P. (1966). Peroxisomes (microbodies and related particles). *Physiol. Rev.* **46**, 323.

Dickinson, D. B., and Preiss, J. (1969). ADP glucose pyrophosphorylase from maize endosperm. *Arch. Biochem. Biophys.* **130**, 119.

Duperon, P., and Duperon, R. (1969). Comparative formation of steroids and sterols liberated during germination of *Phaseolus vulgaris*. Hypothesis on the role of steroids in plants. *C. R. Acad. Sci., Ser. D* **269**, 157.

Earle, F. R., Vanetten, C. H., Clark, T. F., and Wolff, I. A. (1968). Compositional data on sunflower seed. *J. Amer. Oil Chem.* **45**, 876.

Eastwood, D., Tavener, R. J. A., and Laidman, D. L. (1969). Induction of lipase and phytase activities in the eleurone tissue of germinating wheat grains. *Biochem. J.* **113**, 32 p.

Ebert, J. D. (1968). Levels of control—a useful frame of perception. *Curr. Top. Develop. Biol.* **3**, XV.

Edelman, J., Shibko, S. I., and Keys, A. J. (1959). The role of scutellum of cereal seedlings in synthesis and transport of sucrose. *J. Exp. Bot.* **10**, 178.

Effer, W. R., and Ranson, S. L. (1967a). Respiratory metabolism in buckwheat seedlings. *Plant Physiol.* **42**, 1042.

Effer, W. R., and Ranson, S. L. (1967b). Some effects of oxygen concentration on levels of respiratory intermediates in buckwheat seedlings. *Plant Physiol.* **42**, 1052.

Engel, C. (1947a). The distribution of the enzymes in resting cereals. I. The distribution of the saccharogenic amylase in wheat, rye, and barley. *Biochim. Biophys. Acta* **1**, 42.

Engel, C. (1947b). The distribution of enzymes in resting cereals. III. The distribution of esterase in wheat, rye, and barley. *Biochim. Biophys. Acta* **1**, 278.

Engel, C., and Bretschneider, L. H. (1947). The distribution of the enzymes in resting cereals. IV. A comparable investigation of the distribution of enzymes and mitochondria in wheat grains. *Biochim. Biophys. Acta* **1**, 357.

Engel, C., and Heins, J. (1947). The distribution of enzymes in resting cereals. II. The distribution of proteolytic enzymes in wheat, rye, and barley. *Biochim. Biophys Acta* **1**, 190.

Fambrough, D. M., Fujimura, F., and Bonner, J. (1968). Quantitative distribution of histone components in the pea plant. *Biochemistry* **7**, 2575.

Fan, D. F., and Maclachlan, G. A. (1966). Control of cellulase activity by indoleacetic acid. *Can. J. Bot.* **44**, 1025.

Fedeli, E., Favini, F., Camurata, F., and Jacini, G. (1968). Regional differences of lipid composition in morphologically distinct fatty tissues. III. Peanut seeds. *J. Amer. Oil Chem.* **45**, 676.

Fedorcsak, I., Natarajan, A. T., and Ehrenberg, L. (1969). On the extraction of nucleic acids with template activity from barley embryos using diethyl pyrocarbonate as nuclease inhibitor. *Eur. J. Biochem.* **10**, 450.

Ferrari, T. E., and Varner, J. E. (1969). Substrate induction of nitrate reductase in barley aleurone layers. *Plant Physiol.* **44**, 85.

Filner, P., Wray, J. L., and Varner, J. E. (1969). Enzyme induction in higher plants. *Science* **165**, 358.

Firenzuoli, A. M., Vanni, P., Mastronezzi, E., Zanovini, A., and Baccari, V. (1968). Participation of the glyoxylate cycle in the metabolism of germinating seed of *Pinus pinea*. *Life Sci.* **7**, Part II, 1251.

Frederick, S. E., and Newcomb, E. H. (1968). Microbody-like organelles in leaf cells. *Science* **163**, 1353.

Fritz, G. J., Stout, E. R., and Leister, D. E. (1963). Estimation of nicotinamide nucleotide coenzymes in etiolated maize seedlings. *Plant Physiol.* **38**, 642.

Fujisawa, H. (1966). Role of nucleic acid and protein metabolism in the initiation of growth at germination. *Plant Cell Physiol.* **7**, 185.

Galston, A. W., and Davies, P. J. (1969). Hormonal regulation in higher plants. *Science* **163**, 1288.

Garrard, L. A., and Humphreys, T. E. (1969). The effect of D-mannose on sucrose storage in the corn scutellum: Evidence for two sucrose transport mechanisms. *Phytochemistry* **8**, 1065.

Gerhardt, B. P., and Beevers, H. (1969). Occurrence of RNA in glyoxysomes from castor bean endosperm. *Plant Physiol.* **44**, 1475.

Gerhardt, B. P., and Beevers, H. (1970). Developmental studies of glyoxysomes in *Racinus* endosperm. *J. Cell Biol.* **44**, 94.

Ghetie, V. (1966). The mechanism of reserve proteins hydrolysis during germination. *Rev. Roum. Biochim.* **3**, 353.

Gientka-Rychter, A., and Cherry, J. H. (1968). De novo synthesis of isocitratase in peanut (*Arachis hypogaea* L.) cotyledons. *Plant Physiol.* **43**, 653.

Goldberg, M. E., Creighton, T. E., Baldwin, R. I., and Yanofsky, C. (1966). Subunit structure of the tryptophan synthetase of *Escherichia coli*. *J. Mol. Biol.* **21**, 71.

Goo, M. (1951). Water absorption by tree seeds. *Bull. Tokyo Univ. Forests* **39**, 55.

Grabe, D. F. (1964). Glutamic acid decarboxylase activity as a measure of seedling vigor. *Proc. Ass. Off. Seed Anal.* **54**, 100.

Grabe, D. F. (1970). Personal communication.

Groat, J. E., and Briggs, D. E. (1969). Gibberellin and α-amylase formation in germinating barley. *Phytochemistry* **8**, 1615.

Haber, A. H., and Tolbert, N. E. (1959). Metabolism of $C^{14}$-bicarbonate, $P^{32}$-phosphate, or $S^{35}$-sulfate by lettuce seed during germination. *Plant Physiol.* **34**, 376.

Hadwiger, L. A., Badiei, S. E., Waller, G. R., and Gholson, R. K. (1963). Quinolinic acid as a precursor of nicotinic acid and its derivatives in plants. *Biochem. Biophys. Res. Commun.* **13**, 466.

Hall, J. R., and Hodges, T. K. (1966). Phosphorus metabolism of germinating oat seeds. *Plant Physiol.* **41**, 1459.

Halperin, W. (1969). Morphogenesis in cell cultures. *Annu. Rev. Plant Physiol.* **20**, 395.

Hanabusa, K. (1961). High yield crystallization of urease from Jack bean. *Nature (London)* **189**, 551.

Hanson, J. B., Vatter, A. E., Fisher, M. E., and Bils, R. F. (1959). The development of mitochondria in the scutellum of germinating corn. *Agron. J.* **51**, 295.

Harris, P. J. (1969). Relation of fine structure to biochemical changes in developing sea urchin eggs and zygotes. *In* "The Cell Cycle" (G. M. Padilla, G. L. Whitson, and I. L. Cameron, eds.), pp. 315–340. Academic Press, New York.

Henshall, J. D., and Goodwin, T. W. (1964). Amino acid-activating enzymes in germinating pea seedlings. *Phytochemistry* **3**, 677.

Herb, S. F. (1968). Gas-liquid chromatography of lipids, carbohydrates and amino acids. *J. Amer. Oil Chem.* **45**, 784.

Hlynka, L., and Robinson, A. D. (1954). Moisture and its measurement. *In* "Storage of Cereal Grain and Their Products" (J. A. Anderson and A. W. Alcock, eds.), pp. 1–12. Ass. Amer. Cereal Chem., St. Paul, Minnesota.

Hock, B., and Beevers, H. (1966). Development and decline of the glyoxylate cycle enzymes in watermelon seedlings. *Z. Pflanzenphysiol.* **55**, 405.

Hoerz, W., and McCarty, K. S. (1969). Evidence for a proposed initiation complex for protein synthesis in reticulocyte polyribosome profiles. *Proc. Nat. Acad. Sci. U.S.* **63**, 1206.

Horner, H. T., Jr., and Arnott, H. J. (1966). A histochemical and ultrastructural study of pre- and post-germinated Yucca seeds. *Bot. Gaz. (Chicago)* **127**, 48.

Howell, R. W. (1961). Change in metabolic characteristics of mitochondria from soybean cotyledons during germination. *Physiol. Plant.* **14**, 89.

Hsiao, T. C. (1968). Ribonuclease activity associated with ribosomes of *Zea mays*. *Plant Physiol.* **43**, 1355.

Hutson, D. H. (1964). *Trans*-β-glucosylation by alfalfa-seed enzymes. *Biochem. J.* **92**, 142.

Hutton, D., and Stumpf, P. K. (1969). Fat metabolism in higher plants. XXXVII. Characterization of the β-oxidation systems from maturing and germinating castor bean seeds. *Plant Physiol.* **44**, 508.

Ihle, J. N., and Dure, L. S. (1969). Synthesis of protease in germinating cotton cotyledons, catalyzed by mRNA synthesized during embryogenesis. *Biochem. Biophys. Res. Commun.* **36**, 705.

Ihle, J. N., and Dure, L. (1970). Hormonal regulation of translation inhibition requiring RNA synthesis. *Biochem. Biophys. Res. Commun.* **38**, 995.

Ingle, J., and Hageman, R. H. (1965). Metabolic changes associated with the germination of corn. II. Nucleic acid metabolism. *Plant Physiol.* **40**, 48.

Ingle, J., Beevers, L., and Hageman, R. H. (1964). Metabolic changes associated with the germination of corn. I. Changes in weight and metabolites and their redistribution in the embryo axis, scutellum, and endosperm. *Plant Physiol.* **39**, 735.

Irving, G. W., Jr., and Fontaine, T. D. (1945). Purification and properties of Arachain, a newly discovered proteolytic enzyme of peanut. *Arch Biochem. Biophys.* **6**, 351.

Jachymczyk, W. J., and Cherry, J. H. (1968). Studies on messenger RNA from peanut plants: *In vivo* polysome formation and protein synthesis. *Biochim. Biophys. Acta* **157,** 368.

Jacks, T. J., Yatsu, L. Y., and Altschul, A. M. (1967). Isolation and characterization of peanut spherosomes. *Plant Physiol.* **42,** 585.

James, E. (1968). Limitation of glutamic acid decarboxylase activity for estimating viability in beans *(Phaseolus vulgaris)*. *Crop Sci.* **8,** 403.

Jerez, C., Sandoval, A., Allende, J. E., Henes, C., and Ofengand, J. (1969). Specificity of interaction of aminoacyl ribonucleic acid with a protein-guanosine triphosphate complex from wheat embryo. *Biochemistry* **8,** 3006.

Jirgensons, B. (1962). "Natural Organic Macromolecule." Pergamon, Oxford.

Jones, R. L. (1969). Gibberellic acid and the fine structure of barley aleurone cells. I. Changes during the lag phase of α-amylase synthesis. *Planta* **87,** 119.

Jones, R. L., and Price, J. M. (1970). Gibberellic acid and the fine structure of barley aleurone cells. III. Vacuolation of the aleurone cells during the phase of ribonuclease release. *Planta* **94,** 191.

Juliano, B. O., and Varner, J. E. (1969). Enzymic degradation of starch granules in the cotyledons of germinating peas. *Plant Physiol.* **44,** 886.

Karow, H., and Mohr, H. (1967). Changes of activity of isocitratase during photomorphogenesis in mustard seedlings. *Planta* **72,** 170.

Kemp, R. J., Goad, L. J., and Mercer, E. I. (1967). Changes in the levels and composition of the esterified and unesterified sterols of maize seedlings during germination. *Phytochemistry* **6,** 1609.

Kikuchi, M., and Hayashi, K. (1969). Conversion of thiamine into its phosphoric esters in germinating broad beans. *Bot. Mag.* **82,** 28.

Klein, S., and Ben-shaul, Y. (1966). Changes in cell fine structure of lima bean axis during early germination. *Can. J. Bot.* **44,** 331.

Klingmüller, W., and Lane, G. R. (1960). Damaging effect of drying on *Vicia faba* seeds. *Nature (London)* **185,** 699.

Knapp, F. F., Aexel, R. T., and Nicholas, H. J. (1969). Steral biosynthesis in sub-cellular particles of higher plants. *Plant Physiol.* **44,** 442.

Kollöffel, C. (1969). The hisotochemical localization of dehydrogenases in the cotyledons of *Pisum sativum* during germination. *Acta Bot. Neer.* **18,** 406.

Kriedmann, P., and Beevers, H. (1967a). Sugar uptake and translocation in the castor bean seedling. I. Characteristics of transfer in intact and excised seedlings. *Plant Physiol.* **42,** 161.

Kriedmann, P., and Beevers, H. (1967b). Sugar uptake and translocation in the castor bean seedling. II. Sugar transformations during uptake. *Plant Physiol.* **42,** 174.

Lado, P., and Schwendimann, M. (1967). Changes of mitochondrial RNA level during the transition from rest to growth in the endosperm of germinating castor beans. *Life Sci.* **6,** Part II, 1681.

Lado, P., Schwendimann, M., and Marré, E. (1968). Repression of isocitrate lyase synthesis in seeds germinated in the presence of glucose. *Biochim. Biophys. Acta* **157,** 140.

Larson, L. A. (1965). Unpublished data.

Larson, L. A., and Beevers, H. (1965). Amino acid metabolism in young pea seedlings. *Plant Physiol.* **40,** 424.

Latzko, E., and Kotze, J. P. (1965). Glycolysis in germinating sprig barley. *Z. Pflanzenphysiol.* **53,** 377.

Ledoux, L., and Huart, R. (1969). Fate of exogenous bacterial deoxyribonucleic acids in barley seedlings. *J. Mol. Biol.* **43,** 243.

Lee, C. Y., and Shallenberger, R. S. (1969). Changes in free sugar during germination of pea seeds. *Experientia* **25**, 692.

Lee, H. J., Kim, S. J., and Lee, K. B. (1964). Study on the glyoxylate cycle in germinating sesame seed embryo. *Arch. Biochem. Biophys.* **107**, 479.

Lee, K. W., and Roush, A. H. (1964). Allantoinase assays and their application to yeast and soybean allantoinases. *Arch. Biochem. Biophys.* **108**, 460.

Legocki, A. B., and Marcus, A. (1971). Polypeptide synthesis in extracts of wheat germ. Resolution and partial purification of the soluble transfer factors. *J. Biol. Chem.* **245** 2814.

Leloir, L. F., De Feketa, M. A., and Cardini, C. E. (1961). Starch and oligosaccharide synthesis from uridine diphosphate glucose. *J. Biol. Chem.* **236**, 636.

Linko, P. (1961). Simple and rapid monometric method for determining glutamic acid decarboxylase activity as quality index of wheat. *J. Agr. Food Chem.* **9**, 310.

Linko, P., and Milner, M. (1959). Enzyme activation in wheat grains in relation to water content. Glutamic acid-alanine transaminase, and glutamic acid decarboxylase. *Plant Physiol.* **34**, 392.

Linko, P., and Sogn, L. (1960). Relation of viability and storage deterioration to glutamic acid decarboxylase in wheat. *Cereal Chem.* **37**, 489.

Lipmann, F. (1969). Polypeptide chain elongation in protein synthesis. *Science* **164**, 1024.

Loening, U. E. (1968). RNA structure and metabolism. *Annu. Rev. Plant Physiol.* **19**, 37.

Longo, C. P. (1968). Evidence of *de novo* synthesis of isocitratase and malate synthetase in germinating peanut cotyledons. *Plant Physiol.* **43**, 660.

Lui, N. S. T., and Altschul, A. M. (1967). Isolation of globoids from cottonseed aleurone grain. *Arch. Biochem. Biophys.* **121**, 678.

Macey, M. J. K., and Stumpf, P. K. (1968). Fat metabolism in higher plants. XXXVI. Long chain fatty acid synthesis in germinating peas. *Plant Physiol.* **43**, 1637.

McKillicam, M. E. (1966). Lipid changes in maturing oil-bearing plants. IV. Changes in lipid classes in rape and crambe oils. *J. Amer. Oil Chem.* **43**, 461.

MacLeod, A. M. (1969). The utilization of cereal seed reserves. *Sci. Progr. (London)* **57**, 99.

McMahon, V., and Stumpf, P. K. (1966). Fat metabolism in higher plants. XXVI. Biosynthesis of fatty acids in tissues of developing seeds and germinating seedlings of safflower *(Carthamus tinetorius)*. *Plant Physiol.* **41**, 148.

Mahler, H. R., and Cordes, E. H. (1968). "Basic Biological Chemistry." Harper, New York.

Malhotra, S. S., and Spencer, M. (1970). Changes in the respiratory, enzymatic, and swelling and contraction properties of mitochondria from cotyledons of *Phaseolus vulgaris* L. during germination. *Plant Physiol.* **46**, 40.

Marchesini, A., Sequi, P., and Lanzani, G. A. (1968). Peroxidase isoenzymes in wheat during germination. *Atti Simp. Int. Agrochim.* **7**, 203.

Marcus, A. (1969). Seed Germination and the capacity for protein synthesis. *Symp. Soc. Exp. Biol.* **23**, 143.

Marcus, A. (1970a). Tobacco mosaic virus ribonucleic acid-dependent amino acid incorporation in a wheat embryo system *in vitro*. Analysis of the rate-limiting reaction. *J. Biol. Chem.* **245**, 955.

Marcus, A. (1970b). Tobacco mosaic virus ribonucleic acid-dependent amino acid incorporation in a wheat embryo system *in vitro*. Formation of a ribosome-messenger "initiation" complex. *J. Biol. Chem.* **245**, 962.

Marcus, A., and Feeley, J. (1962). Nucleic acid changes in the germinating peanut. *Biochim. Biophys. Acta* **61**, 830.

Marcus, A., and Feeley, J. (1964a). Activation of protein synthesis in the imbibition phase of seed germination. *Proc. Nat. Acad. Sci. U.S.* **51**, 1075.

Marcus, A., and Feeley, J. (1964b). Isocitrate lyase formation in the dissected peanut cotyledon. *Biochim. Biophys. Acta* **89**, 170.

Marcus, A., and Feeley, J. (1965). Protein synthesis in imbibed seeds. II. Polysome formation during imbibition. *J. Biol. Chem.* **240**, 1675.

Marcus, A., and Feeley, J. (1966). Ribosome activation and polysome formation in vitro: Requirement for ATP. *Proc. Nat. Acad. Sci. U.S.* **56**, 1770.

Marcus, A., and Velasco, J. (1960). Enzyme of glyoxylate cycle in germinating peanuts and castor beans. *J. Biol. Chem.* **235**, 563.

Marcus, A., Feeley, J., and Volcani, T. (1966). Protein synthesis in imbibed seeds. III. Kinetics of amino acid incorporation, ribosome activation, and polysome formation. *Plant Physiol.* **41**, 1167.

Marcus, A., Luginbill, B., and Feeley, J. (1968). Polysome formation with tobacco mosaic virus RNA. *Proc. Nat. Acad. Sci. U.S.* **59**, 1247.

Marré, E. (1967). Ribosome and enzyme changes during maturation and germination of the castor bean seeds. *Curr. Top. Develop. Biol.* **II**, 76.

Marré, E., Cornaggia, M. P., Alberghina, F. A. M., and Bianchetti, R. (1965a). Substrate level as a regulating factor of the synthesis of fructokinase, hexokinase, and other carbohydrate-metabolizing enzymes in higher plants. *Biochem. J.* **97**, 20P.

Marré, E., Cocucci, S., and Sturani, E. (1965b). On the development of the ribosomal system in the endosperm of germinating castor bean seeds. *Plant Physiol.* **40**, 1162.

Matheson, N. K., and Strother, S. (1969). The utilization of phytate by germinating wheat. *Phytochemistry* **8**, 1349.

Matsushita, S., Mori, T., and Hata, T. (1966). Enzyme activity associated with ribosomes from soybean seedlings. *Plant Cell Physiol.* **7**, 533.

May, J. T., and Symons, R. H. (1968). Properties and intracellular distribution of cytidine and uridine diphosphokinase of cucumber cotyledons. *Phytochemistry* **7**, 1271.

Mayer, A. M., and Poljakoff-Mayber, A. (1963). "The Germination of Seeds." Macmillan, New York.

Mayer, A. M., and Shain, Y. (1968). Zymogen granules in enzyme liberation and activation in pea seeds. *Science* **162**, 283.

Mayer, A. M., Krishmaro, N., and Poljakoff-Mayber, A. (1968). Isocitric lyase and isocitric dehydrogenase in germinating lettuce. *Physiol. Plant.* **21**, 183.

Mihalyfi, I. P. (1968). Catalase activity in almond seeds. *Ann. Univ. Sci. Budapest. Rolando Eotvos Nominatae, Sect. Biol.* **9–10**, 305.

Mori, T., Ibuki, F., Matsushita, S., and Hata, T. (1968). Occurrence of ribonucleic acids that have template activities in soybean seeds. *Arch. Biochem. Biophys.* **124**, 607.

Mori, T., Ibuki, F., Matsushita, S., and Hata, T. (1969). Characterization of ribonucleic acid contained in the soluble fraction from soybean seeds. *Agr. Biol. Chem.* **33**, 1229.

Morton, R. K., and Raison, J. K. (1964). The separate incorporation of amino acids into storage and soluble protein catalysed by two independent systems isolated from developing wheat endosperm. *Biochem. J.* **91**, 528.

Morton, R. K., Polk, B. A., and Raison, J. K. (1964). Intracellular components associated with protein synthesis in developing wheat endosperm. *Biochem. J.* **91**, 522.

Mounter, L. A., and Mounter, M. E. (1962). Specificity and properties of wheat germ esterase. *Biochem. J.* **85**, 576.

Moustafa, E. (1963). Purification and properties of valine-activating enzyme from wheat germ. *Biochim. Biophys. Acta* **76**, 280.

Mukherji, S., Dey, B., and Sircar, S. M. (1968). Changes in nicotinic acid content and its nucleotide derivatives of rice and wheat seeds during germination. *Physiol. Plant.* **21**, 360.

Muller, M., Hogg, J. F., and de Duve, C. (1968). Distribution of tricarboxylic acid cycle enzymes and glyoxylate cycle enzymes between mitochondria and peroxisomes in *Tetrahymena pyriformis*. *J. Biol. Chem.* **243**, 5385.

Murata, T., Akazawa, T., and Shikiku, F. (1968). Enzymic mechanism of starch breakdown in germinating rice seeds. I. An analytical study. *Plant Physiol.* **43**, 1899.

Nagamachi, K. I., Fuji, M., and Honda, K. (1967). Studies on isocitrate lyase of the germinating castor bean endosperm. A possible metabolic control of NADP-specific isocitrate dehydrogenase and isocitrate lyase. *J. Agr. Chem. Soc. Jap.* **41**, 381.

Nass, M. M. K. (1969). Mitochondrial DNA: Advances, problems, and goals. *Science* **165**, 25.

Newcomb, E. H. (1967). Fine structure of protein-storing plastids in bean root tips. *J. Cell Biol.* **33**, 143.

Nomura, T., Kono, Y., and Akazawa, T. (1969). Enzymic mechanism of starch breakdown in germinating rice seeds. II. Scutellum as the site of sucrose synthesis. *Plant Physiol.* **44**, 765.

Nutile, G. E. (1964). Effect of desiccation on viability of seeds. *Crop Sci.* **4**, 325.

Ofelt, C. W., Smith, A. K., and Mills, J. M. (1955). Proteases of the soybean. *Cereal Chem.* **32**, 53.

Opik, H. (1966). Changes in cell fine structure in the cotyledons of *Phaseolus vulgaris* L. during germination. *J. Exp. Biol.* **17**, 427.

Opik, H. (1968). Development of cotyledon cell structure in ripening *Phaseolus vulgaris* seeds. *J. Exp. Bot.* **19**, 64.

Ory, R. L. (1969). Acid lipase of castor bean. *Lipids* **4**, 177.

Ory, R. L., Yatsu, L. Y., and Kircher, H. W. (1968). Association of lipase activity with the spherosomes of *Racinus communis*. *Arch. Biochem. Biophys.* **123**, 255.

Paleg, L., and Hyde, B. (1964). Physiological effects of gibberellic acid. VII. Electron microscopy of barley aleurone cells. *Plant Physiol.* **39**, 673.

Payne, P. I., and Boulter, D. (1969). Free and membrane bound ribosomes of the cotyledons of *Vicia faba*. *Planta* **87**, 63.

Penner, D., and Ashton, F. M. (1966). Proteolytic enzyme control in squash cotyledons. *Nature (London)* **212**, 935.

Penner, D., and Ashton, F. M. (1967). Hormonal control of proteinase activity in squash cotyledons. *Plant Physiol.* **42**, 791.

Perner, E. (1965). Elektronenmikroskopische Untersuchungen An Zellen von Embryonen in Zustand völliger Samenruhe. *Planta* **65**, 334.

Pinfield, N. J. (1968). The effect of gibberellin on the metabolism of ethanol-soluble constituents in the cotyledons of hazel seeds (*Corylus avellana* L.). *J. Exp. Bot.* **19**, 452.

Pollard, C. J. (1969). A survey of the sequence of some effects of gibberellic acid in the metabolism of cereal grain. *Plant Physiol.* **44**, 1227.

Pollard, C. J., and Singh, B. N. (1968). Early effects of gibberellic acid on barley aleurone layers. *Biochem. Biophys. Res. Commun.* **33**, 321.

Pollock, B. M., and Toole, V. K. (1966). Imbibition period as the critical temperature sensitive stage in germination of lima bean seeds. *Plant Physiol.* **41**, 221.

Pradet, A., and Bomsel, J. L. (1969). *In vivo* control of metabolic activity by "energy charge." *Abstr. Int. Bot. Congr., 11th, 1969* p. 173.

Pradet, A., Armugakannud, N., and Vermeersch, J. (1968). Plant tissue AMP, ADP, and ATP. III. Energy metabolism during the first stages of germination of lettuce seeds. *Bull. Soc. Fr. Physiol. Veg.* **14**, 107.

Preece, I. A., Grav, H. J., and Wadham, A. T. (1960). Studies on phytin. I. The inositol phosphates. *J. Inst. Brewing London* **66**, 487.

Prentice, N., Burger, W. C., and Widerholt, E. (1969). Distribution of acidic and neutral peptidases in germinating barley. *Physiol Plant.* **22,** 157.

Price, C. E., and Murry, A. W. (1969). Purine metabolism in germinating wheat embryo. *Biochem. J.* **115,** 129.

Pridham, J. B., Walter, M. W., and Worth, H. G. J. (1969). Metabolism of raffinose and sucrose in germinating broad bean *(Vicia faba)* seeds. *J. Exp. Bot.* **20,** 317.

Ragland, T. E., and Hackett, D. P. (1965). Radioactive tracer studies of the metabolic fates of intracellular generated NADH and NADPH in higher plant tissues. *Plant Physiol.* **40,** 1191.

Rebeiz, C., and Castelfranco, P. (1964). An extra-mitochondrial enzyme system from peanuts catalyzing the β-oxidation of fatty acids. *Plant Physiol.* **39,** 932.

Rebeiz, C., Castelfranco, P., and Engelbrecht, A. H. (1965a). Fractionation and properties of an extra-mitochondrial enzyme system from peanut catalyzing the β-oxidation of palmitic acid. *Plant Physiol.* **40,** 281.

Reibeiz, C., Castelfranco, P., and Breidenbach, R. W. (1965b). Activation and oxidation of acetic acid 1-$C^{14}$ by cell free homogenate of germinating peanut cotyledons. *Plant Physiol.* **40,** 286.

Ries, S. K., Chmiel, H., Dilley, D. R., and Filner, P. (1967). The increase of nitrate reductase activity and protein content of plants treated with simazine. *Proc. Nat. Acad. Sci. U.S.* **58,** 526.

Roos, A. J., Spronk, A. M., and Cossins, E. A. (1968). 5-Methyltetrahydrofolic acid and other folate derivatives in germinating pea seedlings. *Can. J. Biochem.* **46,** 1533.

Rossi-Fanelli, A., Cavallini, D., Mondovi, B., Wolf, A. M., Sciosca-Santoro, A., and Riva, F. (1965). Studies on cottonseed enzymes. I. Proteolytic and glutamic dehydrogenase activities in cottonseeds. *Arch. Biochem. Biophys.* **110,** 85.

Ryrie, L. J., and Scott, K. J. (1969). Nicotinate, quinolinate and nicotinamide as precursors in the biosynthesis of nicotinamide adenine dinucleotide in barley. *Biochem. J.* **115,** 679.

Sane, P. V., and Zalik, S. (1968a). Metabolism of proline-U-$^{14}$C during germination of barley. *Can. J. Bot.* **46,** 1331.

Sane, P. V., and Zalik, S. (1968b). Amino acids, sugar, and organic acid content during germination of Gateway barley and its chlorophyll mutant. *Can. J. Biochem.* **46,** 1479.

Sarkissian, I. V., and Schmalstieg, F. C. (1969). Citrate synthase of bean seedlings — comparison of activity following *in vitro* and *in vivo* treatments of enzyme. *Life Sci.* **8,** Part II, 933.

Sasaki, S., and Brown, G. N. (1969). Changes in nucleic acid fractions of seed components of red pine *(Pinus resinosa* Ait.) during germination. *Plant Physiol.* **44,** 1729.

Satoh, Y. (1968). Behavior of keto acids during the germination of pumpkin seed. *Sci. Rep. Saitama Univ., Ser. B* **5,** 125.

Scala, J., Patrick, C., and Macbeth, G. (1968). Castor bean FDPase from endosperm tissue: Properties and partial purification. *Life Sci.* **7,** Part II, 407.

Scandalios, J. G. (1969). The genetic regulation of multiple molecular forms of enzymes in plants. *Biochem. Genet.* **3,** 37.

Schweizer, C. J., and Ries, S. K. (1969). Protein content of seed: Increase improves growth and yield. *Science* **165,** 73.

Scolnik, T., Tompkins, R., Caskey, T., and Nirenberg, M. (1968). Release factors differing in specificity for terminator codons. *Proc. Nat. Acad. Sci. U.S.* **61,** 768.

Setterfield, G., Stern, H., and Johnston, F. B. (1959). Fine structure in cells of pea and wheat embryos. *Can. J. Bot.* **37,** 65.

Shain, Y., and Mayer, A. M. (1968). Activation of enzymes during germination: Trypsin-like enzyme in lettuce. *Phytochemistry* **7**, 1491.

Sinha, S. K., and Cossins, E. A. (1965). Pathways for the metabolism of glyoxylate and acetate in germinating fatty seeds. *Can. J. Biochem.* **43**, 1531.

Smith, C. R., Jr., and Wolff, I. A. (1966). Glycolipids of *Briza spicata* seed. *Lipids* **1**, 123.

Sparvoli, E. (1969). Cytoplasmic and nuclear DNA synthesis in the endosperm of germinating castor bean seeds. *Abstr. Int. Bot. Congr., 11th, 1969* p. 206.

Spedding, D. J., and Wilson, A. M. (1968). Studies of the early reactions in the germination of *Sinapis alba* seeds. *Phytochemistry* **7**, 897.

Spirin, A. S. (1966). On "masked" forms of messenger RNA in early embryogenesis and in other differentiating systems. *Curr. Top. Develop. Biol.* **1**, 1.

Splittstoesser, W. E. (1969). Metabolism of arginine by aging and 7-day-old pumpkin seedlings. *Plant Physiol.* **44**, 361.

Spradlin, J. E., Thomas, J. A., and Filmer, D. (1969). Beta amylase thiol groups. Possible regulator sites? *Arch. Biochem. Biophys.* **134**, 262.

Sprent, J. I. (1968). Effects of benzyladenine on cotyledon metabolism and growth of peas. *Planta* **81**, 80.

Srivastava, B. I. S. (1964). The effect of gibberellic acid on ribonuclease and phytase activity of germinating barley seeds. *Can. J. Bot.* **42**, 1303.

Stanley, R. G. (1957). Krebs cycle activity of mitochondria from endosperm of sugar pine seed (*Pinus lambertiana* Dougl.). *Plant Physiol.* **32**, 409.

Stanley, R. G. (1958). Gross respiratory and water uptake patterns in germinating sugar pine seed. *Physiol Plant.* **11**, 503.

Stanley, R. G., and Conn, E. E. (1957). Enzyme activity of mitochondria from germinating seedlings of sugar pine (*Pinus lambertiana* Dougl.). *Plant Physiol.* **32**, 412.

Stewart, C. R., and Beevers, H. (1967). Gluconeogenesis from amino acids in germinating castor bean endosperm and its role in transport to embryo. *Plant Physiol.* **42**, 1587.

Stinson, R. A., and Spencer, M. (1970). Respiratory control, oxidative phosphorylation, respiration rate of ATP hydrolysis, and ethylene evolution in subcellular particulate fractions from cotyledons of germinating seedlings. *Can. J. Biochem.* **48**, 541.

Stumpf, P. K. (1969). Metabolism of fatty acids. *Annu. Rev. Biochem.* **38**, 159.

Sturani, E. (1968). Protein synthesis activity of ribosomes from developing castor bean endosperm. *Life Sci.* **7**, Part II, 527.

Sturani, E., Cocucci, S., and Marré, E. (1968). Hydration dependent polysome-monosome interconversion in germinating castor endosperm. *Plant Cell Physiol.* **9**, 783.

Sund, H. (1968). The pyridine nucleotide coenzymes. *In* "Biological Oxidations" (T. S. Singer, ed.), p. 625. Wiley (Interscience), New York.

Swain, R. R., and Dekker, E. E. (1966a). Seed germination studies. I. Purification and properties of an α-amylase from cotyledons of germinating peas. *Biochim. Biophys. Acta* **122**, 75.

Swain, R. R., and Dekker, E. E. (1966b). Seed germination studies. II. Pathways for starch degradation in germinating pea seedlings. *Biochim. Biophys. Acta* **122**, 87.

Swain, R. R., and Dekker, E. E. (1969). Seed germination studies. III. Properties of a cell-free amino acid incorporating system from pea cotyledons; possible origin of cotyledonary α-amylase. *Plant Physiol.* **44**, 319.

Takeo, K. (1969). Aldolase isozymes of higher plants. *Phytochemistry* **8**, 2127.

Takeuchi, Y., Kushige, T., Kanei, Y., and Iwasa, Y. (1967). The lipid metabolism in castor beans. I. The activity of thiokinase in germinating castor beans. *Bull. Univ. Osaka Prefect., Ser. B* **19**, 19.

Tamaoki, T., Faber, A. J., and Kato, T. (1969). Dehydration of *Excherichia coli* ribosomes in air without loss of activity. *Nature (London)* **221**, 1050.

Tanner, W., and Beevers, H. (1965). The competition between the glyoxylate cycle and the oxidative breakdown of acetate in *Ricinus* endosperm. *Z. Pflanzenphysiol.* **53**, 126.

Tavener, R. J. A., and Laidman, D. L. (1969). Induction of lipase activity in the starchy endosperm of germinating wheat grains. *Biochem. J.* **113**, 32p.

Tezuka, T., and Yamamoto, Y. (1969). NAD kinase and phytochrome. *Bot. Mag. Tokyo* **82**, 130.

Tolbert, N. E., Oeser, A., Kisaki, T., Hageman, R. H., and Yamazaki, R. K. (1968). Peroxisomes from spinach leaves containing enzymes related to glycolate metabolism. *J. Biol. Chem.* **51**, 79.

Tombs, M. P. (1967). Protein bodies of the soybean. *Plant Physiol.* **42**, 797.

Tookey, H. L., and Balls, A. K. (1956). Plant phospholipase D. I. Studies on cotton seeds and cabbage phospholipase C. *J. Biol. Chem.* **218**, 213.

Trewavas, A. J., and Gibson, I. (1968). Ribosomal RNA nucleotide sequence homologies in plants. *Plant Physiol.* **43**, 445.

Tsai, C. Y. and Nelson, O. E. (1969). Two additional phosphorylases I and II of maize endosperm. *Plant Physiol.* **44**, 159.

Tsai, C. Y., and Nelson, O. E. (1968). Phosphorylases I and II of maize endosperm. *Plant Physiol.* **43**, 103.

Umebayashi, M. (1968). Aminoacylase in high plants. VI. Aminoacylase in the seeds of grasses. *Soil Sci. Plant Nutr. (Tokyo)* **14**, 211.

Vold, B. S., and Sypherd, P. S. (1968a). Modification in transfer RNA during the differentiation of wheat seedlings. *Proc. Nat. Acad. Sci. U.S.* **5–9**, 453.

Vold, B. S., and Sypherd, P. S. (1968b). Changes in soluble RNA and ribonuclease activity during germination of wheat. *Plant Physiol.* **43**, 1221.

Waller, G. R., Yang, K. S., Gholson, R. K., Hadwiger, L. A., and Chaykin, S. (1966). The pyridine nucleotide cycle and its role in the biosynthesis of ricinine by *Ricinus communis* L. *J. Biol. Chem.* **241**, 4411.

Walton, D. C., and Soofi, G. S. (1969). Germination of *Phaseolus vulgaris*. III. The role of nucleic acid and protein synthesis in the initiation of axis elongation. *Plant Cell Physiol.* **10**, 307.

Wanka, F., and Walboomers, J. M. (1966). Thymidine kinase and uridine kinase in corn seedlings. *Z. Pflanzenphysiol.* **55**, 458.

Waters, L. C., and Dure, L. S. (1966). Ribonucleic acid synthesis in germinating cotton seeds. *J. Mol. Biol.* **19**, 1.

Weeks, D. P., and Marcus, A. (1969). Polyribosome isolation in the presence of diethyl pyrocarbonate. *Plant Physiol.* **44**, 1291.

West, S. H. (1962). Protein, nucleotides, and nucleic acid metabolism in corn during germination under water stress. *Plant Physiol.* **37**, 565.

Wiley, L., and Ashton, F. M. (1967). Influence of embryonic axis on protein hydrolysis in cotyledons of *Cucurbita maxima*. *Physiol. Plant.* **20**, 688.

Wilson, A. M. (1970). Incorporation of $^{32}$P in seeds at low water potentials: Evaluation of microbiological contamination. *Plant Physiol.* **45**, 524.

Wilson, A. M., and Harris, G. A. (1966). Hexose-, inositol-, and nucleoside phosphate esters in germinating seeds of crested wheatgrass. *Plant Physiol.* **41**, 1416.

Wilson, A. M., and Harris, G. A. (1968). Phosphorylation in crested wheatgrass seeds at low water potentials. *Plant Physiol.* **43**, 61.

Wolfe, F. H., and Kay, C. M. (1967). Physiochemical, chemical and biological studies on wheat embryo ribosomes. *Biochemistry* **6**, 2853.

Wolff, I. A. (1966). Seed lipids. *Science* **154**, 1140.

Woodstock, L. W. (1965). Initial respiration rates and subsequent growth in germinating corn seedlings. *BioScience* **15**, 783.

Woodstock, L. W., and Grabe, D. F. (1967). Relationship between seed respiration during imbibition and subsequent seedling growth in *Zea mays* L. *Plant Physiol.* **42**, 1071.

Woodstock, L. W., and Pollock, B. M. (1965). Physiological predetermination: Imbibition, respiration, and growth in lima bean seeds. *Science* **150**, 1031.

Yamada, M. (1955a). Studies on fat metabolism in germinating castor beans. I. The utilization of bound and free fatty acids *in vivo. Sci. Pap. Coll. Gen. Educ., Univ. Tokyo* **5**, 149.

Yamada, M. (1955b). Studies on fat metabolism in germinating castor beans. II. Different type of respiration in embryonic organs and the conversion of fat to sugars and their translocation. *Sci. Pap. Coll. Gen. Educ., Univ. Tokyo* **5**, 161.

Yamada, M. (1957). Studies on fat metabolism in germinating castor beans. III. Lipase in decotylated embryo tissue. *Sci. Pap. Coll. Gen. Educ., Univ. Tokyo* **40**, 653.

Yamada, M., and Stumpf, P. K. (1965). Fat metabolism in higher plants. XXV. The enzymic degradation of hydroxy long chain fatty acids by extracts of *Racinus communis. Plant Physiol.* **40**, 659.

Yamamoto, Y. (1963). Pyridine nucleotide content in the higher plant. Effect of age of tissue. *Plant Physiol.* **38**, 45.

Yamamoto, Y. (1967). Enzymological studies on *Vigna* seed germination. *In* "Physiology, Ecology, and Biochemistry of Germination" (II. Borriss, ed.), p. 673. Ernst-Moritz-Arndt-Universitaat.

Yamamoto, Y. (1969). Modification of metabolis pattern by variation of nicotinamide adenine dinucleotide phosphate level. *Plant Physiol.* **44**, 407.

Yamamoto, Y., and Beevers, H. (1960). Malate synthetase in higher plants. *Plant Physiol.* **35**, 102.

Yamamoto, Y., and Tezuka, T. (1969). NAD Kinase activity regulated with phytochrome action. *Abstr. Int. Bot. Congr., 11th, 1969* p. 245.

Yatsu, L. Y. (1965). The ultrastructure of cotyledonary tissue from *Gossypium hirsutum* L. seeds. *J. Cell Biol.* **25**, 193.

Yatsu, L. Y., and Jacks, T. J. (1968). Association of lysosomal activity with aleurone grains in plant seeds. *Arch. Biochem. Biophys.* **124**, 466.

Young, J. L., and Varner, J. E. (1959). Enzyme synthesis in the cotyledon of germinating seeds. *Arch. Biochem. Biophys.* **84**, 71.

Young, J. L., Huang, R. C., Venecke, S., Marks, J. D., and Varner, J. E. (1960). Conditions affecting enzyme synthesis in cotyledons of germinating seeds. *Plant Physiol.* **35**, 288.

Zilkey, B. F. (1969). Site of oleic acid biosynthesis in developing castor endosperm. *Abstr. Int. Bot. Congr., 11th, 1969* p. 248.

Zimmerman, D. C., and Klosterman, H. J. (1965). Lipid metabolism in germinating flaxseed. *J. Amer. Oil Chem.* **42**, 58.

# 3

## SEED DORMANCY

### T. A. Villiers

If you can look into the seeds of time
And say which grain will grow, and which will not . . .
*Macbeth*

## I. Introduction

The scientific literature on dormancy is vast, and it is impossible to attempt a comprehensive survey within the scope of one chapter. The works to which reference has been made, therefore, is a purely personal sampling chosen in an attempt to give an adequate description of the phenomenon of seed dormancy. Reference has been made to dormancy in other organs where it was considered necessary to help to explain the problem in seeds.

One of the most important and interesting problems encountered in seed physiology is the failure of otherwise viable seeds to recommence development immediately when supplied with water and oxygen at temperatures recognized as normally favorable for plant growth. This delay may last for a variable period of time under constant conditions and in some cases may continue indefinitely until some special condition is fulfilled.

Such failure to germinate is broadly termed "dormancy" and is a wide-spread phenomenon, especially in the seeds of temperate zone plants, but is also encountered in many subtropical and tropical species. It is considered to be biologically advantageous in adapting the growth cycles of the plant both to seasonal and fortuitous variations in environmental conditions. For example, the requirement for low temperatures shown by the seeds of many temperate zone plants ensures that germination and early seedling development are delayed until the more favorable spring weather prevails. In addition, in many species, seeds from the same harvest, or even the same flower head may possess different dormancy characteristics, so that the period of germination may be spread over many months, or even over several years, thereby increasing the probability of survival of at least some of the seeds. Growing plants, and especially seedlings, are more susceptible to frost and drought than dormant tissues, which may show great resistance to adverse environmental conditions.

Apart from the intrinsic biological interest whereby the study of dormancy may provide valuable information on the control mechanisms of growth and development, a knowledge of the types of dormancy is essential in agriculture, horticulture, and forestry, where special treatments before sowing or during the imbibition of batches of seeds may be necessary either to overcome the dormancy state of the seeds or to prevent the imposition of secondary dormancy in an otherwise normal batch of seeds, giving rise to a "dead sowing" or at the least to the possibility of a long delay before seedling production. One of the biological advantages shown by weed seeds is the possession of efficient dormancy-maintaining mechanisms which enable them to remain dormant but viable in the soil for many years, in a state highly resistant to measures of weed eradication.

Dormancy is not, of course, a condition peculiar to seeds, but is a property shown by other plant organs such as terminal buds, rhizomes, corms, bulbs, and tubers, as well as both the mitospores and meiospores of almost all plants. It also has its counterpart in the spores of bacteria and fungi and in almost all groups of animals, from the deeply dormant encysted stages of protozoa and crustaceans such as the brine shrimp, to the state of diapause in insects and aestivation and hibernation shown by a wide range of animals, including many vertebrates.

## A. Terms Used in Dormancy Studies

Throughout this chapter, the word "seed" will be used purely as a convenient term to describe the dispersal unit of the higher plant propagule, even where the true seed might be further enclosed within layers of the pericarp, which remain attached during dispersal and germination. Thus, such strictly botanical terms as "caryopsis" and "schizocarp" will be avoided and individual seed coat layers described only where this is necessary for an understanding of a particular process.

In its very broadest sense, attempting to include dormancy phenomena in other structures and organisms, the term dormancy may be used to describe any stage in the life cycle, whether this is a regular phase in the developmental process or a fortuitous occurrence in which active growth is suspended for a period of time. This may include the development of special structures in recognizable stages of morphogenetic development, such as seed coats, bud scales, and spore and cyst walls, or merely the suspension of development without obvious morphogenetic changes, which may occur in hibernation. Vegis (1964) states that by the term "dormant condition," he infers "true" dormancy, a condition in which growth, or normal growth, cannot be resumed whatever the external conditions may be. However, while it is possible that the true sense of his meaning has been altered in translation, it is clear that this definition is far too restrictive. *"Whatever the external conditions might be"* logically excludes the application of the term to the physiological state of unchilled embryos of various plants if these germinate when placed in, say, long-day illumination, or if excised from the seed coats, or placed in a solution of a germination-stimulating substance. In this case it is clear that while the conditions of such forced germination may be unnatural, they fall within the strict meaning of Vegis's phrase "whatever the external conditions might be," and we should apparently take his meaning of "true dormancy" as being unable to germinate under the conditions which are natural for the growth and development of that species at that season, and should preclude the use of artificial manipulations of the tissues or the environment.

Koller *et al.* (1962) complain of the indiscriminate use of the term dormancy and specify "primary" dormancy as that condition shown by the ripe dispersal unit at the time of dispersal or harvest, and "secondary" or "induced" dormancy as the type of dormancy appearing in the dispersal unit when imbibed under conditions unfavorable for germination.

Pollock and Toole (1961) use the term dormancy to cover both the condition resulting from an unfavorable environment such as inadequate water supply and also the condition where some internal block to the germination process exists. They appear to favor the use of the term "rest" for the latter state, but go on to give a valuable treatment of the use of the term by discussing actual examples in order to show that dormancy is a relative rather than an absolute state, is highly variable from species to species and even within one sample of seeds.

In their discussion of the terminology of dormancy and germination Sussman and Halvorson (1966) recognize that dormancy relates to a variety of states lying somewhere between the extremes of the active, vegetative condition and the cryptobiotic state "which is imposed by temperatures low enough to vitrify the contents of cells." They describe dormancy as any rest period or reversible interruption of development, and distinguish between "constitutive" dormancy where development is delayed because of an innate property of the dormant organ or organism such as barriers to the penetration of substances, the presence of a metabolic block, or the presence of a self-inhibitor, and "exogenous" dormancy, where development is delayed because of unfavorable environmental conditions.

In this chapter "dormancy" will be reserved to describe the state of arrested development whereby the organ or organism, by virtue of its structure or chemical composition, may possess one or more mechanisms preventing its own germination, while the term "quiescence" will be used to describe a state of arrested development maintained solely by unfavorable environmental conditions such as inadequate water supply. It then becomes proper to apply the term "secondary dormancy" to those cases where dormancy (as used earlier in this paragraph) becomes imposed upon seeds by imbibition under conditions unfavorable to the germination of that species, e.g., at temperatures unfavorable to germination so that seeds become unable to germinate at temperatures normally favorable for germination of that species until some special releaser stimulus is applied.

If the above terms were in general use, much confusion in scientific writing could be avoided. As an example of the confusion resulting from the indiscriminate use of the word dormancy, the fine structure of the dry pea cotyledon (Barker and Rieber, 1967), the dry embryo of *Lactuca sativa* (Paulson and Srivastava, 1968), and other species have been described as investigations into the ultrastructure of *dormant* embryos. On careful consideration, it will be seen that if cytological changes connected with release from dormancy are sought as clues to the cellular site and the control mechanisms involved in *dormancy,* as opposed to the events taking place upon imbibition and germination, it is not sufficient to examine the cytology or biochemistry of a merely *quiescent* embryo, requiring only the availability of water to recommence development. Such clues should obviously be sought by experiments upon dormant seeds or embryos, where the physical process of tissue hydration can be clearly separated in time from the application of the dormancy release mechanism which leads to germination.

In addition, it is obvious that while the ability to withstand severe desiccation is of great interest in itself, it is probable that fundamentally valuable information will emerge from research on the mechanisms by which development is suspended and quiescent seeds prepare for and *enter* the state of quiescence rather than studying changes upon imbibition. Further, the study of dormant as opposed to merely quiescent embryos might give valuable information on control of the growth processes themselves, if the precise molecular effects of the dormancy-releasing stimuli can be explained.

The dormancy phase most often coincides with seasonally unfavorable climatic conditions, and while the state may be induced by fortuitous exposure to such conditions, it is frequently found that an organism normally enters the dormancy state even while conditions are apparently favorable for active growth. Thus seeds produced in a growing season usually fail to continue development and germinate until they have passed through a period of low temperatures experienced at some later date, or until they have been subjected to a particular light regime, or undergone desiccation. Again, the resting buds of many trees are formed during conditions which are apparently conducive to growth. In animals, cyst formation in certain nematodes, diapause in insects, and encystment in certain crustacean embryos are examples of dormancy states which can be entered during apparently favorable environmental conditions, as are the preparations for entry into hibernation by altered metabolic pathways.

It is not surprising to find that the environmental conditions which control the onset of dormancy are not similar to those which relieve the state of dormancy. For example, short photoperiod controls the formation of winter buds in many trees (Wareing, 1954) and entry into diapause in insects (C. M. Williams, 1969), whereas cold treatment relieves these dormancy states. Similarly, the secondary dormancy imposed by high-temperature imbibition in *Lactuca* seeds may be broken by exposure to red light (Pollock and Toole, 1961).

As the dormancy state is considered to be biologically advantageous, it would be expected that there exist evolutionary pressures refining the methods by which dormancy is imposed, and also that parallel evolution has brought about the development of mechanisms imposing the dormancy state based on different anatomic and metabolic characters. When the various mechanisms which are found to impose dormancy are examined it is therefore found that there appears to be no single hypothesis which can entirely explain the phenomenon. On the other hand comparisons can be drawn between certain types of dormancy, for example, the low-temperature requirement for the release of dormancy in seeds and in buds. Even here, however, it is necessary to be cautious, as the same environmental releaser stimulus, although perhaps acting in fundamentally similar ways, may be operating upon different biochemical pathways within the tissues.

### B. Types of Dormancy

Before attempting any explanation of dormancy, it is advantageous to survey and attempt to classify the various mechanisms by which dormancy may be imposed and maintained. A very complete review of these mechanisms has been presented by Nikolaeva (1969) who has suggested a detailed classification where four types of dormancy are established as follows: (*A*) properties of the outer covers of the embryo; (*B*) under-development of the embryo; (*C*) physiological condition of the embryo itself; and (*D*) types of combined dormancy.

Subdivisions of these classes are presented. For example, class *A* is subdivided into physical impermeability of the seed coats to water, inhibiting action of substances within the seed coats, and mechanical resistance to the growth of the embryo. Class *C* is subdivided in a very detailed way according to the combination of low permeability of the seed coats to gases coupled with varying physiological conditions of the embryo itself. For full details of this very complex classification the reader is referred to the original publication. The placing of coat impermeability to gases in class *C* is apparently justified because a lessened availability of oxygen to the embryo is usually connected with physiologically maintained dormancy of the embryo itself, whereas imperviousness to water is not. Class *D,* combined types of dormancy, is distinguished as combinations of the exogenously imposed dormancy types of class *A* with endogenously imposed dormancy classes *B* or *C*. The value of such a detailed classification as that devised by Nikolaeva lies largely in the ease of comparison of the various dormancy-inducing and maintaining mechanisms with the methods used to break dormancy and allow germination.

A more simple, and more generally quoted classification of dormancy types is that of Crocker (1916) who described dormancy as resulting from: (1) immaturity of the embryo; (2) impermeability of the seed coats to water; (3) mechanical resistance of the seed coats to embryo growth; (4) low permeability of the seed coats to gases; (5) dormancy resulting from a metabolic block within the embryo itself [requirement for (a) light, and (b) chilling]; (6) a combination of the above; or (7) secondary dormancy.

This classification has served its purpose well through the years, both as a basis for the investigation of dormancy characteristics in many seed types and for the application of dormancy-breaking measures for the practical needs of agriculture, horticulture, and forestry. The following general descriptions of various types of dormancy therefore follows the classification of Crocker.

## 1.  IMMATURITY OF THE EMBRYO

The embryo of *Ilex opaca* is an undifferentiated mass of cells before germination treatments (Ives, 1923). Similarly, the embryos of *Heracleum sphondylium* are rudimentary and germination is delayed until differentiation is completed (Stokes, 1952). On the other hand, several species possess embryos which may be differentiated when the seed is dispersed, but when reimbibed with water, the embryo continues growth to a larger size before germination can take place. In this group are species of *Fraxinus* (Steinbauer, 1937), species of *Viburnum* (Giersbach, 1937), and *Pinus koraiensis* (Asakawa, 1955).

It is difficult to decide whether this embryo development is part of the final stages of seed development or the initial stages of the germination process. The embryo of *Fraxinus excelsior* is morphologically complete when the seed is dispersed from the tree in an air-dry state, but grows to twice its original length and dry mass after reimbibition. Whereas this growth of the embryo takes place most rapidly at warm temperatures (18–20°C), even when the embryo is fully grown the seed remains dormant unless exposed to low temperatures (approximately 5°C) for several months (Villiers and Wareing, 1964). I have kept seeds of *Fraxinus excelsior* fully imbibed at laboratory temperatures for 6 years, during which time no germination has taken place and the seeds are still viable.

Warm temperatures are not always most effective in embryo development within the seed, as Stokes (1952) has shown that differentiation of the embryos of *Heracleum sphondylium* proceeds more rapidly at 2°C than at higher temperatures. In this case presumably embryo maturation is accomplished and the requirement for chilling is fulfilled at the same time.

## 2. IMPERMEABILITY OF THE SEED COATS TO WATER

This would appear to be one of the simplest but most highly effective means of delaying germination and of spreading the production of seedlings from any particular batch of seeds over a period of time. Agriculturally, seeds rendered dormant by this mechanism are termed "hard seeds" and this dormancy mechanism is exhibited by members of several families, including particularly members of the Leguminosae, also the Malvaceae, Chenopodiaceae, Liliaceae, and Solanaceae.

The uptake of water is prevented by the testa, and rupture of this layer is promptly followed by swelling due to water uptake, and germination commences almost immediately (Fig. 3.1). There is usually much

FIG. 3.1. Seeds of *Bauhinia* sp. after soaking in water for 24 hours showing the increase in volume due to uptake of water following scarification. *Left,* untreated seeds. *Right,* testas damaged by filing. The seeds on the left did not germinate, whereas those on the right germinated fully within 7 days.

variation in the degree of impermeability to water within a single seed batch, such that a few of the seeds are able to become imbibed and germinate almost immediately, while the uptake of water in the remaining seeds is delayed for a variable length of time. The structures of the testa which in many cases prevent water uptake appear to be a layer of palisade-like cells especially thick-walled on their outer surfaces, and having a layer of waxy, cuticular substance external to this.

Hyde (1954) described the structure of the hilum of certain leguminous seeds and suggested that it operates as an hygroscopically-activated valve, allowing water loss from the seed without permitting water uptake. The fissure of the hilum was seen to open rapidly when the seed was placed in relatively dry air, and closed rapidly when in moist air. In experiments using iodine vapor, it was shown that in dry air, staining of internal tissues occurred in the vicinity of the hilum, whereas when the seeds were exposed to iodine vapor in moist air, little or no staining of the tissues occurred. Thus, the anatomic structure of the hilum is such that when closed, even vapors are unable to penetrate within the seed. The impermeability of the testa could be ascribed to its cuticular and thick-walled palisade layers, and the presence of the hilar valve mechanism explains the observation that in conditions where the moisture content fluctuates, the seed itself gradually decreases in moisture content in a series of steps.

In seeds which have lost their water-hardness, it is often assumed that bacterial and/or fungal action has caused surface deterioration and thus broken the cuticular barrier. However, while there can be little doubt that microorganisms are usually concerned in the process, several investigations have been unable to demonstrate structural differences in the coats of swelling and nonswollen seeds of lucerne (Bredeman, 1938).

While selection has possibly favored the exclusion of hardseeded characteristics from many common crop plants, it is found that, as with other types of dormancy mechanisms, even closely related plants may vary widely in the degree of hardness of their seeds. Nikolaeva (1969) cites the case of the European lupine, where the yellow form produces a majority of seeds impermeable to water, while the white lupine produces few or no hard seeds.

Mechanical and chemical treatments of the seed coats are the obviously effective measures taken to release hard seeds from dormancy, but dry heat and boiling water are also effective. For example, the standard method of treatment of the seeds of *Acacia melanoxylon* is to soak the seeds for several minutes in boiling water, and in the field it is found that vast numbers of seedlings are produced when standing trees of this species have been destroyed by fire.

### 3. MECHANICAL RESISTANCE TO EMBRYO GROWTH

Most authors regard mechanical resistance to embryo growth as one type of seed dormancy although in reality it is probably a rare occurrence.

Ferenczy (1955) rejected the view of Steinbauer (1937) that seeds of the genus *Fraxinus* possessing immature embryos have mechanically resistant coats which retard the further development of the embryo once it has become fully grown within the seed. Ferenczy showed that cutting away the testa and endosperm from around the radicle tip did not allow germination. Indeed, when one considers the great pressures generated by growing plant organs, such as the breaking of concrete roadways and lifting of paving stones by plant organs growing beneath them, it is only among those seeds which remain enclosed within an extremely hard endocarp that one would look for cases of dormancy imposed solely by such a mechanism.

In reality, many such reported cases are no doubt caused by factors other than, or in addition to, mechanical constraint. Nevertheless, it is common knowledge that many seeds, or the dispersal units, have an extremely hard and durable structure. In certain rosaceous seeds, such as those of *Rosa* and *Crataegus,* great mechanical pressures are needed to destroy the hard endocarps, but even in these species it has been shown that, in addition to the presence of the stony endocarp, it is probable that the embryo itself is maintained in a state of dormancy by other mechanisms. For example, in addition to any mechanical resistance imposed on the embryo by the stony remains of the pericarp in *Rosa* species, growth inhibitors have been demonstrated to be present (Jackson and Blundell, 1965). Further, secondary dormancy can be induced in *Rosa* embryos by long imbibition periods without chilling, such that embryos will not germinate even when excised from the enclosing structure (Semeniuk *et al.,* 1963).

## 4.   LOW PERMEABILITY OF THE SEED COATS TO GASES

There is good evidence that interference with oxygen uptake can maintain embryos in a state of dormancy. The classic example is the work of Crocker (1906) on seeds of *Xanthium pennsylvanicum,* the fruit of which contains two seeds, an upper, dormant seed and a lower non-dormant one. Not only has it been shown that the removal of the testa of the dormant seed allows germination, but experiments showed that pieces of detached testa constitute a physical barrier to oxygen diffusion. Removal or damage to parts of the seed coats, or increasing the oxygen tension of the surrounding air leads to increases in the rate of embryo respiration in many species of seeds, and frequently results in germination, and the inference often made is that increased oxygen leads to an increase in the availability of energy by oxidation processes.

Germination of seeds of many species of the Gramineae is markedly enhanced by damage to the coats by scarifying, cutting, acid treatment, removal of the coats, or placing in high oxygen tension as in the case of *Avena fatua,* the wild oat (Johnson, 1935). E. Brown *et al.* (1948) showed that in cultivated oats and barley, germination would occur at low temperatures only if the part of the seed coats covering the embryo were removed. Nutile and Woodstock (1967) were able to induce secondary dormancy in *Sorghum* seeds by reducing the moisture content below that of the normal air-dry state. Seeds formerly nondormant were then only able to germinate if the coats were cut and on reimbibition, the dormant seeds had a lower rate of oxygen uptake compared with nondormant seeds. It was suggested that physical changes in the seed coats during the artificial drying restricted the rate of oxygen uptake upon reimbibition.

Similar findings have been reported in other groups of plants. Measurements of the rate of movement of various gases through the seed coats of *Cucurbita pepo* were made by R. Brown (1940), who found the inner seed coat to be more permeable than the outer coat to gases. In addition, the coats were less permeable to oxygen than they were to carbon dioxide. In apple seeds, the coats greatly restrict the oxygen supply, making it insufficient for the high respiratory activity of the embryo when the seeds are maintained imbibed at warm temperatures (Visser, 1954) (Fig. 3.2).

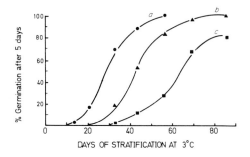

FIG. 3.2. The influence of testa and endosperm on the germination of apple seeds at 25°C. Germination of excised embryos (*a*); germination of embryos with intact endosperm (*b*); and germination of embryos with both endosperm and testa intact (*c*) (from Visser, 1954).

## 5. ENDOGENOUS DORMANCY OF THE EMBRYO

a. REQUIREMENT FOR LIGHT. The study of the light requirement of seeds has been the subject of much research in recent years, partly because seeds are extremely convenient material both for the statistical design and the convenient handling of experiments. The mechanism of the release of dormancy by light has been shown to be similar to that controlling many other morphogenetic stages in the development of plants, and it was a study of the photocontrol of germination in the seeds of lettuce which gave the first clues to the energy receptor system in light-controlled development (Borthwick *et al.,* 1952).

Many species of seeds have been shown to be light sensitive. The germination responses of seeds such as *Betula* spp., *Lepidium virginicum, Nicotiana tabacum,* and some varieties of *Lactuca sativa* are promoted by light, and such seeds are termed "positively photoblastic," while others such as *Phacelia tanacetifolia* and *Nemophila insignis* are inhibited by light, and are termed "negatively photoblastic." The level of illumination required may actually be quite low, and germination depends in general upon the quantity of light received, i.e., the intensity multiplied by the time of exposure, and at high light intensities a very short exposure will suffice. In addition, certain seeds have been shown to exhibit a true photoperiodic effect.

Light-sensitive seeds are only responsive to the stimulus when imbibed with water, and the response is greatly affected by the presence of the seed coats and the prevailing temperature. Gassner (1911) showed that the seed of *Chloris ciliata* required light for germination, but excised embryos would germinate in the dark. However, when excised embryos were wrapped in moist filter paper the light requirement was reimposed. Again, when portions of the seed coat near to the radicle of the light-inhibited seeds of *Phacelia* were removed, the seeds were able to germinate in the light, but if the holes were covered again by wet filter paper the seeds were again light-inhibited (Böhmer, 1928).

The studies of Black and Wareing (1959) showed that seeds of *Betula pubescens* required long day illumination at 15°C, but if the temperature was raised to 20°C good germination was obtained in either long or short days. At temperatures above 25°C germination of *Betula* seeds takes place equally well in darkness (Vaartaja, 1956). Seeds which are promoted by a particular day length at one temperature may be adversely affected at another temperature. For example, seeds of *Tsuga canadensis* germinated best in short days at 17°–20°C but at 27°C they were favored by long days (Stearns and Olson, 1958).

b. Requirement for Chilling. When removed from their enclosing seed coats, the embryos of plants such as *Betula* begin growth immediately to produce apparently normal seedlings, whereas others, such as *Malus* and *Prunus* spp. may remain dormant or grow slowly to form "physiological dwarfs," with little or no elongation of the internodes, and showing chlorosis and wrinkling of the leaves (Table 3.I; Fig. 3.3). The term physiological dwarf is used to differentiate the condition from that caused by genetic dwarfing which is transmitted from generation

**TABLE 3.I**
EXAMPLES OF SPECIES REPORTED TO REQUIRE MOIST CHILLING (STRATIFICATION)
BEFORE GERMINATION[a]

| Species | Temperature (°C) | Range (°C) | Time (days) |
|---|---|---|---|
| *Abies arizonica* | 1 | 1–5 | 30 |
| *Betula* spp. | 5 | 1–10 | 60–70 |
| *Crataegus mollis* | 5 | 5 | 180 |
| *Fraxinus excelsior* | 5 | 1–8 | 150–180 |
| *Gentiana acaulis* | 1 | 1–5 | 60–90 |
| *Juniperus* spp. | 5 | 5 | 100 |
| *Picea canadensis* | 1 | 1–5 | 30–60 |
| *Pinus lambertiana* | 5 | 1–10 | 90 |
| *Pyrus malus* | 5 | 1–5 | 60 |
| *Rosa multiflora* | 5 | 5–8 | 50 |
| *Sorbus aucuparia* | 1 | 1–5 | 60–120 |
| *Vitis vinifera* var. *Concorde* | 5 | 5–10 | 90 |

[a] From Stokes (1965).

FIG. 3.3. Physiological dwarfing in the peach (variety Elberta). *Left*, seedling grown from excised, unchilled embryo. *Right*, seedling grown from embryo excised after chilling for 6 weeks at 5°C.

to generation by heredity. The symptoms may persist in the plants for
many years, but may be relieved by measures taken to overcome the
dormancy of the seed itself, i.e., by chilling, providing a suitable photoperiodic regime, or applying gibberellic acid (GA$_3$). Pollock (1959) has
shown that the temperature maintained at the time of excision is important in determining the severity of the symptoms of this dwarfing
condition. In the peach variety Elberta, he showed that dwarfism resulted
from exposure to temperatures in the range 23°–27°C during the first
seven days of germination of the excised embryos. Figures 3.4 and 3.5
illustrate the effect of moist storage at 4°C on seeds of apple.

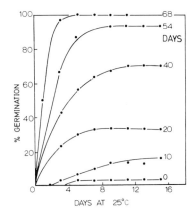

Fig. 3.4. Germination of excised apple embryos placed at 25°C to germinate after increasing periods of moist chilling (from Luckwill, 1952).

Fig. 3.5. Seeds of apple after imbibition at 25°C for 60 days (left), and at 5°C for 60 days
(right).

This type of embryo dormancy has been most frequently associated with the presence of germination-inhibiting substances in the embryos. If such embryos are excised and washed with water, germination often ensues, whereas if they are excised and held in a humid atmosphere without loss of water from the tissues, and where leaching of substances from the embryo is impossible, the embryos will remain dormant. In the case of *Xanthium pennsylvanicum* (Wareing and Foda, 1957) leaching of excised embryos allowed germination, a decrease in inhibitor could be detected within the embryos, and germination inhibitors were detected in the water used for the soaking. That the germination of these leached embryos could not be ascribed merely to increased water uptake was shown by redrying the embryos back to their former wet weight before leaching, when germination still occurred.

Although inhibitors could not be leached from embryos of *Fraxinus excelsior* by soaking, it was found that soaked embryos were capable of germination, and had up to 20% higher water content than those present in fully imbibed seeds. In addition, it was found that when these embryos were redried back to their former weight upon excision, dormancy was reimposed. When excised embryos were held in a water-saturated atmosphere and distilled water was slowly added individually to each embryo so that their final wet weights were equivalent to those of soaked embryos, dormancy was broken (Villiers and Wareing, 1965b). Therefore, if dormancy is imposed upon embryos of *F. excelsior* by inhibiting substances, the effect of soaking must be to dilute the inhibitor within the tissues to a concentration at which germination becomes possible.

Removal of the cotyledons of unchilled, excised red oak acorns prevented physiological dwarfing in the seedlings produced, and Cox (1942) suggested that the cotyledons produced an inhibitor which diffused to the meristem and imposed dormancy. Similarly Knowles and Zalik (1958) correlated the presence of an inhibitor with epicotyl dormancy in *Viburnum trilobum* and showed that this dormancy could be relieved by removal of the cotyledons.

As stated above, this type of embryo dormancy shows many striking similarities to bud dormancy, and it is not surprising that such inhibitors have also been implicated in bud dormancy. Much of the early experimental work was of doubtful value because crude extracts of plant material were usually used. Again, while many substances can be shown to be inhibitors of growth in the conventional biological assays, such as the coleoptile elongation test (Bentley and Housley, 1953), it is preferable to test extracted substances upon the germination or growth process which they are suspected of controlling. It is not sufficient to infer that because inhibitors may vary in activity within the tissues of seeds and buds that they are therefore the agents responsible for the control of germination or growth. On the other hand, while it is difficult to separate cause from effect, much evidence has accumulated in favor of the action of such substances in imposing and maintaining dormancy.

I. D. J. Phillips and Wareing (1958a) showed that while growth inhibitors were present throughout the year in the shoots of *Acer pseudoplatanus,* they increased in activity at the beginning of the dormancy period and decreased towards the end. In a further communication (I. D. J. Phillips and Wareing, 1958b) they showed that upon transfer of seedlings of this species from long to short days, a rise in inhibitor level could be detected after only 2 short-day cycles, and before any morphological modification of the organs in the shoot apex could be detected.

The activity of extracts in conventional growth substance assays is not a specific test for a positive dormancy-inducing substance, and Eagles and Wareing (1964) extracted and partially purified the growth-inhibiting substances produced by short-day treatment of *Betula* seedlings and applied this extract to the leaves of seedlings of the same species but growing in long days. These seedlings, otherwise capable of continued growth, ceased elongation and formed apparently typical winter buds. In addition, an increased inhibitor content was found in the shoot apices of these seedlings following treatment of the leaves, indicating internal transport.

These results are highly significant, especially when the sequence of events in the experimental treatment is compared with the natural sequence. In *Acer pseudoplatanus,* the induction of dormancy in the shoot apex is determined by exposure of the mature *leaves* to short days, whereas the response is given by the shoot *apex* (Wareing, 1954). In the same report, Wareing showed that if dormant *Betula* seedlings are placed in long-day illumination, the buds will resume growth, whereas if the buds are exposed to long days while the leaves are subjected to a short-day regime, the buds remain dormant.

Growth and germination-inhibiting substances have been isolated from almost every plant and plant organ so far investigated and many substances have been isolated, identified, and correlated with stages of dormancy. A comprehensive review of endogenous inhibitors and their possible effects in seed dormancy has been made by Wareing (1965). One of the most frequently studied substances has been the unsaturated lactone coumarin and its derivatives. It occurs widely in plant material as the glycoside, or in a substituted form (Mayer and Poljakoff-Mayber, 1963), and has a strong inhibiting action on seed germination. Evidence for its physiological action is provided by the work of Nutile (1945) who first showed that it has the property of inducing a light-sensitive state of dormancy in varieties of lettuce seeds not normally requiring light for germination. This imposed light requirement can be replaced by gibberellin application as in the case of normal, light-requiring varieties.

Of very wide occurrence in plants is a complex of inhibiting substances known as inhibitor $\beta$, from the original work on the separation of plant extracts by chromatography (Bennet-Clark and Kefford, 1953). Housley and Taylor (1958) showed several substances to be present, including coumarin, scopoletin, and several fatty acids, all of which were inhibiting in the *Avena* coleoptile growth assay. Thus inhibitor $\beta$ is the term used to describe an almost ubiquitous zone of inhibitory activity found on chromatograms of plant extracts.

Working with extracts of the leaves of *Acer pseudoplatanus,* Robinson *et al.* (1963) isolated a single substance from the inhibitor $\beta$ fraction which they called "dormin" and claimed that it constituted the main inhibiting component. Cornforth *et al.* (1965) established that this inhibitor was identical with the substance "abscisin II" which had been earlier isolated as an abscission-accelerating substance from *Gossypium hirsutum* (Ohkuma *et al.,* 1963), and identified as a sesquiterpenoid (Ohkuma *et al.,* 1965). A chemical structure was proposed which has been confirmed by synthesis. The name abscisic acid has been agreed upon, with the reservation that the term "dormin" be still used for the class of substances having the same physiological functions (Addicott and Lyon, 1969) (Fig. 3.6).

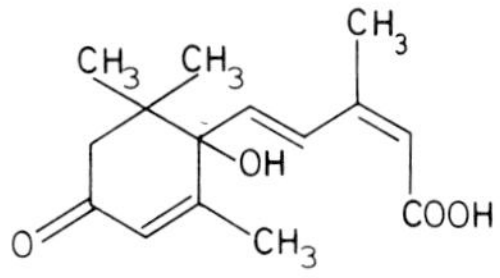

FIG. 3.6. The chemical structure of abscisic acid, a naturally occurring dormancy-inducing hormone.

While it is probable that abscisic acid is generally implicated in the control of dormancy and growth in plants (Lang, 1967) and fulfills all the requirements for the substance controlling dormancy in buds, it is possible that the inhibition of germination in seeds may be controlled by other substances in addition to or in place of abscisic acid. Wareing and Foda (1957) showed that on germination of *Xanthium* embryos the disappearance of the two seed inhibitors coincided with the appearance of an inhibitor having properties more similar to those of inhibitor $\beta$ of the mature plant. Blumenthal-Goldschmidt (1960), and Poljakoff-Mayber *et al.* (1957) showed that similar changes occurred on the germination of lettuce seeds, and C. N. Williams (1959) demonstrated that inhibitor $\beta$, although present in seeds of *Striga lutea,* did not inhibit germination but only affected the growth of the seedlings. Villiers and Wareing (1965b) showed that on germination, *Fraxinus excelsior* embryos showed changes in the properties of the inhibitor content such that the new inhibitor formed was more similar to inhibitor $\beta$ and also to the properties of the inhibitor in the vegetative organs of the mature tree.

More recently, Pammenter and Villiers (1971) have been able to demonstrate the presence of abscisic acid in the leaves and buds of grow- ing *F. excelsior* trees, but found that the seed inhibitor was structurally different from abscisic acid and behaved in certain physiological tests, such as the abscission-accelerating test, in a manner different from ab- scisic acid. On the other hand, Sondheimer *et al.* (1968) have reported the presence of abscisic acid in seeds of *F. americana,* which has, how- ever, different dormancy characteristics from *F. excelsior,* and Jackson and Blundell (1966) have reported the presence of abscisic acid in the pericarp of *Rosa.*

While a great deal of work has been described on the presence and ac- tivities of endogenous inhibitors in seeds, there is much controversy over the question of whether changes in inhibitor levels occur during dor- mancy release. Barton and Solt (1949) found no correlation between the inhibitor content of many seeds, including those of *Sorbus* and *Berberis,* and the stages of after-ripening during chilling, and Flemion and De Silva (1960) were similarly unable to show any connection between the ac- tivities of inhibitors extracted from a variety of seeds and their state of dormancy.

Luckwill (1952) was unable to demonstrate removal of inhibitors from seeds of apple during chilling or even during germination, and concluded that the production of growth stimulators rather than the disappearance of inhibitors was responsible for breaking the dormancy of apple seeds. Similarly, Lasheen and Blackhurst (1956), on the basis of their failure to correlate the germination capacity of excised embryos of blackberry with their inhibitor content, also suggested the possibility that growth promoters were formed before germination could occur.

As described above, seeds of *Betula pubescens* require long-day treat- ment before germination at 15°C, and therefore Black and Wareing (1959) pointed out that if the action of light was to destroy an inhibitor within the seed (Section I,A,5,a), then the total quantity of light received would be more important than a regular light/dark cycle. They suggested there- fore that while the maintenance of dormancy appeared to be connected with the presence of endogenous inhibitor, the action of light might be in the formation of some sort of germination promoting substance.

The evidence for removal of inhibitors from dormant buds during chilling is also contradictory. As one example, Blommaert (1963) showed a direct relationship between the amounts of inhibitor present in buds of peach and the progress of chilling of the buds. On the other hand, the application of inhibitor to the swelling buds did not prevent their release from dormancy. As the inhibitor is present only in the bud scales, and as the ratio of bud scales to enclosed tissues changes during the swelling of the buds, it has been questioned whether the change in inhibitor level does in fact occur (Dennis and Edgerton, 1961).

Inhibitors have therefore been shown to be physiologically active in preventing germination and growth, but may not necessarily be reduced in activity by treatments which promote germination or bud burst. Also leaching, which could be expected to allow germination by removal of inhibitor, leads in many cases to the development of dwarf seedlings. Therefore, while there is much evidence that inhibitors may impose and maintain dormancy, it is possible that their effect is normally overcome by the action of some growth factor produced during chilling as suggested above for the action of light.

It is unlikely that auxins are the substances primarily involved, as it is well known that exogenously applied auxins are ineffective in breaking the dormancy of seeds or buds, and may often impose dormancy. On the other hand gibberellins promote the germination of seeds having a low-temperature, or photoperiodic light requirement, prevent the symptoms of seedling dwarfness when unchilled seeds are forced into germination, and relieve such symptoms in the seedling and also branch rosette growth in the adult plant.

Endogenous gibberellins were therefore studied in relation to the breaking of dormancy in seeds of *Fagus sylvatica* and *Corylus avellana* (Frankland and Wareing, 1962). These seeds require chilling and possess inhibitors which apparently remain unchanged in activity during dormancy-breaking. Gibberellins were apparently absent from unchilled dormant seeds of *Corylus* but became detectable after 6 weeks of chilling, at which time the seeds were able to germinate (Fig. 3.7.). Extracts of seeds of *Fagus sylvatica* showed gibberellin-like activity in both dormant and nondormant embryos, but differences were found in the chromatographic pattern of activity. Similarly, in buds of *Acer* and *Betula,* while the onset of dormancy was postulated to be controlled by inhibitors, the emergence from dormancy was not correlated with a decrease in the quantity of inhibitors, but with an increase in the amount of gibberellin-like substances within the buds (Eagles and Wareing, 1964). In support of this, an interaction between extracted inhibitors and exogenous $GA_3$ was shown in the control of the sprouting of buds on branches of the same species.

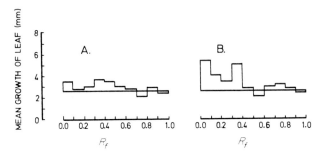

Fɪɢ. 3.7. Test for gibberellin-like substances in unchilled (A) and 6-week-chilled (B) seeds of *Corylus avellana*. Dwarf maize leaf-section assays of chromatographically purified seed extracts (from Frankland and Wareing, 1966).

Villiers and Wareing (1965b) found that a water-soluble fraction of an extract of chilled embryos of *Fraxinus excelsior,* separated by paper chromatography, promoted the rapid germination of unchilled excised embryos of the same species. This germination stimulator also had the property of overcoming the effect of the added inhibitor. It was found to occur only in the embryos and not in the endosperm, and it appeared at the end of the low-temperature after-ripening period. Further, unchilled embryos of *Fraxinus* forced to germinate were stunted, whereas unchilled embryos stimulated to germinate by addition of $GA_3$ or the endogeneous germination stimulator did not appear to be dwarfed.

In general, therefore, while it is possible that dormancy of the embryo itself might be considered to be caused by the absence of an essential germination-stimulating substance, or by the presence of substances actively inhibiting germination, it appears more likely that periods of active growth and dormancy in buds and also the onset of dormancy and its release in seeds are dependent on the balance between growth-promoting and growth-inhibiting substances. Acceptance of this approach to the problem of dormancy makes it unnecessary to postulate removal of inhibitor during or at the end of the treatment which breaks dormancy, and it may be that in some species a change in inhibitor level may not be detectable. What is important is the relative activities of the growth pro-moters and the inhibitors. In a review summarizing work on the inhibitor–promoter balance in seed germination, Amen (1968) goes so far as to state the opinion that *all forms* of seed dormancy are basically concerned with this mechanism, and the apparent differences in dormancy lie merely in the mode of changing this balance to one favorable for growth rather than for rest.

At the present time the substances most likely to be identified as the growth-promoting factors are the gibberellins. These occur widely in plants, especially in seeds and buds, and have been shown to vary with the state of growth activity (Frankland and Wareing, 1966; Eagles and Wareing, 1964). It has been shown that gibberellins increase in amount during chilling of the embryos of *Fraxinus excelsior* (Kentzer, 1966) and may therefore be identical with the germination-promoting substance produced during chilling and shown to counteract the endogenous inhibitors (Villiers and Wareing, 1965b). In addition, gibberellins have been shown to interact with inhibitors, notably abscisic acid, in the control of many physiological growth reactions and biochemical systems. Especially noteworthy is the interaction of gibberellin with abscisic acid in the synthesis of hydrolytic enzymes in the aleurone layers of germinating cereal seeds (Chrispeels and Varner, 1966). This aspect is also discussed in Section IV,E.

Many gibberellins have now been extracted from plants and the different forms have been shown to vary in activity in different systems (Fig. 3.8). For example, a gibberellin sample, shown to be a mixture of gibberel-

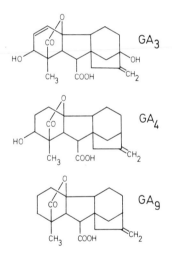

FIG. 3.8. The chemical structure of three naturally occurring gibberellins.

lins $A_4$ and $A_9$, has been shown to be thirty times more active than gibberellic acid ($GA_3$) in stimulating the germination of seeds of *Diplotaxis* and 2000 times more active than $GA_3$ in the germination of *Lepidium* (Borris and Schmidt, 1961).

The great similarities between the mechanisms of dormancy in seeds and buds have been described above. Inhibitors and promoters have been shown to be implicated in controlling the state of dormancy (Fig. 3.9), and $GA_3$ has been shown to replace light and chilling in these

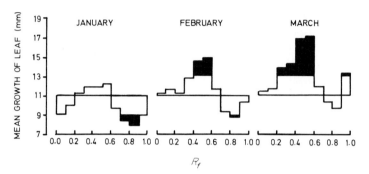

FIG. 3.9. Effect of chilling at 2°C on the endogenous inhibitor and gibberellin levels in buds of black currant. The growth activity was measured by the dwarf maize leaf-section assay (from Wareing, 1969).

phenomena and also to relieve physiological dwarfism and rosette growth. However, although the basic principle of hormonal balance may be the same in both cases, the comparison should not be taken too far, as the causes of the onset of dormancy in the seed and the bud may, in fact, be quite different. In the bud the onset of dormancy may be considered to be controlled hormonally, and Smith and Kefford (1964) have reviewed the process of winter bud induction, concluding that bud formation is a necessary developmental phase in the life cycle of a perennial plant. They point out that dormancy development is a phase of increasing cell division and decreasing cell enlargement, in contrast to the spring burst of growth, which is mainly caused by cell enlargement. The increase in meristematic activity during the induction of dormancy has been recorded many times. For example, Parke (1959) described changes in the shoots of *Abies concolor* which lead to an enlarged apex and a greatly increased rate of production of leaf primordia while the internodal cells remained un-expanded.

In addition, the induction of dormancy involves in many cases changes in the path of development of certain primordia which, instead of de-veloping into vegetative leaves, develop into bud scales which may be very different morphologically, cytologically, and biochemically. De-velopmental changes occurring within the apex of a shoot in response to a stimulus perceived by the mature leaves strongly suggest the production and transport of a substance which would be by definition a hormone, and the use of the term "growth inhibitor" for this type of substance is un-fortunate, as it implies that dormancy is merely a state of temporarily arrested growth. The introduction of the term "dormin" to designate a class of substance concerned in dormancy induction and control shows that workers have become conscious of the negative implications of the growth inhibitor concept, at least when applied to work on bud dormancy.

In the development of the seed we find that the embryo develops without cytoplasmic continuity with the enclosing tissues, upon which it is nutritionally dependent at all stages. While the embryo may be considered to be responsible to a large extent for its own hormonal control of differentiation and development, it must be influenced to some extent by the hormonal balance of the surrounding maternal tissues. In addition, and this may be a decisive factor in the development of the dormancy state, it is at *all stages* enclosed, sometimes very deeply, within maternal tissues and possibly also the endosperm. These tissues must compete with the embryo for oxygen and perhaps maintain a higher $CO_2$ content by interfering with the diffusion gradient of this gas from the embryo. In addition, they may produce other waste products which interfere with metabolism of the embryo.

Further, these maternal tissues, upon terminal differentiation, may provide an evermore efficient barrier impeding the rate of gaseous exchange between the embryo and the environment, and breakdown products resulting from senescence and death of seed-enclosing tissues may further interfere with the metabolic processes of the embryo. The cytological appearance of cells within the dormant tissues of embryos shows evidence of a state of mild anoxia (Villiers, 1971). This chain of events, to which the embryo is subjected from the very inception of its development, may be more than sufficient to divert metabolic pathways so that inhibiting substances are produced and further growth of the embryo is stopped. Ultimately the water supply to the embryo may be cut off and the seed becomes air-dry.

In short, it can be seen that there is no lack of possible causes of the disturbance of metabolism which might favor the balance of promoters and inhibitors on the side of the accumulation of growth-inhibiting substances. Following the above argument, it is significant that the removal and culture of embryos from developing seeds can result in their continued development into seedlings without the intervention of the dormancy stage (Lammaerts, 1942).

## 6. COMBINATIONS OF DORMANCY TYPES

The types of dormancy described are by no means mutually exclusive, and more than one mechanism for the imposition of dormancy may be possessed by seeds. Therefore, while mechanisms breaking dormancy might replace one another, such as the replacement of a chilling requirement by light, more than one treatment may be necessary in order to break dormancy in certain seeds. One example of this is in the relief of dormancy in *Fraxinus excelsior,* chosen because it is most familiar to me. The dispersal unit is an indehiscent samara, where the dry pericarp is extended on one side to form a wing. This species shows a deep dormancy where germination does not occur until the second growing season after the fruit is shed from the tree. A study by Villiers and Wareing (1964) showed that although it was morphologically complete at the time of formation, for germination to occur, the embryo had to grow from approximately half the length of the seed to the full length, and it approximately doubled in dry weight. This growth of the embryo takes place most rapidly at warm temperatures and is impeded by the presence of the indehiscent pericarp, which restricts oxygen supply to the seed. However, even when fully grown, the embryo is still deeply dormant, and must be chilled for a further period of about 4 months before germination can take place. Thus we have restriction of oxygen supply, immaturity of embryo, and a requirement for chilling all contributing to extend the period of dormancy for at least 18 months, and often much longer, depending on the initial rate of decomposition of the pericarp in the forest floor litter and on the total period of chilling achieved in the second winter.

## 7. SECONDARY DORMANCY

Seeds which are normally germinable may be induced to become dormant by maintaining them in unfavorable environmental conditions for a time (Table 3.II). This state of secondary dormancy can be relieved by methods usually used to relieve dormancy, such as chilling, or illumi-

**TABLE 3.II**

IMPOSITION OF SECONDARY DORMANCY ON THE INDEHISCENT FRUITS OF
*Fraxinus americana* BY MAINTENANCE IN IMBIBED STATE AT 20–22°C

| Treatment | | | Final % germination |
|---|---|---|---|
| 1 month 20°C | | | 8 |
| 1 month 5°C | | | 64 |
| 2 months 5°C | | | 100 |
| 1 month 20°C | + | 1 month 5°C | 18 |
| 1 month 20°C | + | 2 months 5°C | 53 |
| 1 month 20°C | + | 3 months 5°C | 95 |
| 2 months 20°C | | | 12 |
| 2 months 20°C | + | 1 month 5°C | 14 |
| 2 months 20°C | + | 2 months 5°C | 43 |
| 2 months 20°C | + | 3 months 5°C | 90 |

nation. The best-known example is that of seeds of *Xanthium,* which may be induced to become dormant when availability of oxygen to the embryo is reduced by embedding seeds in various moist media, at temperatures of 27°C or above (Thornton, 1935). After this treatment, even excision of the embryos fails to induce germination. Nutile and Woodstock (1967) induced secondary dormancy in seeds of *Sorghum vulgare* by forced drying at 46–48°C to about 7% moisture content. They suggested that this secondary dormancy state might be due to physical changes brought about in the seed coats by excessive drying, so that upon later imbibition gas exchange was more restricted.

Lettuce varieties which do not normally require light for germination may be induced to become light-requiring by imbibing at 35°C (Borthwick *et al.,* 1952), and this induced dormancy can also be relieved by $GA_3$. In other cases holding light-requiring seeds in an imbibed state in the dark for long periods can lead to dormancy even when the seeds are placed under illumination, as in the case of *Chloris ciliata* (Gassner, 1911).

An effect which may be similar to that of secondary dormancy induction may be shown by the water sensitivity of seeds of barley, which although otherwise able to germinate, will not do so if placed in an excess amount of water, even though the seeds are not actually submerged in the water (Roberts, 1969).

Secondary dormancy can also be induced in buds, such as in peach, which will not begin extension growth after winter chilling if they are maintained at temperatures higher than the normal growth range. In this case, a further period of chilling is required before bud burst occurs (Vegis, 1964). It is generally found that secondary dormancy can be induced by high temperatures in the presence of structures such as seed coats or bud scales which might be considered to restrict the oxygen supply, or when high temperature is combined with artificial restriction of rate of gas exchange.

## II.  Development of the Dormant State

### A.  *Factors Causing the Onset of Dormancy*

The factors leading to initiation of the dormant state have been studied in great detail in the shoot apices of woody plants. Here, the deciding factor has been shown to be the day length to which the mature leaves are exposed. Transmission of a stimulus to the shoot apices from leaves held in short days has been demonstrated (Wareing, 1954), the seasonal variation in amount of growth-inhibiting substances has been shown (I. D. J. Phillips and Wareing, 1958a), the application of extracted inhibitor to the leaves of plants held in long days has been shown to result in an increase in inhibitor level in the shoot apex, followed by the cessation of extension growth and the formation of a winter bud (Eagles and Wareing, 1964).

The active principle in the extracted growth-inhibiting complex has been isolated, its structure has been elucidated and confirmed by synthesis (Cornforth *et al.*, 1965), and the physiological activities of the synthetic inhibitor have been investigated by several workers (Milborrow, 1966). This substance has been named abscisic acid and has been shown to act in opposition to $GA_3$ in the control of several plant growth phenomena. In short, a whole chain of physiological events has been followed from the environmental stimulus to final hormonal effect, and the problem is now mainly in the province of the biochemist for further work on the fundamental molecular mechanisms involved.

Unfortunately, the picture is neither so clear nor apparently so precise for the control of the onset of dormancy in the seed. The embryo is completely enclosed at all stages from its inception as a fertilized meiospore to the actual time of germination — a period which may be measured in years. During the early stages of its development, the embryo is entirely dependent upon maternal tissues for its nutritional supplies, and must at least be influenced by the hormonal status of the surrounding tissues. In addition, it is at a disadvantage in competition for oxygen with the enclosing tissues, and this must certainly influence its metabolism.

Vegis (1964) has proposed a general theory of dormancy control based on the restriction of gas exchange. A restriction of oxygen supply by the enclosing seed coats at warm temperatures could in fact cause a change in metabolic pathways due to partial anaerobiosis. He suggests that under these conditions lipids accumulate, and that lipid accumulation and especially its association with the cell membranes is generally associated with the dormancy state in a wide range of organisms.

As pointed out above, the developing embryo is at all times enclosed within tissues which in some cases are very extensive and must certainly compete with the embryo for the inwardly diffusing oxygen. However, in the case of dormancy in buds, it is evident that at least the changes leading to dormancy must begin before there are any bud scales or secretions which could restrict the oxygen supply. Thus, cessation of extension growth, increase in mitotic activity, and a redirection of development of the leaf primordia to form bud scales all occur at temperatures and water and oxygen supplies which have supported extension growth and leaf formation until that time. Thus, while the argument could hold in the case of seeds, this cannot be the cause of entry into dormancy in buds.

Again, I have shown (Villiers, 1971) that during the warm imbibition pretreatment of seeds of *Fraxinus excelsior,* the initially high lipid content of the embryos decreases to a very low level from 20% to less than 4% of the dry weight, and the lipid entirely loses its previous association with the plasma membranes of the cells. The embryo, however, is still completely dormant even if it is excised from the seed coats, and still requires chilling before germination can take place.

Thus we see that in buds the restriction of oxygen supply and accumulation of lipids is unlikely to be the cause of the *onset* of dormancy, but could *maintain* it, whereas in seeds the converse is true — that restricted oxygen supply and consequent lipid accumulation could cause entry into dormancy, but not its maintenance.

However, there is evidence that the cause of dormancy in both seeds and buds is a change in the balance between the growth-promoting and dormancy-inducing hormones. In buds the dormancy-inducing substance abscisic acid produced in the leaves is transported to the shoot apex and causes the changes leading to winter bud formation. The ensuing interference with respiratory exchange at the apex could then maintain the apex in a state of dormancy. In seeds, the dormancy-inducing hormone could be produced *in situ* by metabolic pathways favored by the state of oxygen deficit caused by the tissues which enclose the embryo at all times, and this in turn could lead to dormancy. It may be remarked that while entry into dormancy in shoots is a period of intense mitotic activity and organ primordium formation without cell elongation, embryogenesis is similarly notable for its intense mitotic activity, organ primordium formation, and little or no cell elongation.

It then becomes understandable that exactly the same measures are capable of relieving dormancy in both seeds and buds and result in a change in the promoter/dormin balance toward the production of growth promoter. Thus, chilling and light may break dormancy naturally, while the application of gibberellins may bypass these requirements. Further, both bud break and germination are concerned with the initiation of cell elongation, and the forcing of buds and embryos into growth without the chilling requirement being fulfilled results in the remarkably similar symptoms of rosette growth and physiological dwarfism, which again can be relieved by exactly similar treatments.

## B. Stages of Dormancy Development

Wareing (1956) divides the process of entry into dormancy in buds into two stages; the first is reversible, and the second, which occurs after the plant has received its minimal number of short-day cycles, is irreversible.

Plants such as *Quercus* spp. which have more than one flush of growth during the same season appear to enter only the reversible part of dormancy induction between the flushes of growth, although this state of suspended extension growth is obviously not caused by changes in day length but has been ascribed to the effects of high temperature (Vegis, 1964). Thus, it is possible to differentiate between "summer" dormancy, in which there is only a temporary suspension of active growth, and "winter" dormancy which may be irreversible even by removal of the conditions causing the development of the dormant state.

The seeds of certain woody species which normally become dormant can be induced to continue their development without pause and to germinate if they are collected and sown without being allowed to become dry. Lammaerts (1942) found that seed dormancy in certain deciduous fruit trees could be avoided and the breeding cycle could be shortened by culturing embryos excised from fruits before they became fully ripened on the parent tree. Again, Pope and Brown (1943) were able to bypass the dormancy stage in barley seeds by supplying the maturing seeds with water from a direct source, thus preventing their drying out on the plant.

This phenomenon may be compared with the germination of cereal grains during a wet harvest while they are still on the parent plant. It has been shown that cold rain is necessary for the induction of this premature germination (Lang, 1965), and as it is well known that higher temperatures are especially liable to impose deep dormancy upon freshly harvested seeds otherwise only slightly dormant (Barton, 1965), it would seem that the temperatures prevailing during maturation of the seed or fruit may determine its degree of dormancy. This could explain the variability which is shown by certain types of seeds from year to year and from place to place within the same year.

In a similar way, woody plants transferred from southern to northern latitudes usually do not enter winter dormancy but continue active extension growth into the autumn and are killed by frost (Lang, 1965), and it would thus appear that in these southern-latitude plants entry into dormancy of both seeds and buds is controlled by high temperature rather than photoperiod, as suggested to be the cause of the induction of summar dormancy in temperate climate woody plants which grow in a series of flushes (Vegis, 1964).

It could therefore be suggested that two stages exist in the induction of both seed and bud dormancy in which a state of reversible relative dormancy is followed by a state of deep dormancy which is not reversible by merely removing the conditions imposing dormancy, but requires completely different conditions for its release. In support of this, it has been found that in cereal grains, three stages of germinability exist. As the embryo develops there is at first a gradual increase in the number of seeds which will germinate, followed by a decrease during the final stages of maturation of the grain. This dormancy is then relieved either by a period of dry storage or by moist chilling after which the germination capacity rises again (Fuchs, 1941).

Most of the work connected with growth- and germination-inhibiting substances in seeds has been concerned with the possible removal of such substances during after-ripening and the breaking of dormancy rather than in the induction of dormancy. Villiers and Wareing (1965a) investigated the growth-substance changes during maturation, after-ripening, and germination of fruits and seeds of *Fraxinus excelsior*. They found that germination-inhibiting substances did not become detectable until after the fruits had become air-dried and then were reimbibed, and that these inhibitors were apparently metabolically produced during the reimbibition process. In this respect, it is interesting that *F. excelsior* seeds will germinate without an intervening period of dormancy if gathered and sown "green," before they have dried out.

### III. Physiology of Dormancy

#### A. Germination Capacity

Lang (1965) has pointed out that a common error in seed germination tests is to assume that seeds germinating rapidly under given conditions are nondormant. He points out that during treatments which break dormancy, seeds do not become nondormant abruptly, but pass through a period of relative dormancy in which their potential activity is sufficiently high to permit germination within certain limits of, for example, temperature. He states that completely nondormant seeds would have "a uniform maximal germination capacity under almost any condition under which they can germinate at all." Differing results will be due to differing speeds of germination, but not to differing germination capacities. It is required therefore to show that the "germination capacity of the seed material is identical under all the conditions which are to be tested" before germination tests can be used to measure the responses of seeds to given conditions.

As stated above (Section II), dormancy develops during seed maturation and is frequently enhanced by higher temperatures, and in this respect is probably similar to the development of secondary dormancy (Section I, B, 7). The first gradual increase in the number of seeds which will germinate during development of the seed is probably due to the attainment by the embryo of a metabolic status which is capable of functioning separately from the parent plant. This is followed by a period of decreasing germinability during which the seed enters a dormancy state which is then relieved by a period of dry storage, or moist chilling, during which the germination capacity rises again (Fuchs, 1941).

Dormancy states in various organs and organisms change in such a way that the initial stages may be looked upon as a state of relative dormancy (Section II, B) gradually deepening until the fully dormant condition is reached. This change is usually accompanied by a progressive narrowing of the range of temperatures within which the dormancy can be reversed. During the stages in the relief of dormancy, resting organs such as seeds will germinate only within a very narrow range of temperatures initially, but this range gradually widens until germination is possible over a wide range of temperatures (Vegis, 1964). During maturation of cereal seeds, the temperatures at which germination can take place gradually fall until germination is possible only at temperatures in the region of 10°C, and during after-ripening, become germinable at successively higher temperatures once more. Seeds of *Chenopodium album* exhibit a raising of the minimum accompanied by a lowering of the maximum germination temperatures as dormancy ensues, but this tendency is reversed as dormancy progresses.

### B. Seed Storage Conditions

#### 1. NECESSITY FOR DRY STORAGE AS A DORMANCY RELEASE MECHANISM

While the problem of seed storage is discussed at length in Chapter 3 of Volume III, it is of concern to dormancy studies in that dormancy in many types of seed is frequently overcome by a period of storage while air-dry, an effect termed "after-ripening" in dry storage (Fig. 3.10). Barton (1965) lists forty-eight species showing this type of dormancy, in which seeds will not germinate if sown immediately after harvesting, but will do so if stored in an air-dry condition for a period (Table 3.III). It is probable that at least some of the cases of dormancy cited are subject to the criticisms of Lang (1965) quoted in Section III,A above. For example, barley seeds which are freshly harvested may not germinate at room temperatures until after a period of dry storage, but will germinate even when fresh at 10°C. During dry storage, the temperatures at which germination is possible become gradually higher.

This type of dormancy may last for long periods in many wild plants, but has largely been lost in cultivated forms owing to selection for strains showing less severe dormancy characteristics. Roberts (1969) points

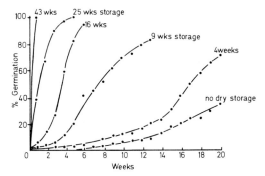

FIG. 3.10. Germination tests at 25°C on seeds of *Impatiens balsamina* after dry storage for the periods indicated (from Crocker, 1949).

**TABLE 3.III**

EXAMPLES OF SPECIES REPORTED TO UNDERGO AFTER-RIPENING IN DRY STORAGE, WITH TREATMENT RECOMMENDED FOR THE RELIEF OF THE DORMANCY STATE[a]

| Species | Period of dry storage | Treatment bypassing dormancy |
|---|---|---|
| *Ambrosia trifida* | 1–2 years (approx.) | Chilling, 3 months |
| *Cyperus rotundus* | 7 years | $H_2SO_4$, 15 minutes |
| *Festuca rubra* | 1–2 months | Chilling, 7 days |
| *Gossypium hirsutum* | 1 month | Thorough drying |
| *Hordeum* spp. | 1½–9 months | Removing hulls |
| *Impatiens balsamina* | 4–6 months | Chilling, 2 weeks |
| *Lactuca sativa* var. | | |
| Grand Rapids | 3–9 months | Exposing to light |
| *Lepidium virginicum* | 2 weeks | Light or $KNO_3$ |
| *Oenothera odorata* | 7 months | $KNO_3$ |
| *Streptanthus arizonicus* | 1–2 years | Alternating temperature |
| *Triticum* spp. | 1–2 months | Pricking coats |

[a] From Barton (1965).

out, however, that at least some degree of dormancy is advantageous agriculturally to prevent germination of the seed in the field before harvest, and is thus frequently a character selected for in the breeding of cereals.

## 2.  DELETERIOUS CYTOLOGICAL CHANGES OCCURRING DURING DORMANCY

The changes which take place during dry storage and the deleterious effects of fluctuating temperatures and high moisture content are dealt with in Chapter 3 of Volume III, and the physiological and biochemical manifestations of these changes are discussed in Chapter 4 of this volume.

However, this section discusses briefly some of the observed changes in cellular organization which appear to take place when seeds are held in an imbibed but dormant condition without germination being possible, i.e., when the required dormancy-releasing agency is withheld.

Seeds and excised embryos of *Fraxinus excelsior* will not germinate if held at laboratory temperatures, and require either chilling at 5°C for 4–6 months or the application of gibberellic acid (GA$_3$). Seeds of this species have been held imbibed at laboratory temperatures for a period of 6 years, and samples of embryos have been used at intervals for germination tests in the presence of GA$_3$ and for examination of their cytological organization in the electron microscope.

Within the first year the changes noted appeared to be mainly in the disposition of various cellular organelles in relation to the endoplasmic reticulum (ER). Plastids containing stored starch grains, and also mitochondria and lipid bodies became closely associated and apparently completely enclosed by layers of ER. The ER itself became dilated in places to form many vacuole-like structures connected by apparently normal sheets of unchanged ER.

ER-enclosed structures were shown to possess acid hydrolase activity and were remarkably similar in structure to the cytolysomes of animal cells in various pathological states (Villiers, 1967) (Fig. 3.11). In animal cells these structures are presumed to be concerned in the sequestration and removal of damaged cell organelles or in the selective resorption of cellular structures during periods of stress or starvation, when hydrolysis of the organelles can take place without damage to the remainder of the cytoplasm.

Longer periods of imbibition while withholding the chilling releaser stimulus results in a progressive loss of organelles, leaving areas of apparently blank cytoplasm and visible damage to the remaining organelles. Nuclear envelopes become greatly dilated and the unit membranes eventually appear to burst. Vesicles with electron-dense cores appear in the cytoplasm and also in association with the plasma membrane. Plastid envelopes swell and burst, and heavy electron-dense deposits are found in the vacuoles. Mitochondria may become enlarged to enormous sizes and present a very contorted appearance, and eventually the endoplasmic reticulum breaks up into countless small vesicles (Villiers, 1972).

A surprising fact is that even after 6 years of such imbibed warm-temperature storage, the embryos may recover and germinate if a suitable stimulus such as gibberellic acid is applied, or if they are exposed to chilling temperatures. In this case a progressive reversal of the degenerative appearance of the cytoplasm is observed. Very similar degeneracy of the cytoplasm has been found in embryos from seeds of maize stored

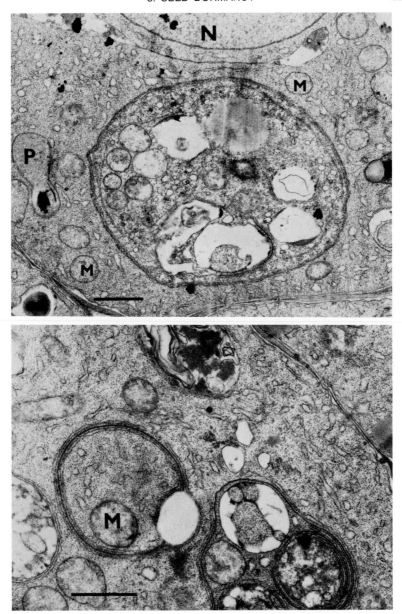

FIG. 3.11. Formation of cytolysomes in embryo radicle cells of *Fraxinus excelsior* during long storage in a moist condition at 20°C. No germination occurred during a period of 6 years in this germination test as the chilling releaser stimulus was withheld. Cytolysomes are considered to be concerned with the resorption of cellular organelles. Magnifications: top, × 14,000; below, × 17,500. (N, nucleus; M, mitochondrion; P, plastid.)

for long periods in an air-dry condition (Berjak and Villiers, 1971). In this case it has also been found that the aggregation of the ribosomes into polysomes (the protein-synthetic units), which usually occurs during seed imbibition in normal embryos, does not take place in embryos stored for prolonged periods of time.

The observations of the accumulation of electron-dense material (which has the appearance of dense precipitates) in the vacuoles are possibly of significance, as the protein reserves of the embryo of *Fraxinus* appear to be in association with bodies similar to small vacuoles, and it has been suggested several times that coagulation of proteins may be one of the causes of loss of viability in stored seeds (e.g., Barton, 1965).

## C.  The Imbibition Phase and Seed Dormancy

### 1.  TEMPERATURE DURING WATER UPTAKE

Imbibition in seeds has been shown to occur in two phases, the first of which appears to be a purely physical, temperature-independent process, while the second phase is affected by temperature and therefore probably a metabolic, energy-requiring process. For example, Marcus *et al.* (1966) have shown that in dry embryos of wheat placed in water, there is an initial phase of rapid uptake during which the fresh weight of the embryos approximately doubles during 1 hour. This is followed by a period of 5–6 hours during which no further increase occurs, after which water uptake begins again at a steadily increasing rate until eventually there are visible signs of germination. They suggest that the lag period between the two phases of water uptake be named the "germination phase," because the *visible* signs of germination are caused either by cell division or cell extension, or both, and cannot be distinguished from normal growth processes.

It becomes important to discover what metabolic processes are being initiated during this lag, or germination phase, both for the elucidation of the changes leading to germination and growth of the embryo, and also because dormant seeds presumably encounter their particular metabolic block during this phase. It has also been shown above (Section I, A, 7) that the temperatures experienced by the seed during this period may be important, and may on occasion determine whether or not the seed will be induced to become secondarily dormant.

Thus, light-requiring varieties of lettuce seeds may be induced to become deeply dormant by imbibition at 35°C. This state of secondary dormancy cannot be relieved by light but only by chilling or the application of gibberellic acid (Mayer and Poljakoff-Mayber, 1963). It is possible that imbibition at high temperatures promotes a rapid respiratory rate which creates an oxygen deficit within the seed. The structure of the seed does not favor a high rate of gaseous exchange and therefore a state of partial anaerobiosis may result in the embryo tissues, in which metabolic pathways might be altered in some way to favor the production of germination-inhibiting substances, the removal of which might require the dormancy-breaking measures of chilling or an artificial alteration of the promoter/inhibitor balance by the addition of gibberellic acid. Roberts (1969) is of the opinion that conditions which favor the operation of respiratory pathways other than the pentose phosphate pathway tend to maintain dormancy.

In a series of investigations on possible causes of the pattern of seasonal changes in the flora of the California deserts, Went (1949) has shown that there exists an interaction between temperature and imbibition resulting in the germination of different species of seeds at different seasons of the year. In these deserts the main rainfall takes place in the winter, with a lighter rainfall period during the summer. During both of these periods, seeds in the soil become imbibed, but the temperature requirements of the desert annuals are such that different species germinate and grow following the winter and the summer rains. For example, seeds of *Pectis papposa* germinate at temperatures in excess of 20°C and therefore appear following summer rains. After production of seed, the plants die but seed germination does not generally occur in the winter rainfall season following because of the lower temperatures prevailing. On the other hand, seeds such as those of *Baeria chrysostoma* germinate at temperatures from below 5°C up to 20°C, and are therefore favored by the winter rains rather than those occurring in the summer.

## 2. GASEOUS EXCHANGE DURING THE PERIOD OF WATER UPTAKE

It has been shown in Section I, B, 4 that the available oxygen may have a profound effect on germination. Many species of seeds, including those of rice and many other water plants can imbibe and germinate in the absence of oxygen. On the other hand, seeds placed in moist sand so that water may be taken up with as little restriction of gas exchange as possible, have an initial RQ value which rises to 1.5 or higher and then falls again. This high RQ value apparently results from restriction of gas exchange by the film of water over the seed coupled with the presence of the husks, and very high RQ values have been measured in barley seeds placed in an excess of water (James and James, 1940).

This water-sensitivity in barley seeds (see also Section I, B, 7) usually leads to damage of the embryo if the seeds are exposed to these conditions for more than a few hours, and the embryos may die, possibly because of the accumulation of lactic acid within the tissues (J. W.

Phillips, 1947). Conversely, the removal of the seed coats, or the maintenance of high oxygen tensions in the atmosphere during the period of imbibition causes germination in barley seeds without an intervening period of dormancy. However, entry into dormancy in this species when seeds are imbibed intact in normal air cannot solely be attributed to seed coat interference with gas exchange, as it has been shown that the oxygen uptake is actually *higher* in dormant than in nondormant barley seeds in early imbibition. In addition, certain respiratory inhibitors may *promote* germination (Major and Roberts, 1968). The effect of these inhibitors is considered to be due to the switching of respiratory pathways towards the pentose phosphate pathway (Section I, B, 4). Thus in barley it can be seen that the availability of oxygen during the period of imbibition can result in promotion of germination, entry into dormancy, or embryo damage and death.

Periods of soaking seeds by complete submersion in water, while leading to more rapid imbibition, can have harmful effects if extended. Generally, presoaking causes more injury to normally nondormant seeds than it does to dormant seeds. The harmful effect of extended pre-soaking would at first sight be expected to be caused by restriction of oxygen supply during some critical metabolic stage of germination, but while aeration of the water during soaking might reduce the harmful effects slightly in some cases, in certain cases (e.g., in cereal and bean seeds) it actually increases the apparent damage (Barton and MacNab, 1956). Further, the soaking injury to beans which is increased by the presence of oxygen is reversed by the presence of excess carbon dioxide. The nature of the soaking injury is therefore difficult to understand and although several suggestions have been made, for example that substances essential to growth are leached out from the seed, no satisfactory explanation has yet been offered.

The most recent attempt at an explanation known to me is that of Orphanos and Heydecker (1968) who ascribe the soaking injury to a critical *early* stage of germination when oxygen supply is limited and the cavity between the cotyledons is flooded with an excess of water.

### D. Gas Exchange in Imbibed Dormant Seeds

In seeds of lettuce varieties which require light for germination, it is found that imbibition in the dark is accompanied by an increase in rate of respiration. This later falls to a low level, but if the seeds are illuminated it increases again and germination ensues (Evenari, 1957). In studies of gas exchange in seeds of various woody species, a similar initial rise in respiration rate was found during and immediately following imbibition.

When the seeds were kept dormant at laboratory temperatures, this rate fell again to a constant low level. However, when placed at low temperatures in order to break dormancy, the respiration rate of most of the species rose gradually during cold treatment (Nikolaeva, 1969).

Similarly, in studies of the respiration rate of cherry seeds, Pollock and Olney (1959) found that dinitrophenol, which acts as an uncoupler of oxidative phosphorylation, stimulated the respiratory rate in dormant seeds, but that this effect diminished as cold treatment proceeded. They suggested that in dormant seeds respiration is limited by the availability of adenosine diphosphate (ADP) and that this becomes available during chilling (Olney and Pollock, 1960).

It is generally regarded as significant if germination is improved when the seed coats are damaged in such a way that their mechanical integrity is maintained largely unimpaired, but that their rate of gas exchange is increased (Table 3.IV). Thus in the case of *Betula pubescens* it is only necessary to cut or even scratch the coats in order to allow germination, expecially if the oxygen tension of the surrounding atmosphere is increased (Black and Wareing, 1959).

**TABLE 3.IV**

GERMINATION OF INDEHISCENT FRUITS AND SEEDS OF *Fraxinus americana* WITH INTACT AND DAMAGED COATS

| Organ and treatment | % Germination (days) | | | |
|---|---|---|---|---|
| | (2) | (4) | (6) | (8) |
| Intact fruits in air | – | 6 | 10 | 10 |
| Fruits with pericarps cut | – | 12 | 28 | 38 |
| Naked seeds in air | – | 14 | 30 | 42 |
| Seeds with testa pricked | 14 | 45 | 53 | 53 |
| Seeds in oxygen | 40 | 56 | 72 | 78 |
| Seeds pricked, in oxygen | 25 | 62 | 70 | 84 |

As emphasized by Nikolaeva (1969) in her classification of dormancy types, the restricting effect of the seed coats on gas exchange is usually accompanied by one or more factors imposing dormancy on the seed, and this observation is supported by the range of dormancy-releasing stimuli which are effective in these cases, including exposure to light and certain photoperiodic regimes, chilling temperatures, and the application of gibberellic acid and dormancy-breaking chemicals such as thiourea, nitrites, and nitrates. The original conclusions referred to above, that restriction of oxygen limits the availability of energy for germination and growth, may be valid in very few cases and in fact the oxygen requirement may be for a different purpose.

In this respect Wareing and Foda (1957) demonstrated that the oxygen requirement of dormant seeds of *Xanthium,* described above, is necessary for the enzymic oxidation of endogenous germination-inhibiting substances in the embryo tissues. Similarly, Black and Wareing (1959) have shown that lack of oxygen itself does not prevent germination of embryos of *Betula,* but lack of oxygen in the presence of seed coat inhibitors does so, and it was therefore concluded that oxygen may be required for the enzymic destruction of the inhibitor as in the case of *Xanthium.* Apart from damage to the seed coats, chilling as well as exposure to light and photoperiodic treatments are effective in promoting germination of intact *Betula* seeds, and it was suggested that the effect of these treatments is possibly to produce germination-stimulating substances within the embryo which would counteract the effect of the inhibitor, rather than remove inhibitor in the manner postulated for seed coat damage and increased oxygen tensions.

Again, though high oxygen tensions and seed coat treatments break dormancy in barley, dormancy cannot be attributed solely to low oxygen tensions as it has been shown that the rates of gas exchange during early imbibition stages of intact barley are actually higher in dormant than in nondormant seeds. Further, although increased oxygen tensions stimulate barley germination, it was found that application of certain respiratory *inhibitors* will also promote germination (Major and Roberts, 1968). Of the various respiratory inhibitors which were used, it was found that those inhibitors which block the Krebs cycle, glycolysis, and terminal oxidation were all effective in breaking dormancy. In addition, any treatment which tended to divert the respiratory activities so that the pentose phosphate pathway predominated, tended to break dormancy.

In this respect it is interesting that chilling has been shown to lead to an increase in the activity of the pentose phosphate pathway in seeds of *Prunus cerasus* (LaCroix and Jaswal, 1967). Roberts (1969) presents and discusses a large body of evidence suggesting (a) that a state of dormancy may be maintained by a restriction of the pentose pathway, (b) that in the early stages of germination, glycolysis, the Krebs cycle and the cytochrome oxidase system are unimportant, whereas the pentose phosphate pathway is operative, and (c) that this tendency becomes reversed in the later seedling stages. The common observation and apparent paradox that high oxygen tensions also break dormancy is explained by the fact that high oxygen tensions tend to favor the pentose phosphate pathway and block the Krebs cycle, and low oxygen tensions favor glycolytic pathways.

That dormancy is not always connected with the accumulation of carbon dioxide within seed tissues is shown by the fact that in some species high $CO_2$ concentrations may break dormancy. For example, Thornton (1935) showed that the deep dormancy induced by imbibing lettuce seeds at 35°C can be broken by an atmosphere containing more than 40% $CO_2$. For a presentation of the large number of examples of dormancy-breaking by $CO_2$, the reader is referred to Carr (1961).

It is frequently suggested that one of the main causes of the non-germination of buried weed seeds is the higher $CO_2$ content of the soil presumed to be caused by microbial activity. However, in those cases where $CO_2$ has been shown to inhibit germination, the concentration required is in most cases far higher than is found to occur in the soil. It is possible that in these cases other factors (such as light requirement) or a combination of factors might operate to prevent germination.

### E.    *Thermoperiodism*

In the field, temperatures fluctuate both diurnally and seasonally, and germination conditions are thus often very different from those used in the laboratory. Went (1957) has clearly shown the beneficial effects of daily alternations of temperature on plant growth, and has shown that the effect is correlated with the relative length of day.

The seasonal change in temperature is clearly linked with the requirement for chilling before germination, and is discussed more fully in Section I, B, 5. However, it may be mentioned here that not all the tissues of the same embryo are sensitive to this change. For example, several species of *Lilium* and *Viburnum* show a condition known as epicotyl dormancy, and while they begin germination in the same season by extension of the radicle into the soil, the growth of the epicotyl does not take place until after the winter chilling (Barton and Crocker, 1948).

Daily alternations of temperature are frequently effective in regulating germination in seeds which may otherwise remain dormant at constant temperatures. Asakawa (1956) found that the percentage germination of *Fraxinus mandshurica* seeds was very low in constant temperatures, whereas rapid, high germination could be brought about by alternating the temperature between 8°C (20 hours) and 25°C (4 hours) each day. Seeds of *Pinus polita* germinated at constant temperatures in the light, but required alternating temperatures in the dark (Asakawa, 1959).

Possible explanations for the thermoperiodic behavior of seeds are presented by Koller *et al.* (1962) as follows: (*a*) endogenous rhythms may exist in seeds, and synchronization of different components of the cycle may be necessary for germination and growth to occur, (*b*) the differential sensitivity of the various enzymes or their precursors to temperature, (c) the creation of a balance of respiratory intermediates at higher temperatures which may be unfavorable to germination but which may be promoting at low temperatures. These suggestions have little evidence for support, and it may be seen that the basis for the effect of alternating temperatures is far from being understood.

### F. Light in Dormancy Control

The effective wavelength of light for the promotion of seed germination has been shown to be the red region of the spectrum (660 nm) and it was also shown that far-red irradiation at 730 nm inhibits the germination of light-promoted seeds (Borthwick *et al.,* 1952). If seeds are irradiated successively with these two wavelengths, whether the seeds are actually promoted to germinate or not depends upon the wavelength to which they are finally exposed. Borthwick *et al.* (1952) postulated that a pigment is converted from one energy state to another, and that when seeds are exposed to red radiation, it is converted from $P_{660}$ to the $P_{730}$ form, and in this form is able to initiate reactions which cause germination. On the other hand, exposure to far-red irradiation converts the pigment back to the $P_{660}$ form, which is probably of a lower energy status and cannot participate in the reactions leading to germination. In the case of *Nemophila insignis,* blue light also inhibits germination in addition to far-red in this negatively photoblastic seed (Black and Wareing, 1960).

If time is allowed to lapse between successive irradiations with red and far-red, it is eventually found that far-red cannot reimpose dormancy. Evidently, with increase in time $P_{730}$ has initiated the reactions leading to germination, and these reactions are not themselves reversed by far-red.

It usually happens that seeds which are light-requiring when newly formed on the plant gradually become insensitive to light when stored in an air-dry condition. However, if the seeds are maintained imbibed under unsuitable conditions of illumination, or are kept at high temperatures, they may become more deeply dormant (see Section I,B,7). Conversely, if imbibed within a narrow range of temperatures, which varies with the species, seeds may have no light requirement. Thus, Grand Rapids lettuce seeds which have lost their original dormancy state may become deeply dormant again by imbibition at 35°C. At temperatures in excess of 25°C ordinarily dormant Grand Rapids seeds are light-requiring, but the percent germination is gradually increased in the dark as the temperature is lowered, and complete germination is obtained in the dark at 10°C. Koller (1962) has shown that the climatic conditions under which a particular batch of seeds has matured on the plant might be important in imposing a greater or lesser degree of light-requirement in lettuce.

White light is as effective as red light illumination in the promotion of germination as the reaction converting the pigment system from $P_{660}$ (inactive) to $P_{730}$ (active) requires only about one-quarter of the energy of the reconversion by far-red. At the end of a period of normal daylight illumination most of the pigment is therefore held in the active form.

It is evident that the amount of the active form of the pigment required for germination-promotion varies according to the condition of the seed, such as the state of hydration, the presence of intact coats, the presence of inhibiting or promoting substances, and treatment with solutions having a high osmotic pressure. Black (1969) has described this by concluding that light is important in the relief of the embryo or seed from conditions of stress.

In the case of seeds which are inhibited by light, or by long photoperiods, blue and far-red light apparently act together (Black, 1969). Long periods of illumination are usually necessary for this effect to be detected, compared with the very short periods of illumination in the case of light-stimulated seeds. The presence of seed coats is also a determining factor in this type of dormancy.

Light sensitivity can be induced by maintaining seeds in unfavorable conditions. It has been shown that in seeds of tomato continuous irradiation with far-red for 18 hours from the beginning of imbibition causes this normally light-insensitive seed to require red light for germination. This has been suggested to be caused by the synthesis of a far-red absorbing pigment over the period of irradiation (Mancinelli *et al.,* 1966).

The time between illumination and germination in lettuce seeds is about 7 hours at the minimum. Also, the time which can elapse before far-red illumination ceases to be effective in reversing the reaction is about 4 hours, and it is difficult to discover which particular type of reaction is important during this long period of time. Black (1969) has summarized the possible effects of light stimulation of seed germination as promoting the following: (*a*) lipid metabolism, (*b*) respiratory control, (*c*) enzyme synthesis, (*d*) hormone synthesis (e.g., gibberellin), and (*e*) membrane permeability effects. Little real evidence can be presented for a role of phytochrome in fat metabolism or in respiration. It has been suggested, however, that one of the very early effects in the process must be the synthesis of mRNA by the genome (Rissland and Mohr, 1967).

The application of gibberellic acid ($GA_3$) can replace both the light and chilling requirements of seeds and buds, and the action of light may therefore be to produce gibberellins which act by initiating the reactions necessary to germination (Fig. 3.12). Brian (1958) suggested that during the ripening of seeds, gibberellin-like hormone may be converted into substances from which the active form may be regenerated by exposure to

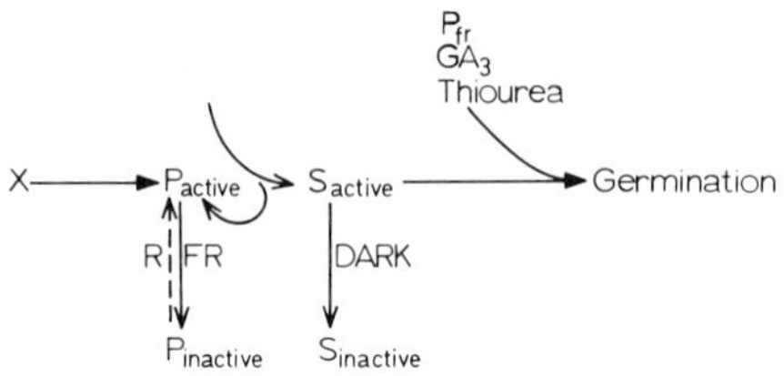

FIG. 3.12. Model showing the possible interactions of the far-red sensitive pigment system. X, Pigment precursor; P, pigment; S, substrate; R, FR, red, far-red light; $GA_3$, gibberellic acid. (From Black, 1969).

light or by low temperature treatment, leading to an alteration of the balance of inhibitors and promoters. Addition of exogenous gibberellin would then bypass the necessity for light or chilling.

It was later shown that $GA_3$ can reverse the dormancy imposed upon lettuce seeds by coumarin (Mayer, 1959), and Ikuma and Thimann (1960) have shown that the time of maximum sensitivity to $GA_3$ coincides with the time of maximum sensitivity to red light, and that the application of $GA_3$ can be reversed by heavy doses of far-red radiation. On the other hand, although they demonstrated that gibberellin-like substances were present in lettuce seeds, they failed to show that these increased in amount after exposure to red light.

A striking synergistic effect of red light and $GA_3$ was shown when suboptimal doses of each were used for lettuce seed germination (Bewley et al., 1958). In these experiments, the use of subthreshold amounts of $GA_3$ showed that after $P_{730}$ had been allowed to act for only 5 minutes, significant changes had already taken place in the seeds even though such a short time would not itself allow germination. These workers therefore suggest that gibberellins are not produced by the rapid action of $P_{730}$ during this short time, since in such case, longer periods of activity of the $P_{730}$ such as 1 hour should induce a high germination percentage, whereas in fact after 1 hour of activity only about 7% germination occurred.

On the other hand, the use of inhibitors of gibberellin synthesis, such as Phosphon D, can reverse the effect of light on germination of seeds of *Verbascum* (McDonough, 1965) which could indicate that gibberellin synthesis is involved somewhere in the light-stimulated pathways in seed germination control, even if it is not one of the direct effects.

The effect of light in the control of germination under natural conditions has been the subject of much discussion. In a review on light control of germination, Evenari (1965) suggested that reaction to light is an important mechanism in the survival of desert plants. The results of Black and Wareing (1959) suggest that, as unchilled seeds of *Betula pubescens* require long day illumination for dormancy release, they would tend to remain dormant until the spring or summer following their production on the tree. However, as chilled seeds will germinate in darkness, even seeds buried in the forest litter would germinate in spring after the winter chilling.

It has been stated above that the relative amounts of red and far-red light in ordinary illumination are important in the control of light-induced dormancy, red light being more active in most cases than far-red. However, it has been shown that under leaf cover the light which filters through to the ground below has altered spectral qualities such that far-red light actually predominates. In this case, seeds might be held dormant under leaf cover until such time as either they are chilled and therefore not sensitive to light, or until the seed coats are decayed and rendered less effective in the restriction of gas exchange. It has been suggested that the control of seed germination in *Chenopodium* could be effected in this way (Cumming, 1963).

### IV.  Biochemistry of Dormancy

#### A.  *Metabolism during Dormancy*

The respiration rates of seeds during imbibition, dormancy, and germination have been frequently studied as an index of the rate of metabolism, and also as an indicator of the depth of dormancy. On imbibition of a nondormant seed, hydrolysis of reserve materials begins almost at once, part of the products being used in the respiratory rise which ensues, while part is made available for the formation of new materials in preparation for and during germination and growth. The products of hydrolysis must frequently be translocated from the storage tissues of endosperm, or of the cotyledons, to the embryonic axis.

While changes in metabolism during dormancy have been studied by a number of workers, little is known about the connection between metabolism and the actual cause of dormancy, as it is very difficult to separate cause from effect. In this respect many of the studies on changes in enzyme activity during dormancy, such as in catalase, peroxidase, and lipase are of doubtful significance. In the context of modern theories on the control of growth, it appears less likely that a gradual build-up of hydrolytic enzymes could be responsible for the relief of dormancy. However, studies on the hormonal control of amylase production suggest that changes probably do occur in the *capacity* to produce certain key enzymes (see Chapter 2, this volume).

Studies in respiratory metabolism during dormancy have shown that in general respiration rates rise during treatments (such as chilling) with a sharp rise during breaking of dormancy. Vegis (1964) has presented a comprehensive scheme to account for the control of respiratory pathways by seed coats and bud scales. He suggests that due to interference with gaseous exchange at warm temperatures, acetyl coenzyme A is not oxidized by normal pathways, but is diverted to the production of lipids. There are physiological, cytological, and anatomic objections to this theory which have been discussed fully (Section II, A). However, while it appears impossible to accept that bud scales are a cause of the onset of dormancy, the tissues external to the embryo could well compete with the embryo for oxygen, and therefore induce dormancy by a shift in metabolic pathways.

## B. Reserve Food Changes

Stokes (1952) showed that reserve food materials were transferred from the endosperm to the embryo of *Heracleum sphondylium* during chilling at 2°C. The embryo grew to twice the volume and twelve times its original dry weight during 9 weeks of chilling, with a concomitant loss of dry matter from the endosperm. If the seeds were stored moist at 15°C, however, these changes did not occur and the seeds remained dormant (Fig. 3.13). Warm temperature interruption did not reverse the progress of dormancy-breaking at any stage, and the total chilling periods were additive. Because of this, it was concluded that chilling did not remove an inhibitor, but was merely concerned with the mobilization of food reserves and their transfer to the embryo.

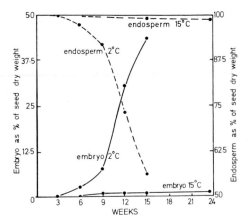

FIG. 3.13. The effect of temperature during moist storage on the dry weights of embryos and endosperms of *Heracleum sphondylium*. (From Stokes, 1952).

Steinbauer (1937) similarly concluded that dormancy-breaking in embryos of *Fraxinus nigra* was concerned with food mobilization, and suggested that the accumulation of starch, soluble carbohydrates, and proteins by the embryo during chilling resulted in the development of sufficient osmotic and imbibitional forces to enable the embryo to overcome the mechanical resistance of the seed coats. However, Villiers and Wareing (1965b) found that the processes of reserve food accumulation and of chilling were separable in time, and further that the excised fully developed embryo of *F. excelsior,* which has precisely similar dormancy characteristics to *F. nigra,* was completely dormant unless chilled or leached with water. They concluded that an inhibitor of germination was present in the embryos, and they demonstrated its presence and physiological activity.

The reserve food materials within the embryo itself can also change during the chilling treatment. The freshly imbibed embryos of *F. excelsior* store mainly lipid, and starch is almost undetectable. The lipid decreases during dormancy from 20% to less than 4% of the dry weight, much starch appears in the tissues and at the same time the protein content increases from 14% to 24% of the dry weight. Varasova (1956, cited by Nikolaeva, 1969), found that protein hydrolysis occurring in the endosperm was accompanied by an increase in protein of the embryo of *F. excelsior* during warm moist storage at warm temperatures, and that if the seed was stored at chilling temperatures immediately after imbibition, this mobilization of nitrogenous substances did not occur. This explains the requirement for a period of warm temperature storage before chilling treatment in this species.

The mobilization of reserve food materials does not therefore appear to be an essential stage in the breaking of dormancy, and Nikolaeva (1969) concluded that it appears that lipid breakdown, starch formation, and increase of sugar can all take place even if the seed is held at temperatures not conducive to dormancy-breaking.

### C. *Phosphate Metabolism*

Phosphorus-containing compounds in the seed include nucleotides, nucleic acids, phospholipids, sugar phosphates, and phytin. Olney and Pollock (1960) found that during the after-ripening of cherry seeds by moist chilling at 5°C, growth occurred in the embryonic axis and was accompanied by a transfer of reserve materials from the cotyledons to the axis. In addition, there was an increase in respiration rate of the axis accompanied by a greater efficiency of the respiratory enzyme system, suggesting that an increase in the supply of available phosphate or phosphate acceptors occurred. Accordingly, they measured the changes in phosphate and its fractions during the after-ripening process. While the total phosphate of the axis increased at the expense of the cotyle-donary reserve, the increase was not uniform in all fractions. In addition, the rate of translocation was found to be in excess of the rate of cell division in the axis, causing an increase in the concentration per cell. At chilling temperatures phosphate accumulated through intermediates such as sugar phosphates and high-energy nucleotides into nucleic acids, whereas in seeds held at laboratory temperatures the normal pathways of metabolism and synthesis did not appear to function, and inorganic phosphate accumulated in the cells. They tentatively concluded that a block in phosphate metabolism appeared to be associated with the dormant condition.

The seeds of *Corylus avellana* require a short period of chilling at 5°C before germination, and Bradbeer and Colman (1967) found that at 5°C feeding radioactive adenine-8-$^{14}$C to cotyledon preparations caused radioactivity to appear in adenosine and adenosine mono- and diphos-phate. As chilling progressed adenosine mono- and diphosphate was formed at the expense of the adenine and adenosine. In addition, they found that there appeared to be differences in the pattern of activity of the tricarboxylic acid cycle during chilling in different batches of seeds, and concluded that the activity of the TCA cycle does not appear to be critical during chilling. On the other hand, the activity of the pentose phosphate pathway decreased relative to the glycolytic pathway in all batches of seeds examined.

Thus, as has been suggested many times from the results of purely physiological experiments, the breaking of dormancy appears to be associated with phosphate metabolism and an increased supply of available energy to the growing centers of the embryo. However, as with so many experiments of this type it cannot be said whether this is the primary block in dormancy or whether such changes follow some more fundamental mechanism.

### D. Nucleic Acid and Protein Synthesis

As stated in Section IV, C, there appears to be a net synthesis of nucleic acid during the dormancy period. Of the nucleic acids, DNA is well established to be the information-containing genetic material, while the various types of RNA are directly concerned in the control of protein synthesis. As the metabolic status depends upon enzymes, which are proteins, it obviously follows that a convenient control of dormancy would be the repression or derepression of the activities of the DNA, or an activation of the RNA in the processes of protein synthesis. For an explanation of the processes concerned in protein synthesis, the reader is referred to Chapter 2, this volume.

Tuan and Bonner (1964) in a study of the process of RNA synthesis in dormant buds of potato, found little RNA synthesis in dormant buds, but a stimulation of RNA synthesis occurred in buds treated in order to break dormancy. Extracted DNA from dormant buds did not support RNA synthesis, but DNA from nondormant buds could do so. RNA production in nondormant buds was inhibited by actinomycin D which is a metabolic inhibitor of the production of messenger RNA (i.e., RNA complementary to the DNA of the cell, and therefore concerned in the transmission of the species-specific genetic information). Therefore, breaking of dormancy in the buds of potato appears to be associated with the derepression of the genetic material and the synthesis of informational mRNA.

Several studies with nondormant seeds such as wheat have shown that it appears likely that the mRNA necessary for at least the initial stages of germination is already present in the dry seed, as actinomycin D does not prevent protein synthesis or the initial stages of germination in im-bibing seeds, even though it does inhibit RNA synthesis, and therefore informational RNA must already exist in the dry seed (Dure and Waters, 1965; Marcus, 1969).

Kahn (1966) suggested that in *Xanthium* seeds protein synthesis itself was not involved in the release from dormancy, but rather that a reversal of inhibition of messenger RNA synthesis occurs because of an en-dogenous inhibitor taking part in the repression of the genetic material. Although kinetin was used in his experiments to overcome the effect of endogenous inhibitors he suggested than an interplay of endogenous inhibitors and promoters, acting together with physical factors such as light and temperature regulate dormancy by repression or derepression of the genetic material.

Increasing times of cold treatment of pear embryos showed a pro-gressively increasing capacity for nucleic acid synthesis (Kahn *et al.,* 1968). All nucleic acid fractions increased, but especially transfer RNA and a fraction considered to be a hybrid RNA-DNA molecule. They sug-gested that this hybrid fraction might be important in dormancy-breaking.

Jarvis *et al.* (1968a) found that during dormancy-breaking in *Corylus avellana* embryos RNA synthesis increased, measured by the incorpora-tion of $^{32}$P into RNA in the tissues (Fig. 3.14). The earliest change was in DNA template availability (Fig. 3.15) which was found to be followed by an increase in RNA polymerase activity (Fig. 3.16). They suggested that a similar mechanism for the control of dormancy operates in both seeds

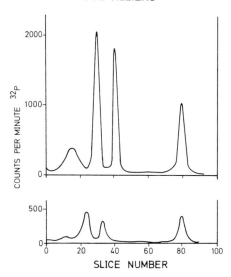

FIG. 3.14. RNA formation in excised embryos of *Corylus avellana* in gibberellic acid (top) and water (below). Polyacrylamide gel electrophoresis of RNA labeled by incorporation of [32]P into embryos excised 24 hours. (From Jarvis *et al.*, 1968a).

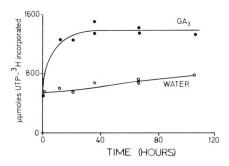

FIG. 3.15. DNA template activity of chromatin from embryos of *Corylus avellana* in gibberellic acid and in water, measured as uridine-[3]H triphosphate incorporated into RNA per 100 μg DNA (From Jarvis *et al.*, 1968b).

and buds, and that the breaking of dormancy is accompanied by an increased capacity of the DNA to support RNA synthesis.

Thus, it appears that in nondormant species, such as wheat, germination does not appear to depend upon the synthesis of new mRNA, but only on activation of preexisting mRNA. This activity is apparently permitted upon hydration of the seed tissues, and germination ensues. In dormant species, however, the absence of mRNA or at least of certain types of mRNA could provide a convenient method for the control of germination at the level of transcription of the genome.

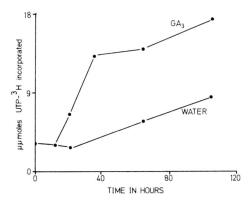

FIG. 3.16. RNA polymerase activity of chromatin from embryos of *Corylus avellana* in gibberellic acid and in water. Measured as uridine triphosphate-³H incorporated per 100 μg DNA. (From Jarvis *et al.,* 1968b).

In completely dormant but imbibed seeds, as well as in those seeds undergoing pretreatment in order to stimulate germination, synthesis of a very wide variety of proteins, both structural and enzymic, can take place. mRNA must therefore be present and active, and if dormancy is due to repression of genetic activity in the synthesis of mRNA, then it must be in the synthesis of certain rather specific types of mRNA. As it might be difficult to detect such a small fraction, it may be that the increases in mRNA shown upon breaking dormancy include RNA produced as a result of the chain of activity set up as a consequence of the more specific derepression mechanism.

### E.   *Effects of Seed Hormones*

As the auxins have not been seriously implicated in the control of dormancy, they will not be considered here. However, substances inhibiting germination are known to occur in seeds and have been suggested to act after the manner of hormones. Abscisic acid is well established to be the cause of entry into dormancy of the shoots of woody plants as described in Section II, A, and is known to antagonize the effects of the gibberellins, which in their turn are well established in the release from dormancy of seeds such as barley.

It is becoming generally accepted that dormancy in higher plants is due to a balance of endogenously occurring growth inhibitors and promoters (Amen, 1968), and dormancy may be considered to be due to the presence of growth inhibitors, the absence of growth promoters, or a combination of both, and it is therefore necessary to determine how the levels of such endogenous hormones exert their effects, and are in turn themselves controlled by the environmental stimuli such as light and temperature.

Section IV, D outlines a method by which the synthesis of certain proteins essential to the process of germination and extension growth could be controlled through the absence or possibly the sequestration of mRNA. Such a scheme would require intermediary substances or hormones which would mediate between the receptor stimuli and the protein synthetic mechanism. In trees, the role of abscisic acid in inducing the formation of the winter bud is well established. In addition, abscisic acid has been shown to act in opposition to the growth-promoting hormone systems of the plant, in particular, the gibberellins.

The close similarities between the rosette growth of unchilled branches of deciduous trees and the physiological dwarfing of seedlings produced from unchilled plants has already been described above, and the factors relieving these symptoms are identical. Again, the action of gibberellic acid in stimulating the *de novo* synthesis of α-amylase in barley germination is well established, and there seem to be excellent grounds for the general assumption that abscisic acid and the gibberellins form the inhibitor–promoter complex, the balance of which is concerned in the control of dormancy.

While inhibitor levels did not decrease during chilling in seeds of *Corylus avellana,* endogenous gibberellins appeared and were shown to increase (Frankland and Wareing, 1962) (Fig. 3.7). Application of $GA_3$ breaks dormancy in this species (Fig. 3.17), and has been shown to cause

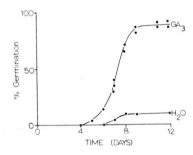

Fig. 3.17. The effect of gibberellic acid on the germination of seeds of *Corylus avellana* at 25°C. (From Jarvis *et al.,* 1968a).

a rapid increase in RNA synthesis (Jarvis *et al.,* 1968a) (Fig. 3.14). The earliest change was found to be in DNA-directed RNA synthesis (Fig. 3.15), followed by increased activity of RNA polymerase activity (Fig. 3.16), and both preceded observable changes in germination by 48 hours.

Several studies have implicated gibberellins in the regulation of protein and nucleic acid synthesis. The work on the production of α-amylase by barley aleurone is now classical. The breakdown of starch reserves in the endosperm of barley is brought about by amylases, and $\alpha$-amylase becomes active after dormancy-breaking but before germination. The presence of the embryo had long been known to be essential for the initiation of amylase activity and it was found that the action of the embryo was in the production of gibberellins which had their effect in the aleurone layer (Paleg, 1965). It has been shown that the $\alpha$-amylase produced by the action of gibberellin is synthesized before germination and that its production is prevented by actinomycin D, and is therefore dependent on mRNA synthesis (Varner and Chandra, 1964).

GA$_3$ has also been shown to increase RNA and DNA synthesis in isolated nuclei from dwarf peas (Johri and Varner, 1968) and in barley aleurone cells (Chandra and Varner, 1965). From the results of physiological experiments, it has been shown that GA$_3$ acts in opposition to abscisic acid (e.g., Aspinall et al., 1967).

In an autoradiographic study at the electron microscope level on the effects of abscisic acid and GA$_3$ on the incorporation of nucleic acid and protein precursors, Villiers (1968) found that abscisic acid maintained embryos of *Fraxinus excelsior* in a state of dormancy. The incorporation of uridine-$^3$H and thymidine-$^3$H were strongly inhibited, but not incorporation of leucine-$^3$H (Fig. 3.18). Therefore, the action of abscisic acid

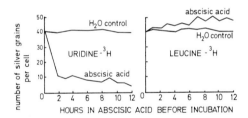

FIG. 3.18. Uptake of uridine-$^3$H and leucine-$^3$H in germinating embryos of *Fraxinus excelsior* placed in abscisic acid or water, and samples removed for isotope incorporation experiments after hourly intervals. Data from silver grain counts over autoradiographs of radicle meristem cells (From Villiers, 1968).

was found to be at the level of nucleic acid synthesis, but to have no direct effect on protein synthesis. On the other hand, GA$_3$ reversed the effect of abscisic acid on the germination of the embryos, greatly promoted the formation of RNA and DNA, but did not increase protein synthesis over the controls during the time of the experiment (2 hours). It was concluded that abscisic acid and gibberellin have opposing effects in controlling the state of dormancy of the embryo, and that their primary effect was upon RNA synthesis in the nuclei (Fig. 3.19). Similar results have been reported for the action of GA$_3$ and abscisic acid in dormant potato buds (Shih and Rappaport, 1970).

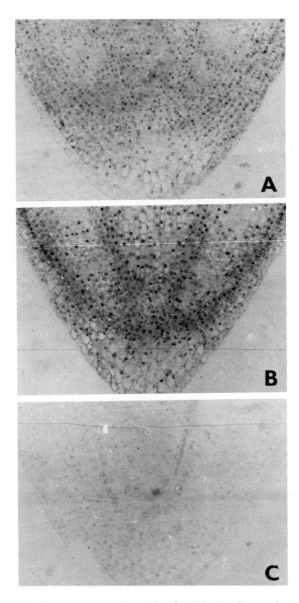

Fig. 3.19. Light microscope autoradiographs of radicle tips from embryos of *Fraxinus excelsior* incubated in 20 $\mu$c/ml uridine-$^3$H for 1 hour followed by a "chase" of nonradioactive uridine for 1 hour. The following incubation media were used: (A) isotope in water, (B) isotope in 100 mg/liter gibberellic acid, (C) isotope in 2 mg/liter abscisic acid. *Note:* The sections are deliberately presented unstained in order to show the distribution of photographic silver grains. (From Villiers, 1968).

From the results of their work on *Corylus* embryos, Jarvis *et al.* (1968a) suggested that $GA_3$ does not act directly upon the DNA in its stimulation of DNA-directed RNA synthesis, but requires another factor. They postulated that $GA_3$ plus this factor could then act upon one or more sites on the DNA equivalent to an "operator" gene, overcoming its repression and thereby enabling the transcription of a number of "structural" genes after the manner of the model described by Jacob and Monod (1961). They suggested that in low-temperature after-ripening, events occur which allow gibberellin synthesis at the higher temperatures of germination thus leading to operator gene derepression and the synthesis of several types of mRNA (Fig. 3.20).

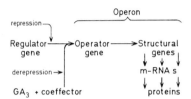

FIG. 3.20. A possible mechanism for the action of gibberellic acid at the level of the genome. (From Jarvis *et al.*, 1968a).

In Section III, F the physiology of light induction of germination is discussed, and the conclusion is reached that activation of the phytochrome system is in some way responsible for production of a germination-promoting substance, suggested by several workers to be a gibberellin-type hormone.

In this type of dormancy we see the possibility of a complete system triggered by a precise environmental stimulus and having a known receptor site, which is not the case with the rather diffuse, dormancy-breaking effect of chilling.

Illumination by red light converts $P_{660}$ to $P_{730}$, which has a higher energy status and is possibly an enzyme involved in some way in the production of gibberellin. The gibberellin acts with a cofactor on a repressor system releasing an operator gene controlling the activity of several structural genes which in turn produce the several types of mRNA responsible for the production of a sequence of enzymes which can now direct one or more metabolic pathways.

## V.   Cytology of Dormant Embryos

Several studies have been made of the changes in cytological organization in the cells of embryos and endosperms of various seeds during imbibition and germination (e.g., Klein and Ben-Shaul, 1966; Opik, 1966; Paulson and Srivastava, 1968; Jones, 1969). However, while some authors have described their material as being dormant, the species chosen were actually in a state of quiescence, merely requiring water before resumption of the growth of the embryo in germination.

In a study of the cytological events following imbibition in embryos of *Fraxinus excelsior* Villiers (1971) has shown that many changes can take place in the cytology of the radicle tip cells even while the embryo is fully dormant, in the true sense of the word. In this species, germination will not take place until the seed has been chilled at 5°C for several months.

During imbibition at 20°C, the disappearance of lipid (from 20% to 4% of the dry weight) was the most obvious change. Whereas upon imbibition the lipid was in close association with the cell membrane, this association was completely lost after about 3 months at 20°C (Fig. 3.21). The significance of this is discussed in Section II, A in connection with the theory of Vegis (1964). Accompanying the loss of lipid there was an associated rise in the starch content of the embryo axis, the proplastids developing into amyloplasts containing starch.

Golgi bodies were absent when the tissues were freshly imbibed, but they appeared and were apparently in an active state, producing many vesicles. In addition, a system of complex multivesicular bodies appeared, usually in association with the cell walls, and it was thought that the Golgi activity and multivesicular bodies are connected with wall synthesis as the embryo grows in size within the seed.

Endoplasmic reticulum appeared and developed into very complex systems of parallel membranes (Fig. 3.22) with countless attached ribosomes, and could have been a preparation for the intensive burst of protein synthesis which is known to occur during germination. Mitochondria increased in number and became very complex in internal structure, and microbodies possessing stored crystalline protein cores appeared. Protein increased from 14% to 24% of the dry weight, possibly partly due to stored protein but partly due to the extensive proliferation of cellular organelles.

During the chilling treatment which is necessary to break dormancy, the most conspicuous changes occurred in the nuclei, where the nucleoli became very large, and autoradiographic studies showed that much RNA synthesis was taking place. Abscisic acid, which is known to be a powerful inhibitor of germination, completely stopped this RNA production, and it is significant that excised embryos stimulated to germinate could be caused to become dormant again by application of abscisic acid, and later released from this dormancy by soaking in water, which presumably re-

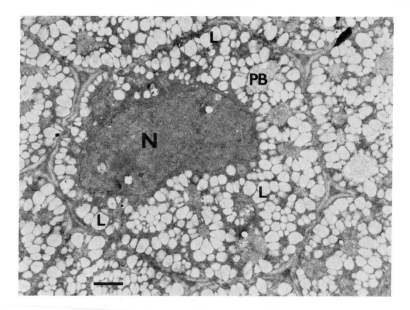

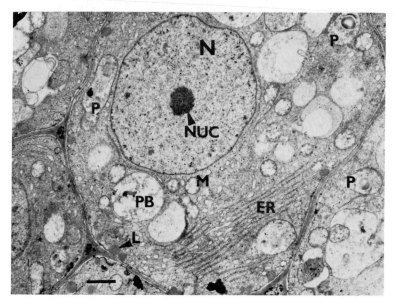

Fɪɢ. 3.21. Electron micrographs showing the loss of stored lipid food material in *Fraxinus excelsior* during maturation of the embryo. The lipid decreases from 20% of the dry weight (top micrograph, × 8000), to less than 4% of the dry weight (bottom micrograph, × 7500). (Villiers, 1971). NUC, nucleolus; N, nucleus; L, lipid body; P, plastid; PB, protein body; M, mitochondrion; ER, endoplasmic reticulum.

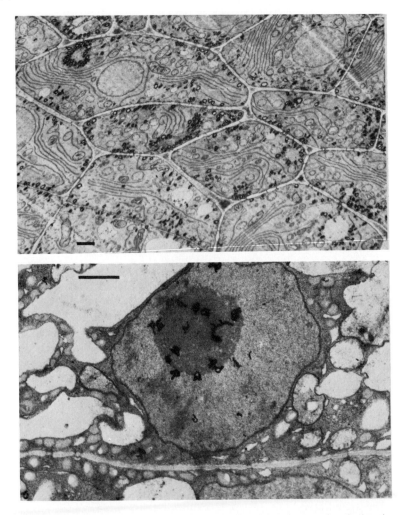

F<small>IG</small>. 3.22. Electron micrographs showing extensive development of endoplasmic reticulum in radicle meristem cells during maturation of the embryo of *Fraxinus excelsior* at 20°C (top micrograph, × 3800), and the incorporation of uridine-³H into the nucleolus during chilling of the embryo at 5°C in the relief of dormancy (bottom micrograph, × 11,000). (Villiers, 1971).

moved the inhibitor from the tissues (Villiers, 1968). It has been shown that abscisic acid occurs in the vegetative organs of this species (Milborrow, 1966).

While food interconversions, wall development, and organelle proliferation can all take place in dormant embryos, the changes taking place during release from dormancy appeared to be mainly connected with

RNA synthesis in the nuclei and nucleoli. This would prepare the embryos for the period of intense protein synthetic activity on germination.

## VI.  Inheritance of Dormancy

Dormant seeds appear to be more common in certain groups of plants than in others. The best example is the wide occurrence of dormancy due to the impermeability of the seed coats in the Leguminosae. Again, all members of the Gesneriaceae appear to be light-requiring while members of the Rosaceae have a chilling requirement (Barton, 1965). Certain varieties of species of cultivated plants may show dormancy, whereas other varieties consistently do not, and the actual degree of dormancy appears to depend to a great extent upon the climatic conditions during seed maturation and also on the conditions of storage. The dormancy characteristics therefore differ from year to year within the same locality and from one area to another within the same year.

On the other hand, several investigations have shown that the possession of seed dormancy characteristics seems to vary according to the geographical spread of the species or genus. Nikolaeva (1969) cites a study of the genus *Fraxinus,* in which she could distinguish five groups of species by their germination behavior, and found that these groups correlated well with the geographical distribution of the species studied.

Similarly, Kramer and Kozlowski (1960) state that Douglas fir, and ponderosa, lodgepole, and loblolly pines have evolved distinct geographic races which differ physiologically from one another, and Stearns and Olson (1958) showed that the germination of seeds of *Tsuga canadensis* was affected by latitudinal origin, where seeds from Quebec province showed a greater moist chilling requirement and lower optimal temperatures for germination than seed from Tennessee.

Lebedeff (1947), in an investigation of hard-seededness in *Phaseolus vulgaris,* made experimental crosses among five strains which differed in their dormancy characteristics, and came to the conclusion that while it is obvious that the ability to develop the seed coat structures responsible for seed-hardness must be a genetic characteristic, a particular strain of seeds would only show a similar degree of hardness under given environmental conditions, and therefore dormancy in this species was a potential property, and the degree of its expression was controlled by the environment.

However, even within the same flower head, seeds frequently vary in their degree of dormancy, and it is well known that in the Compositae, the germination characteristics of the seeds produced from the ray and from the disk florets are different. Seed samples gathered from the same plant of *Rumex crispus* have also been shown to differ in their degree of light- and temperature-dependence. When samples are mixed and seeds withdrawn for germination tests, these variations are obscured by the presentation of average percentage germination figures. However, these apparently small variations are obviously of great ecological and adaptive significance in spreading the period of germination over a period of time.

## REFERENCES

Addicott, F. T., and Lyon, J. L. (1969). Physiology of abscisic acid and related substances. *Annu. Rev. Plant Physiol.* **20,** 139.

Amen, R. D. (1968). A model of seed dormancy. *Bot. Rev.* **34,** 1.

Asakawa, S. (1955). Hastening the germination of *Pinus koraiensis. J. Jap. Forest. Soc.* **37,** 127.

Asakawa, S. (1956). Thermoperiodic control of germination of *Fraxinus mandshurica* var. *japonica* seeds. *J. Jap. Forest. Soc.* **38,** 269.

Asakawa, S. (1959). Germination behaviour of several coniferous seeds. *J. Jap. Forest. Soc.* **41,** 430.

Aspinall, D., Paleg, L. G., and Addicott, F. T. (1967). Abscisin II and some hormone-regulated plant responses. *Aust. J. Biol. Sci.* **20,** 869.

Barker, G. R., and Rieber, M. (1967). The development of polysomes in the seed of *Pisum arvense. Biochem. J.* **105,** 1195.

Barton, L. V. (1965). Seed dormancy: General survey of dormancy types in seeds, and dormancy imposed by external agents. *In* "Handbuch der Pflanzenphysiologie" (W. Ruhland, ed.), Vol. 15, Part 2, p. 699. Springer-Verlag, Berlin and New York.

Barton, L. V., and Crocker, W. (1948). "Twenty Years of Seed Research." Faber & Faber, London.

Barton, L. V., and MacNab, J. (1956). Relation of different gases to the soaking injury of seeds. III. Some chemical aspects. *Contrib. Boyce Thompson Inst.* **18,** 339.

Barton, L. V., and Solt, M. L. (1949). Growth inhibitors in seeds. *Contrib. Boyce Thompson Inst.* **15,** 259.

Bennet-Clark, T. A., and Kefford, N. P. (1953). Chromatography of the growth substances in plant extracts. *Nature (London)* **171,** 645.

Bentley, J. A., and Housley, S. (1953). Growth of *Avena* coleoptile sections in solutions of 3-indolylacetic acid and 3-indolylacetonitrile. *Physiol. Plant.* **6,** 480.

Berjak, P., and Villiers, T. A. (1970). Ageing in plant embryos. *New Phytol.* **69,** 929.

Bewley, J. D., Negbi, M., and Black, M. (1968). Immediate phytochrome action in lettuce seeds and its interaction with gibberellins and other germination promoters. *Planta* **78,** 351.

Black, M. (1969). Light-controlled germination of seeds. *Symp. Soc. Exp. Biol.* **23,** 193.

Black, M., and Wareing, P. F. (1959). The role of germination inhibitors and oxygen in the dormancy of the light-sensitive seeds of *Betula* spp. *J. Exp. Bot.* **10,** 134.

Black, M., and Wareing, P. F. (1960). Photoperiodism in the light-inhibited seed of *Nemophila insignis. J. Exp. Bot.* **11,** 28.

Blommaert, K. L. J. (1963). Winter rest of deciduous fruit trees in relation to the problem of delayed foliation. *S. Afr. J. Sci.* **59,** 316.

Blumenthal-Goldschmidt, S. (1960). Changes in the content of germination regulating substances in lettuce seeds during imbibition. *Bull. Res. Counc. Isr., Sect. D* **9,** 93.

Böhmer, M. (1928). Die bedeutung der samenteile für die lichtwirking und die wechsel-beziehung von Licht und Saverstoff bei der keimung lichtempfindlicher samen. *Jahrb. Wiss. Bot.* **68,** 544.

Borris, H., and Schmidt, S. (1961). Die bildung keimungsfordernder stoffwechselprodukte in submerskulturen von *Gibberella fujikuroi Naturwissenschaften* **48,** 483.

Borthwick, H. A., Hendricks, S. B., Parker, M. W., Toole, E. H., and Toole, V. (1952). A reversible photoreaction controlling seed germination. *Proc. Nat. Acad. Sci. U.S.* **38,** 662.

Bradbeer, J. W., and Colman, B. (1967). Studies in seed dormancy. I. The metabolism of $2\text{-}^{14}C$ acetate by chilled seeds of *Corylus avellana*. New Phytol. **66,** 5.

Bredeman, G. (1938). Zur frage der bewertung hartschaliger Luzernesamen. *Proc. Int. Seed Test. Ass.* **10,** 1 (cited in Nikolaeva, 1969).

Brian, P. W. (1958). Effects of gibberellins on plant growth and development. *Biol. Rev.* **34,** 37.

Brown, E., Stanton, T. R., Wiebe, G. A., and Martin, J. H. (1948). Dormancy and the effect of storage on oats, barley and sorghum. *U.S. Dep. Agr., Tech. Bull.* **953.**

Brown, R. (1940). An experimental study of the permeability to gases of the seed-coat membranes of *Cucurbita pepo*. *Ann. Bot. (London)* [N.S.] **4,** 379.

Carr, D. J. (1961). Chemical influences of the environment. *In* "Handbuch der Pflanzen-physiologie" (W. Ruhland, ed.), Vol. 16, p. 737. Springer-Verlag, Berlin and New York.

Chandra, G. R., and Varner, J. E. (1965). Gibberellic acid controlled metabolism of ribo-nucleic acid in aleurone cells of barley. *Biochim. Biophys. Acta* **108,** 583.

Chrispeels, M. J., and Varner, J. E. (1966). Inhibition of gibberellic acid induced formation of α-amylase by abscisin II. *Nature (London)* **212,** 1066.

Cornforth, J. W., Milborrow, B. V., Ryback, G., and Wareing, P. F. (1965). Identity of Sycamore 'Dormin' with Abscisin II. *Nature (London)* **205,** 1269.

Cox, L. G. (1942). "A Physiological Study of Embryo Dormancy in the Seed of Native Hardwoods and Iris." Ph.D. Dissertation, Cornell University, Ithaca, New York.

Crocker, W. (1906). Role of seed coats in delayed germination. *Bot. Gaz.* **42,** 265.

Crocker, W. (1916). Mechanisms of dormancy in seeds. *Amer. J. Bot.* **3,** 99.

Crocker, W. (1949). "The Growth of Plants." Van Nostrand-Reinhold, Princeton, New Jersey.

Cumming, B. G. (1963). The dependence of germination on photoperiod light quality and temperature in *Chenopodium* spp. *Can. J. Bot.* **41,** 1211.

Dennis, F. G., and Edgerton, L. J. (1961). The relationship between an inhibitor and rest in peach flower buds. *Proc. Amer. Soc. Hort. Sci.* **77,** 107.

Dure, L., and Waters, L. (1965). Long-lived messenger RNA: Evidence from cotton seed germination. *Science* **147,** 410.

Eagles, C. F., and Wareing, P. F. (1964). The role of growth substances in the regulation of bud dormancy. *Physiol. Plant.* **17,** 697.

Evenari, M. (1957). The physiological action and biological importance of germination inhibitors. *Symp. Soc. Exp. Biol.* **11,** 21.

Evenari, M. (1965). Light and seed dormancy. *In* "Handbuch der Pflanzenphysiologie" (W. Ruhland, ed.), Vol. 15, Part 2, p. 804.

Ferenczy, L. (1955). The dormancy and germination of seeds of *Fraxinus excelsior* L. *Acta Biol. (Szeged)* **1,** 17.

Flemion, F., and De Silva, D. S. (1960). Bioassay and biochemical studies of extracts of peach seeds in various stages of dormancy. *Contrib. Boyce Thompson Inst.* **20,** 365.

Frankland B., and Wareing, P. F. (1962). Changes in endogenous gibberellins in relation to chilling of dormant seeds. *Nature (London)* **194,** 313.

Frankland, B., and Wareing, P. F. (1966). Hormonal regulation of seed dormancy in hazel (*Corylus avellana* L.) and beech (*Fagus sylvatica* L.). *J. Exp. Bot.* **17**, 596.

Fuchs, W. H. (1941). Keimungstundien an Getreide. I. Keimungstemperatur und Reifazustand. *Z. Pflanzenzuecht.* **24**, 165.

Gassner, G. (1911). Vorläufig mitteilung neuerer Ergebnisse einer Keimungsuntersuchungen mit *Chloris ciliata*. *Ber Deut. Bot. Ges.* **29**, 708.

Giersbach, J. (1937). Germination and seedling production of species of *Viburnum*. *Contrib. Boyce Thompson Inst.* **9**, 79.

Housley, S., and Taylor, W. C. (1958). Studies on plant growth hormones. VI. The nature of inhibitor $\beta$ in potato. *J. Exp. Bot.* **6**, 129.

Hyde, E. O. C. (1954). The function of the hilum in some Papilionaceae in relation to the ripening of the seed. *Ann. Bot. (London)* [N.S.] **18**, 241.

Ikuma, H., and Thimann, K. V. (1960). Action of GA on lettuce seed germination. *Plant Physiol.* **35**, 557.

Ives, S. A. (1923). Maturation and germination of seeds of *Ilex opaca*. *Bot. Gaz.* **76**, 60.

Jackson, G. A. D., and Blundell, J. B. (1965). Germination of *Rosa arvensis*. *Nature (London)* **205**, 518.

Jackson, G. A. D., and Blundell, J. B. (1966). Effect of dormin on fruit-set in *Rosa*. *Nature (London)* **212**, 1470.

Jacob, F., and Monod, J. (1961). On the regulation of gene activity. *Cold Spring Harbor Symp. Quant. Biol.* **26**, 193.

James, G. M., and James, W. O. (1940). The formation of pyruvic acid in barley respiration. *New Phytol.* **39**, 266.

Jarvis, B. C., Frankland, B., and Cherry, J. H. (1968a). Increased nucleic acid synthesis in relation to the breaking of dormancy of hazel seed by gibberellic acid. *Planta* **83**, 257.

Jarvis, B. C., Frankland, B., and Cherry, J. H. (1968b). Increased DNA template and RNA polymerase associated with the breaking of seed dormancy. *Plant Physiol.* **43**, 1734.

Johnson, L. V. P. (1935). General preliminary studies in the physiology of delayed germination in *Avena fatua*. *Can. J. Res.* **13**, 283.

Johri, M. M., and Varner, J. E. (1968). Enhancement of RNA synthesis in isolated pea nuclei by gibberellic acid. *Proc. Nat. Acad. Sci. U.S.* **59**, 269.

Jones, R. L. (1969). The fine structure of barley aleurone cells. *Planta* **85**, 359.

Kahn, A. A. (1966). Breaking of dormancy in *Xanthium* seeds by kinetin mediated by light and DNA-dependent RNA synthesis. *Physiol. Plant.* **19**, 869.

Kahn, A. A., Heit, C. E., and Lippold, P. C. (1968). Increase in nucleic acid synthesizing capacity during cold treatment of dormant pear embryos. *Biochem. Biophys. Res. Commun.* **33**, 391.

Kentzer, T. (1966). Gibberellin-like substances and growth inhibitors in relation to the dormancy and after-ripening of ash seed (*Fraxinus excelsior* L.). *Acta Soc. Bot. Pol.* **35**, 575.

Klein, S., and Ben-Shaul, Y. (1966). Changes in cell fine structure of lima bean axes during early germination. *Can. J. Bot.* **44**, 331.

Knowles, R. H., and Zalik, S. (1958). Dormancy in *Viburnum trilobum*. *Can. J. Bot.* **36**, 561.

Koller, D. (1962). Preconditioning of germination in lettuce at time of fruit ripening. *Amer. J. Bot.* **49**, 841.

Koller, D., Mayer, A. M., Poljakoff-Mayber, A., and Klein, S. (1962). Seed germination. *Annu. Rev. Plant Physiol.* **13**, 437.

Kramer, P. J., and Kozlowski, T. T. (1960). "Physiology of Trees." McGraw-Hill, New York.

LaCroix, L. J., and Jaswal, A. S. (1967). Metabolic changes in after-ripening seed of *Prunus cerasus*. *Plant Physiol.* **42**, 479.

Lammaerts, W. E. (1942). Embryo culture: An effective technique for shortening the breeding cycle of deciduous trees and increasing germination of hybrid seed. *Amer. J. Bot.* **29**, 166.

Lang, A. (1965). Effects of some internal and external conditions on seed germination. *In* "Handbuch der Pflanzenphysiologie" (W. Ruhland, ed.), Vol. 15, Part 2, p. 848. Springer-Verlag, Berlin and New York.

Lang, A. (1967). Plant growth regulation. *Science* **157**, 589.

Lasheen, A. M., and Blackhurst, H. T. (1956). Biochemical changes associated with dormancy and after-ripening of blackberry seed. *Proc. Amer. Soc. Hort. Sci.* **67**, 331.

Lebedeff, G. A. (1947). Studies on the inheritance of hard seeds in beans. *J. Agr. Res.* **74**, 205.

Luckwill, L. C. (1952). Growth-inhibiting and growth-promoting substances in relation to the dormancy and after-ripening of apple seeds. *J. Hort. Sci.* **27**, 53.

McDonough, W. T. (1965). Some effects of phosphon on germination induced by red radiation and gibberellic acid in seeds of *Verbascum thapsus*. *Plant Physiol.* **40**, 575.

Major, W., and Roberts, E. H. (1968). Dormancy in cereal seeds. I. The effects of oxygen and respiratory inhibitors. *J. Exp. Bot.* **19**, 77.

Mancinelli, A. L., Borthwick, H. A., and Hendricks, S. B. (1966). Phytochrome action in tomato-seed germination. *Bot. Gaz.* **127**, 1.

Marcus, A. (1969). Seed germination and the capacity for protein synthesis. *Symp. Soc. Exp. Biol.* **23**, 143.

Marcus, A., Feeley, J., and Volcani, T. (1966). Protein synthesis in imbibed seeds. III. Kinetics of amino acid incorporation ribosome activation, and polysome formation. *Plant Physiol.* **41**, 1167.

Mayer, A. M. (1959). Joint action of gibberellic acid and coumarin in germination. *Nature (London)* **184**, 826.

Mayer, A. M., and Poljakoff-Mayber, A. (1963). "The Germination of Seeds." Pergamon, Oxford.

Milborrow, B. V. (1966). The effects of synthetic *dl*-Dormin (Abscisin II) on the growth of the Oat Mesocotyl. *Planta* **70**, 155.

Nikolaeva, M. G. (1969). "Physiology of Deep Dormancy in Seeds." Nat. Sci. Found., Washington, D.C.

Nutile, G. E. (1945). Inducing dormancy in lettuce seeds with coumarin. *Plant Physiol.* **20**, 433

Nutile, G. E., and Woodstock, L. W. (1967). The influence of dormancy-inducing desiccation treatments on the respiration and germination of Sorghum. *Physiol. Plant.* **20**, 554.

Ohkuma, K., Lyon, J. L., Addicott, F. T., and Smith, O. E. (1963). Abscisin II, an abscission-accelerating substance from young cotton fruit. *Science* **142**, 1592.

Ohkuma, K., Addicott, F. T., Smith, O. E., and Thilsen, W. E. (1965). The structure of Abscisin II. *Tetrahedron Lett.* **29**, 2529.

Olney, H. O., and Pollock, B. M. (1960). Studies of rest period. II. Nitrogen and phosphorus changes in embryonic organs of after-ripening cherry seed. *Plant Physiol.* **35**, 970.

Opik, H. (1966). Changes in cell fine structure in the cotyledons of *Phaseolus vulgaris* L. during germination. *J. Exp. Bot.* **17**, 427.

Orphanos, P. I., and Heydecker, W. (1968). On the nature of the soaking injury of *Phaseolus vulgaris* seeds. *J. Exp. Bot.* **19**, 770.

Paleg, L. G. (1965). Physiological effects of the gibberellins. *Annu. Rev. Plant Physiol.* **16**, 291.

Pammenter, N., and Villiers, T. A. (1971). Unpublished data.

Parke, R. V. (1959). Growth periodicity and the shoot tip of *Abies concolor*. *Amer. J. Bot.* **46**, 110.

Paulson, R. E., and Srivastava, L. M. (1968). The fine structure of the embryo of *Lactuca sativa*. I. Dry embryo. *Can. J. Bot.* **46**, 1437.

Phillips, I. D. J., and Wareing, P. F. (1958a). Studies in dormancy of sycamore. I. Seasonal changes in the growth-substance content of the shoot. *J. Exp. Bot.* **9**, 350.

Phillips, I. D. J., and Wareing, P. F. (1958b). Effect of photoperiodic condition on the level of growth inhibitors in *Acer pseudoplatanus*. *Naturwissenschaffen.* **45**, 317.

Phillips, J. W. (1947). Studies on fermentation in rice and barley. *Amer. J. Bot.* **34**, 62.

Poljakoff-Mayber, A., Blumenthal-Goldschmidt, S., and Evenari, M. (1957). The growth substances content of germinating lettuce seeds. *Physiol. Plant.* **10**, 14.

Pollock, B. M. (1959). Temperature control of physiological dwarfing in peach seedlings. *Nature (London)* **183**, 1687.

Pollock, B. M., and Olney, H. O. (1959). Studies of the rest period. I. Growth, translocation and respiratory changes in the embryonic organs of after-ripening cherry seeds. *Plant Physiol.* **34**, 131.

Pollock, B. M., and Toole, V. K. (1961). After-ripening, rest period and dormancy. *Yearb. Agr. (U.S. Dep. Agr.)* p. 106.

Pope, M. N., and Brown, E. (1943). Induced vivipary in three varieties of barley possessing extreme dormancy. *J. Amer. Soc. Agron.* **35**, 161.

Rissland, I., and Mohr, H. (1967). Phytochrom-induzierte Enzymbildung (Phenylalanin-desaminase, ein schnell ablaufender Prozess). *Planta* **77**, 239.

Roberts, E. H. (1969). Seed dormancy and oxidation processes. *Symp. Soc. Exp. Biol.* **23**, 161.

Robinson, P. M., Wareing, P. F., and Thomas, T. H. (1963). Isolation of the inhibitor varying with photoperiod in *Acer pseudoplatanus*. *Nature (London)* **199**, 875.

Semeniuk, P., Stewart, R. N., and Uhring, J. (1963). Induced secondary dormancy of rose embryos. *Proc. Amer. Soc. Hort. Sci.* **83**, 825.

Shih, C. Y., and Rappaport, L. (1970). Regulation of bud rest in tubers of potato. VII. Effect of abscisic acid and gibberellic acid on nucleic acid synthesis in excised buds. *Plant Physiol.* **45**, 33.

Smith, H., and Kefford, N. P. (1964). The chemical regulation of the dormancy phases of bud development. *Amer. J. Bot.* **51**, 1002.

Sondheimer, E., Tzou, D. S., and Galson, E. C. (1968). Abscisic acid levels and seed dormancy. *Plant Physiol.* **43**, 1443.

Stearns, F., and Olson, J. (1958). Interaction of photoperiod and temperature affecting seed germination in *Tsuga canadensis*. *Amer. J. Bot.* **45**, 53.

Steinbauer, G. P. (1937). Dormancy and germination of *Fraxinus* seeds. *Plant Physiol.* **12**, 813.

Stokes, P. (1952). A physiological study of embryo development in *Heracleum sphondylium*. I. Effect of temperature on embryo development. *Ann. Bot. (London)* [N.S.] **16**, 441.

Stokes, P. (1965). Temperature and seed dormancy. "Handbuch der Pflanzenphysiologie" (W. Ruhland, ed.), Vol. 15, Part 2, p. 746. Springer-Verlag, Berlin and New York.

Sussman, A. S., and Halvorson, H. O. (1966). "Spores: Their Dormancy and Germination." Harper, New York.

Thornton, N. C. (1935). Factors influencing germination and development of dormancy in cocklebur seeds. *Contrib. Boyce Thompson Inst.* **7**, 477.

Tuan, D. T. H., and Bonner, J. (1964). Dormancy associated with repression of genetic activity. *Plant Physiol.* **39**, 768.

Vaartaja, O. (1956). Photoperiodic response in the germination of seeds of certain trees. *Can. J. Bot.* **34,** 377.

Varner, J. E., and Chandra, G. R. (1964). Hormonal control of enzyme synthesis in barley endosperm. *Proc. Nat. Acad. Sci. U.S.* **52,** 100.

Vegis, A. (1964). Dormancy in higher plants. *Annu. Rev. Plant Physiol.* **15,** 185.

Villiers, T. A. (1967). Cytolysomes in long-dormant plant embryo cells. *Nature (London)* **214,** 1356.

Villiers, T. A. (1968). An autoradiographic study of the effect of the plant hormone abscisic acid on nucleic acid and protein metabolism. *Planta* **82,** 342.

Villiers, T. A. (1971). Cytological studies in dormancy. I. Embryo maturation during dormancy in *Fraxinus excelsior.* New Phytol. **70,** 751.

Villiers, T. A. (1972). Cytological studies in dormancy. II. Pathological ageing changes occurring during prolonged dormancy. *New Phytol.* **71,** 145.

Villiers, T. A., and Wareing, P. F. (1964). Dormancy in fruits of *Fraxinus excelsior. J. Exp. Bot.* **15,** 359.

Villiers, T. A., and Wareing, P. F. (1965a). The growth substance content of dormant fruits of *Fraxinus excelsior. J. Exp. Bot.* **16,** 533.

Villiers, T. A., and Wareing, P. F. (1965b). The possible role of low temperature in breaking the dormancy of seeds of *Fraxinus excelsior. J. Exp. Bot.* **16,** 519.

Visser, T. (1954). After-ripening and germination of apple seeds in relation to the seed coats. *Proc., Kon. Ned. Akad. Wetensch., Ser. B* **57,** 175.

Wareing, P. F. (1954). Growth studies in woody species. VI. The locus of photoperiodic perception in relation to dormancy. *Physiol. Plant.* **7,** 261.

Wareing, P. F. (1956). Photoperiodism in woody plants. *Annu. Rev. Plant Physiol.* **7,** 191.

Wareing, P. F. (1965). Endogenous inhibitors in seed germination and dormancy. *In* "Handbuch der Pflanzenphysiologie" (W. Ruhland, ed.), Vol. 15, Part 2, p. 909. Springer-Verlag, Berlin and New York.

Wareing, P. F. (1969). Germination and dormancy. *In* "Physiology of Plant Growth and Development" (M. B. Wilkins, ed.), p. 605. McGraw-Hill, New York.

Wareing, P. F., and Foda, H. A. (1957). Growth inhibitors and dormancy in *Xanthium* seeds. *Physiol. Plant.* **10,** 266.

Went, F. W. (1949). Ecology of desert plants. II. The effect of rain and temperature on germination and growth. *Ecology* **30,** 1.

Went, F. W. (1957). "The Experimental Control of Plant Growth." Chronica Botanica, Waltham, Massachusetts.

Williams, C. M. (1969). Photoperiodism and the endocrine aspects of insect diapause. *Symp. Soc. Exp. Biol.* **23,** 285.

Williams, C. N. (1959). Action of inhibitor $\beta$ on the growth of *Striga* seedlings. *Nature (London)* **184,** 1577.

# 4

PHYSIOLOGICAL AND BIOCHEMICAL

DETERIORATION OF SEEDS

*Aref A. Abdul-Baki and James D. Anderson*

## I. Introduction

The realization that deterioration occurs in all forms of life and leads ultimately to death is not at all comforting. Through the ages, experience and information have accumulated which now allow man to slow up but not to stop these degenerative, aging processes. Because the rate of deterioration of organisms is influenced by a variety of agents or conditions, an understanding of the mechanisms which control that rate might provide ways of curbing deterioration and senescence, and could thereby extend the lives of organisms.

Gove (1965) defined deterioration as "the falling from a higher to a lower level in quality, character, or vitality. It implies the impairment of vigor or usefulness." In the following discussion, the term "seed deterioration" implies an irreversible degenerative change in the quality of a seed after it has reached its maximum quality level. The highest quality level for each species is the theoretical maximum attained under that complex of conditions evoking the most favorable interactions between the genetic make-up of a seed and the environment under which it is produced, harvested, processed, and stored. Many factors can prevent attainment of the theoretical maximum during seed development and maturation. Mineral deficiences or presence of toxic elements in the soil, infestation of a crop by diseases or insects, and crop damage by heavy rain or frost are only a few of the factors that can prevent seeds from reaching their theoretical maximum quality. Such factors as mechanical damage during harvesting, heat damage during curing, and poor storage conditions lower the quality attained and, therefore, are among the major inducers of seed deterioration. Some of these factors will be discussed briefly.

Helmer *et al.* (1962) suggested that seeds reach maximum quality at physiological maturity, and that from that time until planting only degenerative changes occur, the rate being dependent upon the degree of deviation from optimum conditions. Physiological maturity may be interpreted as a condition which must be attained before the optimum stage for harvesting the seed can begin. This condition normally coincides with the stage of maximum quality. Thus, in the process of producing high quality seed, cultural practices applied before the seeds reach full physiological maturity help to bring their quality close to the theoretical maximum, whereas proper harvesting, drying, and storing slow deterioration by maintaining the quality as close as possible to the highest, that is, the level that was attained at the beginning of physiological maturity.

Our knowledge of physiology and biochemistry of seed deterioration is still elementary (Barton, 1961). The bibliographies compiled by E. James (1961, 1963) attest to the fact that in the years covered, most investigations were concerned with establishing optimum conditions for producing, handling, storing, and determining the longevity of seeds under various conditions, with very little having been accomplished on mechanisms of deterioration. This review emphasizes advances made during the past decade on physiology and biochemistry of seed deterioration and presents evidence to support some interpretations that we offer.

## II.  Indices for Measuring Deterioration

Indices for measuring seed deterioration have been arbitrary and far from standardized. The difficulty stems from the lack of a stable reference for each plant species or variety against which a given seed lot can be evaluated at any desired time. A search for reliable indices revealed that, regardless of the causal agents, changes in some quantifiable traits occur when seeds deteriorate, and some of these changes have been used to estimate deterioration. Perhaps the most widely accepted and useful index of seed deterioration is reduction in viability, which is often accompanied by reduced seedling growth. Recently, increasing emphasis has been placed on biochemical or physiological changes such as quantitative and qualitative changes in specific enzymes, respiratory metabolism, synthesis of proteins and carbohydrates, leaching of inorganic and organic material, and degradation of storage compounds. The role of each of these important changes in reducing seed quality is discussed separately.

## III.  Physiological Manifestations of Seed Deterioration

As seeds age, they maintain germinability for some time. Subsequently they enter a period of rapid decline during which some seeds completely fail to germinate while others germinate and grow normally. Kearns and Toole (1939) related differences in viability among seeds of the same age to heterogeneity of individual seeds within a seed lot. This interpretation implies that significant differences in quality level exist among individual seeds within a lot. As the seed lot as a whole deteriorates, those seeds which initially had the lowest quality level lose viability first.

The time which marks the first detectable decline in germinability does not coincide with the actual beginning of deterioration. Changes in the major biochemical processes that are associated with deterioration occur in seeds before germinability declines. This is illustrated by the great decline in synthesis of carbohydrates and proteins by seeds while their germinability still remains unchanged (Abdul-Baki, 1969b).

Among the many physiological manifestations of seed deterioration are changes in seed color, delayed germination, decreased tolerance to suboptimal environmental conditions during germination, lowered tolerance to adverse storage conditions, higher sensitivity to radiation treatments, reduced growth of seedlings, reduced germinability, and increased number of abnormal seedlings. Of these responses, reduced germinability has been the most widely accepted single criterion of seed deterioration. All these indications may or may not appear at one time in a single sample of seed, and some are peculiar to seeds of only certain species of plants (Toole *et al.,* 1948).

Delay in the full expression of germination is usually the earliest detectable physiological sign of quality loss (Kearns and Toole, 1939; Schwemmle, 1940; Toole *et al.,* 1948; Parkinson, 1948; Filutowics and Bejnar, 1954). Toole *et al.* (1948) and Parkinson (1948) used delayed expression of full germinability to measure vigor because it provided an easily used tool, particularly when employed on seed lots of known history. Very low seed moisture (Whymper and Bradley, 1934; Nutile, 1964), dormancy, or low permeability of seed coats (Hyde, 1954) also can produce a similar effect. While certain chemical or mechanical treatments promote germination of dormant seeds and of seeds with hard or impermeable seed coats, no physical or biochemical treatment that we know of can bring deteriorated seeds back to their original quality level.

Associated with reduced germination is the change in one or more environmental requirement (Fig. 4.1). Larsen (1965) stored samples of vigorous (95% germination) alfalfa seeds at temperatures and moisture levels for durations which produced three lots with 95%, 85%, and 75% germination. He observed that the emergence curve for the first lot (most vigorous) had a broader temperature base as a result of the ability of these seeds to germinate at lower temperatures than seeds from the other two lots. Furthermore, whereas all three lots had equal germinability at temperatures above optima, they differed significantly at temperatures below optima. In addition, the temperature which produced maximum germination in the most vigorous seed lot shifted from 25°C to 26°C and 27°C for the less and least vigorous lot, respectively. Freshly harvested seeds which have high-temperature dormancy (Kearns and Toole, 1939; Hutchinson *et al.,* 1948; George, 1967) exhibit higher percent germination at low than at high temperatures. After overcoming such dormancy, the seeds are capable of germinating over a wide range of temperatures. The failure of these dormant seeds to germinate at high temperatures should not be interpreted as deterioration.

Deteriorated seeds, if they germinate at all often produce seedlings which grow slowly (Kearns and Toole, 1939; Parkinson, 1948; Toole and Toole, 1953; Toole *et al.,* 1957). However, reductions in seedling growth which either precede or accompany loss of germinability do not necessarily occur in every case of seed deterioration (J. D. Anderson, 1970a). This observation suggests that germinability and seedling growth, though closely related, are regulated by two mechanisms which seem to operate independently during deterioration.

Generally, when the percent germination of a seed lot declines many of the seedlings obtained are abnormal (De Vries, 1901; Nawashin, 1933a,b;

Toole *et al.,* 1948) and are not capable of surviving to maturity (D'Amato, 1954). Surviving seedlings may show poorly developed roots and/or shoots (Toole *et al.,* 1948), have necrotic root meristems (Nawashin, 1933b; Toole *et al.,* 1948; D'Amato, 1954), or they may mature into

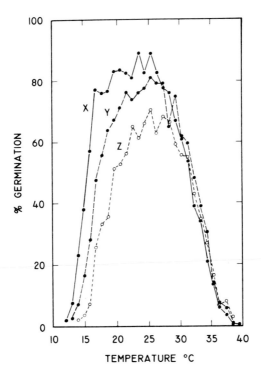

FIG. 4.1. Percent germination of high (X), medium (Y), and low (Z) germinating alfalfa seed as a function of temperature. Germination counts were made 6 days after planting. Each point represents percent germination of 180 seeds. From Larsen (1965).

plants having reduced pollen fertility (Avery and Blakeslee, 1936; Schwarnikow, 1937; Abdalla and Roberts, 1968). Seeds of carrot, celery, and pepper are exceptions and do not show large increases in production of abnormal seedlings during deterioration (Toole *et al.,* 1948).

Mechanical injuries during harvesting and processing cause many seed abnormalities. Large-seeded legumes are among the most susceptible species. Borthwick (1932) listed several types of abnormalities in seeds of machine-harvested lima beans. These included damaged seed coats, broken hypocotyls, and cotyledons that had become disconnected from the embryonic axis. Seed desiccation (Nutile, 1964), and rapid changes in temperature during storage of very dry seeds (Toole and Toole, 1946) also led to the development of a higher percentage of abnormal seedlings from these seeds.

Changes in seed color occur during deterioration. Grain seeds that are damaged by heat (Zeleney, 1954) or by fungi during storage (Christensen and Kaufmann, 1969) lose their natural luster. In lima beans (Pollock and Toole, 1966; Wester, 1970), snap beans (Toole et al., 1948), alfalfa, and clover (Whitcomb, 1923, 1942; West and Harris, 1963), seeds which changed color had low germinability both in the laboratory and field. West and Harris (1963) evaluated some of the physiological and biochemical differences between normal and off-color alfalfa seeds. They found that off-color seeds had lower respiration rates, lower rates of root growth, and higher shoot-root ratios than normal seeds. In addition, aqueous extracts from off-color seeds were more acidic than those from normal seeds and they contained certain compounds which absorbed light in the ultraviolet range. These extracts from off-color seeds also slowed the germination rate of normal seeds of the same species.

Other physiological manifestations of seed deterioration include low tolerance to adverse storage conditions (Burns, 1957; Helmer et al., 1962; Clark et al., 1968; Delouche et al., 1968; J. D. Anderson, 1970a), increased susceptibility to fungal invasion (Dickson and Holbert, 1926; Isely, 1957) and high frequency of chromosomal damage in response to gamma or x-radiation (Gustafsson, 1947; Nilan and Gunthardt, 1956; Sax and Sax, 1962).

## IV.  Biochemical Manifestations of Seed Deterioration

Many biochemical changes have been detected in seeds as they deteriorate. These include changes in enzymic activity, respiratory and synthetic pathways, membranes, storage compounds, and chromosomes. Yet, the primary causes of seed deterioration, the sites and the nature and sequence of biochemical reactions involved remain unknown. Efforts have been made to correlate biochemical changes with more visible manifestations of deterioration, such as reductions in germinability, seedling growth, or yield. Comparisons have also been made between the rates at which biochemical changes proceed, and the earliest detectable physiological changes. The relationship between biochemical and physiological changes, which is cause and which is result, might be resolved in time-course studies in which both physiological and biochemical changes would be measured periodically.

Our knowledge of biochemical deterioration of seeds is still limited and therefore does not permit making firm conclusions. Therefore, in discussing some of the major biochemical changes that are observed in seeds as they deteriorate, allowance will be made for the possibility that most of these changes are results rather than causes.

## A. *Changes in Respiration*

Of the biochemical changes which take place in deteriorating seeds, none has been investigated as extensively as the alteration in respiration. Respiratory metabolism results in oxidation of food reserves, production of a large number of intermediates which serve as building blocks for synthesis of protoplasmic components (proteins, nucleic acids, and lipids), and production of biochemical energy in the form of adenosine triphosphate for these reactions.

Early work suggested that a certain specific amount of heat is produced per unit weight of healthy seed as a product of respiratory metabolism (Darsie *et al.,* 1914). Excess heat indicated infection with microorganisms, and subnormal heat suggested low vigor. Subsequent, studies on respiration have been carried out by using standard manometric procedures for measuring oxygen ($O_2$) uptake, carbon dioxide ($CO_2$) evolution, or both. Seeds of many species have been used to study changes in respiration induced by natural aging (Woodstock and Grabe, 1967; Kittock and Law, 1968; Abdul-Baki, 1969a; J. D. Anderson, 1970a), accelerated aging (Woodstock and Feeley, 1965; Abdul-Baki, 1969a), chilling injury (Throneberry and Smith, 1954; Woodstock and Pollock, 1965; Woodstock and Feeley, 1965), and irradiation (Woodstock and Combs, 1965; Woodstock and Justice, 1967). In most cases, germination and growth tests were made on seeds from the same lot in an attempt to correlate respiratory to physiological changes. In general, correlations between $O_2$ uptake by germinating seeds and their germinability and seedling growth were positive and significant. An exception to this conclusion is the stimulation of $O_2$ uptake by chilled cacao seeds (Ibanez, 1964; Woodstock *et al.,* 1967). Mechanically injured seeds would probably exhibit higher respiratory rates than sound seeds.

High respiratory quotient (RQ) values are often observed in deteriorated seeds (Woodstock and Grabe, 1967; J. D. Anderson, 1970a). Such RQ values (1.5 or higher) are due either to increases in $CO_2$ evolution, reductions in $O_2$ uptake, or both. In certain cases increases in $CO_2$ production occur before germinability declines (J. D. Anderson, 1970a, 1970b), whereas, reductions in $O_2$ uptake appear primarily during advanced deterioration (Throneberry and Smith, 1955; Woodstock and Grabe, 1967). In these situations where $CO_2$ production increased, the source of the extra $CO_2$ has not been identified. An increase in activity of one or more decarboxylases could bring about this change. However, investigations of changes in activity of glutamic acid decarboxylase (the only decarboxylase investigated in seeds with reference to deterioration) indicate that activity of this enzyme is reduced in deteriorated seeds of many species (Linko and Sogn, 1960; Linko, 1961; Bautista and Linko, 1962; Bautista *et al.,* 1964; Grabe, 1964).

Our work on barley and wheat indicates that increased $CO_2$ production by deteriorated seeds during early germination does not seem to result from glucose oxidation (Abdul-Baki, 1969a,b; J. D. Anderson, 1970b). Evidence supporting this conclusion comes from experiments in which seeds of different quality levels are incubated in a medium containing glucose-$^{14}C$. Seeds which underwent some deterioration "took in" the same amount of glucose but utilized it at a much lower rate, and ultimately produced less $^{14}CO_2$ than nondeteriorated seeds. Yet, the former produced more total $CO_2$ which was not radioactive. On the basis of the significant differences in specific radioactivity of $^{14}CO_2$ released by the two types of seeds we conclude that most of the $CO_2$ produced by deteriored seeds is not due to glucose oxidation.

Lack of $O_2$ diffusion into seeds which are germinated in air affects $O_2$ utilization and therefore leads to higher RQ values (W. O. James and James, 1940; Merry and Goddard, 1941). Seeds germinated in 100% $O_2$ consume more $O_2$ than those germinated in air (Woodstock and Grabe, 1967; Woodstock *et al.,* 1967). However, $O_2$ limitation appears in both vigorous and deteriorated seeds as indicated by stimulation of $O_2$ utilization by both types of seeds when imbibed under high $O_2$ tension. It should be noted that this difference in $O_2$ consumption continues to exist even when the seeds are germinated in 100% $O_2$. Throneberry and Smith (1955) suggested that such differences in $O_2$ uptake might be due, in part, to less active cytochrome oxidase.

The question as to whether reduction in respiration of deteriorated seeds is limited by the activity of one or more enzymes in the glucose utilizing pathways (glycolytic, pentose phosphate and tricarboxylic acid pathways), by the low demand for energy production, or by both, remains unresolved. It is also highly important to determine whether deteriorated and sound seeds have similar efficiencies in producing chemical energy from equal amounts of oxidizable substrate.

## B. *Changes in Enzymes*

The isolation and crystallization of the first enzyme (urease), from dry Jack bean seed rather than from any other plant or animal tissue shows an interest of early chemists and biologists in seed enzymes (Sumner, 1926). Numerous attempts to correlate loss of seed viability with decreased activity of certain enzymes were made several decades before urease was crystallized. Table 4.I summarizes some of the enzymes which were in-

**TABLE 4.I.**
ENZYMES INVESTIGATED WITH RESPECT TO SEED VIABILITY

| Enzyme | Seed | Reference |
|---|---|---|
| Alcohol dehydrogenase | Corn | Throneberry and Smith (1954, 1955) |
| Amylase | Rice | Sreenivasan (1939); Rao *et al.* (1954) |
| | Wheat | Fleming *et al.* (1960) |
| | Barley | J. D. Anderson (1970a) |
| Catalase | Timothy | Esbo (1959); Nemeč and Duchoň (1923) |
| | Johnson grass, Sudan grass, and *Amaranthus* | Crocker and Harrington (1918) |
| | Wheat | Marotta and Kaminka (1924) |
| Cellulase | Timothy | Esbo (1959) |
| Cytochrome oxidase | Corn | Throneberry and Smith (1954, 1955) |
| Diastase | Timothy | Esbo (1959) |
| Glutamic acid decarboxylase | Wheat | Linko and Sogn (1960); Linko (1961) |
| | Corn | Grabe (1964); Woodstock and Grabe (1967) |
| | Beans | E. James (1968) |
| Malic dehydrogenase | Corn | Throneberry and Smith (1954, 1955) |
| Peroxidase | Corn, hemp, tomato, oat, cow pea, lettuce, soybean, and castor bean | McHargue (1920) |
| Phenolase | Oats, wheat, cucumber, and barley | Davis (1931) |

vestigated. Oxidases (catalase, peroxidase, and phenolase) were the first enzymes investigated with the objective of establishing a correlation between their activities and seed viability. These oxidases were among the few enzymes which were known and for which simple assay procedures were established. More recently, changes in activity of amylase, cytochrome oxidase, glutamic acid decarboxylase, and dehydrogenases have been investigated in deteriorating seeds (Throneberry and Smith, 1955; Fleming *et al.,* 1960; Grabe, 1964). These enzymes play important roles in mobilizing and utilizing food reserves during germination.

The results of these limited studies suggest a relationship between activities of certain enzymes and seed viability. McHargue (1920) proposed that seeds of nine species of plants could be separated into high, medium, and low viability classes on the basis of peroxidase activity.

Throneberry and Smith (1955) also reported positive correlations between germinability of corn seeds and activity of cytochrome oxidase, malic dehydrogenase, and alcohol dehydrogenase. Grabe (1964) noted a positive correlation between activity of glutamic acid decarboxylase and vigor of corn seedlings. In contrast to these findings are reports of high activity of peroxidase in very old wheat seeds that failed to germinate (Brocq-Rosseu and Gain, 1908), high activity of glutamic acid decarboxylase in nonviable bean seeds (E. James, 1968), and high activity of dehydrogenases in heat-damaged barley seeds (MacLeod, 1952).

In view of such conflicting findings on the relation of enzymic activity to seed deterioration, we tend to favor the conclusions of Palladin (1926) and Hibbard and Miller (1928), namely that enzymes are not necessarily to be taken as indices of life in an organism.

## C. Changes in Food Reserves

The view that loss of seed viability resulted from depletion of food reserves was questioned because when seeds lose their capacity to germinate they still have a large amount of stored food in their tissues (Barton, 1961). Harrington (1967) suggested that depletion of available oxidizable material in meristematic cells (presumably of the embryonic axis) might cause deterioration, even when adjacent tissues such as cotyledons and endosperms still contained abundant food. In this case, the lack of mobilization of food in dry seeds would lead to starvation of these cells.

Significant changes in major food reserves take place as seeds deteriorate. For literature on these changes as they affect food quality, we refer the reader to the excellent reviews of J. A. Anderson and Alcock (1954), and Christensen and Kaufmann (1969). This chapter will discuss only those changes which affect seed viability.

One of the changes associated with deterioration of seeds in general, and of oily seeds in particular, is increase in their acidity (Zeleney and Coleman, 1938, 1939; Milner and Geddes, 1946; Milner et al., 1947; Sorger-Domenigg et al., 1955; and others). Zeleney and Coleman showed that these acids consist of: (a) free fatty acids produced by action of lipases on lipids; (b) acid phosphates resulting from hydrolysis of phytin by phytase; and (c) amino acids produced by hydrolysis of proteins by proteases. Of all three groups, the largest and earliest increases occur in the free fatty acids. Milner et al. (1947) demonstrated that this specific type of deterioration was caused by microorganisms. They showed that increased fat acidity and loss of germination accompanied growth of molds on wheat. They stated that when wheat seeds were stored in a nitrogen atmosphere which inhibited fungal growth, neither fat acidity nor germinability changed. The role of storage fungi in producing fatty acids has been well documented (Christensen and Kaufmann, 1969).

Of significance is the 50-fold increase in the level of lactic acid in deteriorated alfalfa seeds (Wyttenbach, 1955). This compound which is an intermediary product of the glycolytic pathway of carbohydrate oxidation is a potent mutagen (Woll, 1953; D'Amato and Hoffmann-Ostenhof, 1956). By comparison, citric acid content of deteriorated cereal seeds was very low (Täufel and Pohlaudek-Fabini, 1955).

Along with the increase in fat acidity there is a decrease in total, neutral, and polar lipids as seeds deteriorate (Pomeranz, 1966; Koostra and Harrington, 1969). But of all the lipoidal changes reported, the phospholipids decreased the fastest (Table 4.II), and this decrease seems to occur only

TABLE 4.II.

CHANGES IN PERCENT EMERGENCE OF RADICLES, TOTAL LIPIDS, AND POLAR LIPIDS OF NATURALLY AGED AND ARTIFICIALLY AGED CUCUMBER SEEDS[a]

| Naturally aged | | | | Artificially aged | | | |
|---|---|---|---|---|---|---|---|
| | | LIPID (% OF DRY WT.) | | | | LIPID (% OF DRY WT.) | |
| Age (years) | Radicle Emergence (%) | Total | Polar | Age (weeks) | Radicle Emergence (%) | Total | Polar |
| 1 | 99 | 28.6 | 1.4 | 0 | 99 | 23.3 | 1.3 |
| 2 | 100 | 26.2 | 1.1 | 1 | 83 | 28.6 | 1.4 |
| 3 | 98 | 31.1 | 1.1 | 2 | 81 | 27.8 | 1.1 |
| 6 | 96 | 24.7 | 1.0 | 3 | 49 | 26.1 | 0.6 |
| 11 | 70 | 22.4 | 1.2 | 4 | 2 | 13.4 | 0.5 |

[a]From Koostra and Harrington (1969).

when deterioration is brought about by high seed temperature and moisture. Should these polar lipids be predominantly part of cellular membranes, as has been suggested (Koostra and Harrington, 1969), then it is reasonable to conclude that rapid aging induced more damage to membranes than slower aging since no change occurred in these compounds when viability decreased slowly. However, polar lipids are contituents of other protoplasmic components in addition to membranes and one of their many functions is to serve as binding agents between polar and nonpolar cellular components (Robinson, 1964). Since these data on changes in phospholipids only suggest membrane deterioration, they should be carefully interpreted. Changes in lipids will be discussed further (see Section IV,D).

High concentrations of inorganic phosphates were found in nongerminable wheat which deteriorated under anaerobic conditions (Lynch *et al.*, 1962) and in crimson clover (Ching and Schoolcraft, 1968). However, germinable and nongerminable seeds of perennial ryegrass and wheat had similar concentrations of inorganic phosphates. These changes and the origin of the inorganic phosphates have been attributed to increases in activity of phosphatases and phytase (Ching and Schoolcraft, 1968). However, no direct evidence was presented and changes in activity of these enzymes in seeds were not determined.

Jones *et al.* (1942) investigated changes in proteins of wheat seeds stored for 24 months under conditions which induced different levels of deterioration. They observed three major changes in proteins of deteriorated seeds: (a) decreased water solubility; (b) partial breakdown of protein molecules as reflected by the smaller amount of proteins precipitable in trichloroacetic acid; and (c) decreased digestibility of proteins by the proteolytic enzymes pepsin and trypsin. They also noted a slight increase in amino acids, presumably resulting from protein degradation. Likewise, Ching and Schoolcraft (1968) reported reductions in seed proteins of crimson clover and perennial ryegrass with concomitant increases in amino acids of deteriorated seeds. In both investigations, the rates of protein loss depended on the severity of storage conditions. The data of Ching and Schoolcraft indicate that significant reductions in protein and increases in amino acids occurred only after seeds had lost viability. Jones *et al.* (1942) did not furnish data on germinability of the wheat seeds they used.

Starch, the primary storage material in most seeds, underwent little or no reduction in deteriorated seeds of crimson clover, perennial ryegrass (Ching and Schoolcraft, 1968), and rice (Kondo and Okamura, 1934). In contrast, amounts of total soluble sugars of deteriorated wheat (Lynch *et al.*, 1962), crimson clover, and perennial ryegrass (Ching and Schoolcraft, 1968) were lower in nondeteriorated seeds. Reducing sugars of wheat which deteriorated in air did not change; however, under anaerobic conditions they increased. Nonreducing sugars decreased under both aerobic and anaerobic conditions.

The relationship between the reduction in polymeric storage compounds (lipids, proteins, and carbohydrates) and the increase in their subunits (fatty acids, amino acids, and sugars) is not quantitative. In general, hydrolysis of these compounds takes place under unfavorable storage conditions of high moisture and temperature. Such conditions are conducive to rapid fungal growth and high metabolic activity by seed. The products of hydrolysis of lipids, proteins, and carbohydrates serve as substrates for growth of fungi and for increased metabolic processes in seed, and their utilization explains the lack of stoichiometry between the polymers lost and the hydrolyzed products formed.

## D. Changes in Membranes and Barriers

It has been known for over a century that electrical conductivity of a solution in which plant tissues are immersed increases as the tissues die (Weber, 1836; Ranke, 1865). Such increase in conductivity is caused by increased membrane permeability of the dying tissue. For example, as unicellular organisms such as yeast, bacteria (Brooks, 1923), and *Nitella* (Osterhout, 1918) progress towards death, their permeability increases and allows more leaching of inorganic and organic salts into the surrounding water, thus increasing its electrical conductivity.

Using conductivity measurements, Hibbard and Miller (1928) showed that in most cases the germinability of corn, wheat, peas, and timothy seeds was related to resistance of the solution in which the seeds were immersed. They concluded that the decrease in resistance of the imbibing solution was caused by increased permeability of membranes which allowed more leaching of salts from deteriorated seeds. Later studies on seeds that were naturally aged, sun bleached, and mechanically injured confirmed the observation by Hibbard and Miller, in that more materials leach out of deteriorated and injured seeds than from vigorous and sound seeds (Pollock and Toole, 1966; Matthews and Bradnock, 1967, 1968; Ching and Schoolcraft, 1968; Takayanagi and Murakami, 1968; Abdul-Baki and Anderson, 1970; Bradnock and Matthews, 1970; and others). Figure 4.2, chosen from the work of Matthews and Bradnock (1968), illustrates the general relationship between leaching of

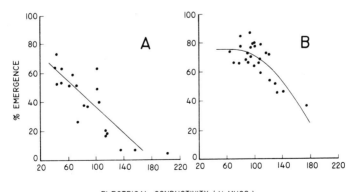

ELECTRICAL CONDUCTIVITY ( μ MHOS )

FIG. 4.2. Relationship between field emergence and electrical conductivity of seed-steep water for peas (A), and French beans (B). From Matthews and Bradnock (1968).

salts from pea and French bean seeds and their field emergence. Likewise, Fig. 4.3 demonstrates the leaching of sugars from barley seeds with

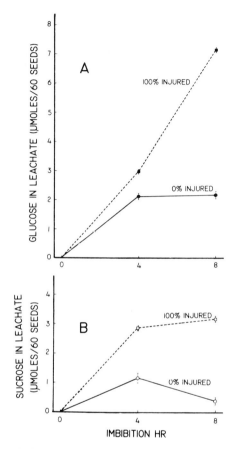

FIG. 4.3. Leaching of glucose (A) and sucrose (B) from sound and mechanically injured barley seeds during germination. Percent germination for sound and injured seeds are 85% and 84%, respectively. From Abdul-Baki and Anderson (1970).

mechanically injured endosperms. Like Hibbard and Miller (1928), some investigators linked the greater leaching to increased membrane permeability (Pollock and Toole, 1966; Ching and Schoolcraft, 1968).

On the basis of results we obtained by following leaching of glucose from barley seeds which had reached different levels of deterioration, we questioned that increased leaching of material from seeds was the result of changes in membrane permeability (Abdul-Baki and Anderson, 1970). Results of these time-course experiments (Fig. 4.4) suggest that the initial rates of glucose accumulation in the imbibing media of seeds with high,

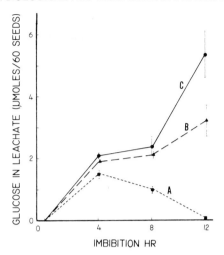

FIG. 4.4. Leaching of glucose from barley seeds of high (A), medium (B), and low (C) viability during germination. Germination percentages are 91%, 84%, 0%, respectively. From Abdul-Baki and Anderson (1970).

medium, and low germinability are about the same. Later, glucose concentration decreases in the imbibing media of seeds of the first type while it continues to increase in the media of seeds of the two other types. We interpreted this pattern of glucose accumulation in the immersion liquid as passive diffusion and concluded that the amount of glucose leached depended on glucose concentration in the seed and the rate at which it was utilized in metabolic processes. The faster the rate of glucose utilization, the lower was the concentration of glucose in the imbibing medium. This interpretation is further supported by the findings that when seeds of high, medium, and zero germinability are placed in a solution of radioactive glucose for a short time (1–2 hours), all three types of seeds take up similar amounts of radioactive glucose (Abdul-Baki, 1969a,b). Possibly the same holds true for other small-molecular-weight compounds. Other observations were reported on seed lots which leached equal amounts of salts and sugars, yet showed great differences in germinability or field emergence (Matthews and Bradnock, 1967; Ching and Schoolcraft, 1968). Furthermore, more sugar was leached from barley seeds with mechanically injured endosperms, yet this injury did not reduce their germinability (Fig. 4.3). Such injury, which is very common in machine-harvested seeds (Bradbury et al., 1956; Toole and Toole, 1960), should be viewed as physical breaking of tissues and cells rather than membrane deterioration. Leaching of sugars from seed due to mechanical injury is very significant, and should not be confused with leaching caused by seed deterioration which results from aging.

All leachates of seeds such as inorganic salts, amino acids, nucleotides, and simple sugars, are small molecules, and their movement across the membrane from a highly concentrated medium to one of very low concentration, rather than against a concentration gradient, does not necessarily imply that membrane permeability has changed. In fact, the most viable seeds allow for leaching of sugars and salts into the immersion medium. The results would be interpreted differently had the imbibing medium contained large molecules such as polypeptides, polysaccharides, or polynucleotides whose movement across intact membranes is normally restricted. A notable exception is the active secretion of amylase, ribonuclease, and protease by aleurone cells of germinating barley seeds into the starchy endosperm (Chrispeels and Varner, 1967; Jacobsen and Varner, 1967).

Before attributing leaching of salts and sugars to changes in membranes and correlating these with seed deterioration, it becomes necessary to determine the part of the seed from which these leachates are lost and the particular membrane which undergoes change. Should the relationship between seed deterioration and leaching of solutes be a direct one, then leaching of material from the embryo alone, rather than from the whole seed, would be the phenomenon of primary importance. Comparative studies on leachates from isolated endosperms and embryos of vigorous barley seeds (Table 4.III) indicated that most sugars were

**TABLE 4.III.**

GLUCOSE AND SUCROSE CONTENTS OF LEACHATES AND EXTRACTS
OF ISOLATED BARLEY ENDOSPERMS AND EMBRYOS IMBIBED AT 25°C[a]

| Seed part | Imbibition (hr) | Leached by 60 seeds ($\mu$moles) | | Extracted from 60 seeds ($\mu$moles) | |
| --- | --- | --- | --- | --- | --- |
| | | Glucose | Sucrose | Glucose | Sucrose |
| Endosperm | 4 | 1.95 | 1.07 | 4.75 | 38.34 |
| | 8 | 5.40 | 0.29 | 5.86 | 42.34 |
| | 12 | 7.36 | 0 | 6.24 | 31.98 |
| Embryo | 4 | 0.84 | 2.58 | 0.62 | 9.78 |
| | 8 | 1.88 | 2.47 | 0.89 | 6.40 |
| | 12 | 2.53 | 1.12 | 1.45 | 2.84 |

[a]From Abdul-Baki and Anderson (1970).

leached from the endosperm which contained the highest concentrations of seed sugars.

Little is known about the changes which seed membranes undergo as seeds deteriorate, and the part of these changes which is associated with deterioration. Notable exceptions are studies on mitochondria isolated from 4-day-old etiolated seedlings of new and aged soybean seeds (Abu-Shakra and Ching, 1967). In these studies, mitochondria from seedlings of aged seeds had lower oxidative phosphorylation efficiencies and higher $O_2$ uptake per milligram of mitochondrial nitrogen than those from new seeds. From these observations the authors concluded that mitochondria from aged seeds seemed to be endogenously uncoupled. However, electron micrographs showed that membranes of mitochondria which were isolated from seedlings of new and old soybean seeds appeared to be similar. These investigators did not indicate whether their preparations contained mainly mitochondria which were present in the embryonic axes of quiescent seeds before germination or whether they represented newly formed mitochondria. It is more likely that the preexisting mitochondria would be more uncoupled than the recently formed ones. This speculation is in accord with results of studies on changes in activity of mitochondria isolated from embryos of developing and mature barley seeds (Abdul-Baki and Baker, 1969) which suggest that significant changes occur in these organelles during seed maturation. Mitochondria from embryos of mature barley seeds show less respiratory activity and ultrastructural organization than those isolated from embryos of immature seeds. The latter had more cristae and denser matrices. Nonetheless, preexisting mitochondria are functional enough to power synthesis of new mitochondria as well as many energy-requiring reactions during germination.

The finding of Koostra and Harrington (1969) that polar lipids, which are essential components of cellular membranes, do not undergo quantitative changes in naturally aged cucumber seeds after viability decreased (Table 4.II), is in accord with observations derived from electron micrographs. These micrographs show that mitochondrial membranes of 4-day-old embryonic axes from new and aged soybean seeds were equally intact. On the other hand reduction in polar lipids of artificially aged cucumber seeds suggests that the biochemical changes brought about by natural aging are indeed different from those caused by artificial aging and that the processes which bring about these changes are not understood. Degradation of seed lipids by storage fungi and concomitant increases in the free fatty acids of seeds infected with these organisms (see Section IV,C) are definitely associated with seed deterioration.

Bain and Mercer (1966) reported that during the phase of rapid synthesis of storage reserves (18 to 28 days from fertilization) in peas, the increase in the size of starch grains in plastids of cotyledons was more rapid than was plastid elongation. This caused disruption of plastid structure. Increase in size of starch grains during the maturation phase (28 to 45 days from fertilization) further disrupted plastids and ultimately led to disappearance of stromas and lamellae. The endoplasmic reticulum and the Golgi bodies also disintegrated at this stage. Likewise, Abdul-Baki and Baker (1970) observed rapid changes in organelles of endosperm cells during maturation of barley seed while organelles in cells of the embryo exhibited much less structural disorganization. These seeds germinated perfectly, suggesting that the described changes in membranes of barley endosperm and pea cotyledons which occurred during seed maturation were not associated with deterioration. On the other hand, many varieties of lima bean which produce seeds with green cotyledons become bleached when exposed to sunlight during seed maturation and dryness. Bleached seeds which lose chlorophyll from their cotyledons are less vigorous than nonbleached seeds (Wester, 1965, 1970; Pollock and Toole, 1966). Whether bleaching and loss of vigor are associated with disruption of membranes has not been established.

Certain seed tissues such as seed coats, lemmas, paleas, and pericarps are several cells thick, and changes in these tissues should not be interpreted as alterations in the classical lipoprotein membranes which envelope cellular organelles. These tissues also play a significant role in protecting seeds from mechanical damage and attack by microorganisms and insects. They also influence gas exchange and water movement into and out of seeds.

### E. Changes in Synthetic Rates

Physiological expressions of deterioration, such as reduced germinability and seedling growth, suggest low rates of synthetic reactions in these seeds. Detection of deterioration through reduction in synthetic rates of macromolecules is an attractive approach for two reasons. First, the reactions involved are energy-requiring and the products (proteins, nucleic acids, and structural polysaccharides) are polymers of biological activity and function. These reactions are therefore part of growth processes. Second, the degree of deterioration in seeds can be determined at an early stage of germination even before shoots and roots emerge, because synthesis of these compounds begins soon after germination is initiated by hydrating the seed (Marcus and Feeley, 1964; Marcus et al., 1966; Jachymczyk and Cherry, 1968; Marcus, 1969; Abdul-Baki, 1969a,b).

Exposure of seeds to conditions which favor deterioration reduces the synthetic rates of many seeds. French (1959) reported that exposure of dry barley seeds to 85°C for only 30 minutes significantly reduced the ability of the embryos to synthesize starch during germination. In these studies endogenous sugars were converted into newly synthesized starch.

Synthesis of starch, cellulose, hemicelluloses, water-soluble polysaccharides, and proteins by wheat and barley seeds during early germination was followed after these seeds had reached different levels of deterioration as a result of natural or rapid aging (Abdul-Baki, 1969a,b). This was accomplished by imbibing seeds for 1 to 2 hours in $^{14}C$-labeled glucose, collecting the $^{14}CO_2$ during incubation, and extracting proteins and the various polysaccharides. Significant reductions were found in synthesis of proteins and polysaccharides in deteriorated seeds. The relationships between reduced germinability and rates of synthesis of proteins and carbohydrates are illustrated in Tables 4.IV and 4.V. Wood-

**TABLE 4.IV.**

Utilization of Glucose-$^{14}C$ into Various Components of Germinating Barley Seeds from Different Crop Years[a]

| Extracted component | Harvested crop results by year[b] | | | |
|---|---|---|---|---|
| | 1966 | 1963 | 1961 | 1958 |
| $^{14}CO_2$ (in $BaCO_3$) | 7710 | 3680 | 730 | 100 |
| Proteins (acid precipitable) | 860 | 140 | 60 | 50 |
| Water-sol, EtOH-insol polysaccharides | 1450 | 650 | 350 | 220 |
| Hemicelluloses and starch | 6590 | 1270 | 130 | 50 |
| Cellulose | 390 | 50 | 10 | 0 |
| Deproteinized EtOH-soluble | 149,980 | 163,540 | 251,580 | 219,550 |
| Total uptake | 166,640 | 169,290 | 252,850 | 219,970 |
| Total utilization | 16,660 | 5750 | 1270 | 420 |
| % Utilization | 10.18 | 3.43 | 0.51 | 0.19 |
| % Germination | 91.0 | 84.0 | 64.0 | 0 |

[a]From Abdul-Baki (1969b).
[b]Results in dpm/20 seeds; measurements were made in 1968.

stock (1969) reported a similar decline in rate of incorporation of leucine-$^{14}C$ into trichloroacetic acid–insoluble material (presumably proteins) by naturally aged barley seeds which had reached different levels of vigor.

Of significance is the observation that reductions in synthesis of carbohydrates and proteins occur before any reductions in germinability, seedling growth, or seed respiration are evident (Figs. 4.5, 4.6). In many cases, reductions of 50% or more in rates of carbohydrate and protein synthesis occur before germinability or seed respiration begin to decline. Whether synthesis of these polymers is restricted by reduction in the activity of one or more enzymes in the biosynthetic pathways, or whether it is due to disorganization of the Golgi apparatus and the slow formation of functional polysomes is yet to be demonstrated.

TABLE 4.V.

**TABLE 4.V.**
EFFECT OF ACCELERATED AGING ON UTILIZATION OF
GLUCOSE-$^{14}$C INTO VARIOUS COMPONENTS OF GERMINATING BARLEY[a]

| Extracted component | Duration of accelerated aging, days (dpm/20 seeds) | | | | | |
|---|---|---|---|---|---|---|
| | 0 | 3 | 6 | 9 | 12 | 15 |
| $^{14}CO_2$ (in $BaCO_3$) | 3290 | 2100 | 1900 | 1760 | 270 | 220 |
| Soluble proteins (acid precipitable) | 130 | 80 | 20 | 30 | 30 | 50 |
| Water-sol, EtOH-insol | | | | | | |
| polysaccharides | 1380 | 920 | 590 | 380 | 210 | 310 |
| Hemicelluloses, starch, | | | | | | |
| and cellulose | 1230 | 970 | 510 | 430 | 100 | 130 |
| Deproteinized EtOH-soluble | 188,440 | 172,180 | 170,050 | 165,440 | 164,510 | 172,220 |
| Total uptake | 194,430 | 176,250 | 173,070 | 168,040 | 165,120 | 172,930 |
| Total utilization | 6030 | 4070 | 3020 | 2600 | 610 | 710 |
| % Utilization | 3.10 | 2.31 | 1.75 | 1.55 | 0.37 | 0.41 |
| % Germination | 99.00 | 98.00 | 97.00 | 89.00 | 66.00 | 30.00 |

[a]From Abdul-Baki (1969b).

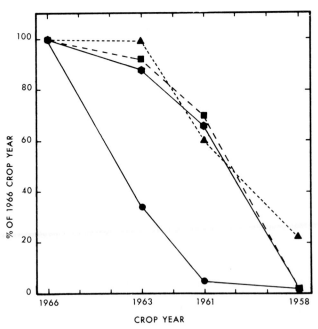

FIG. 4.5. Glucose utilization (●—●), % germination (■—■), seedling shoot growth (□—□), and $O_2$ uptake (Δ—Δ) by germinating barley seeds from different crop years. Absolute values for the 1966 crop are: 10% glucose utilization, 91% germination, 44 mm shoot length and 36 $\mu$l $O_2$ uptake. Experiments were conducted in 1967. From Abdul-Baki (1969b).

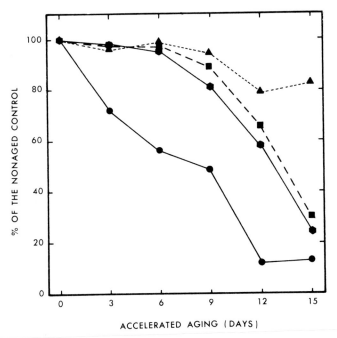

FIG. 4.6. Effects of accelerated aging of barley seed on glucose utilization (●—●), % germination (■—■), seedling shoot growth (◐—◐), and $O_2$ uptake (▲—▲). Absolute values for nonaged control are: 16.5% glucose utilization, 99% germination, 54 mm shoot length and 48 μl $O_2$ uptake. From Abdul-Baki (1969b).

## F. Chromosomal Damage

The observation by De Vries (1901) that higher percentages of abnormal plants are produced by old *Oenothera lamarckiana* seeds than by new seeds marked the beginning of mutational studies in seeds in relation to their deterioration. Studies made 30 years later confirmed De Vries' observation that the occurrence of abnormalities was more frequent in seedlings from old seeds than from new ones (Nilsson, 1931; Nawashin, 1933a; Cartledge and Blakeslee, 1934; Gerassimova, 1935; Schwarnikow, 1937; Nichols, 1941; Gunthardt *et al.*, 1953; and others).

Some of the common chromosomal abnormalities in root tips of aged barley, broad bean, and pea seeds during anaphase of the first mitoses are shown in Fig. 4.7 (Abdalla and Roberts, 1968). These abnormalities include bridges, fragments, and rings which result from breakage of chromosomes rather than chromatids. These chromosomal aberrations are most common in meristematic cells of seedling roots and to a less extent in the meristematic cells of shoots, and in pollen cells (Nawashin and Gerassimova, 1936a,b; Gunthardt *et al.*, 1953). Chromosomal abnor-

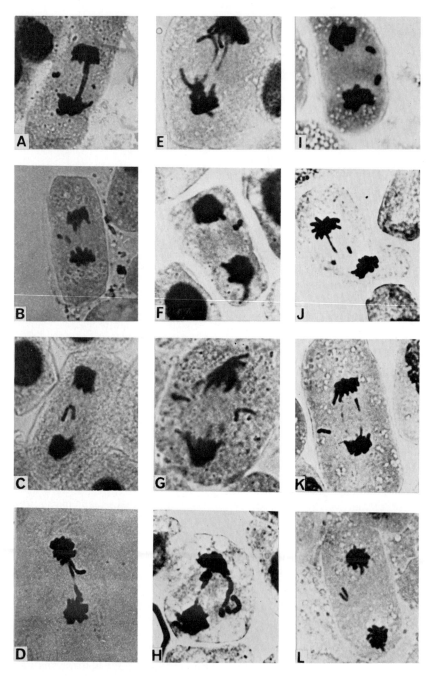

FIG. 4.7. Examples of abnormal anaphase configurations during the first mitoses in the radicles of aged seeds. A–D, barley; E–H, broad bean; I–L, pea. Magnification: C and D, × 900; remainder × 1000. From Abdalla and Roberts (1968).

malities which are highest in meristematic cells during the first mitotic division become less frequent as a seedling develops, presumably due to preferential growth of normal cells over those with chromosomal damage (Peto, 1933; Nichols, 1941). However, some abnormalities persist up to the time of flowering and cause pollen abortion as reported in *Datura* and barley (Cartledge and Blakeslee, 1933, 1934; Abdalla and Roberts, 1969).

Initial reports attributed the high incidence of chromosomal aberrations to seed aging alone. Subsequent studies confirmed that some old seeds produced seedlings with high rates of chromosomal aberrations. Cartledge and Blakeslee (1934) suggested that factors other than seed age or viability determined the frequency of pollen abortion. In several instances high negative correlations existed between the number of chromosomal aberrations in root tips during the first anaphase and percent germination (Nawashin *et al.,* 1940; Nichols, 1941; Roberts *et al.,* 1967; Abdalla and Roberts, 1968). Correlations were low between chromosomal aberrations and germinability in cases where either control seeds showed low germination capacity or aberrations were induced by x-radiation treatments which did not affect viability (Harrison and McLeish, 1954; Sax and Sax, 1962). Great variation has been reported in the number of aberrations in seeds of the same age within the same variety as well as among different varieties (Nichols, 1941).

Mechanisms of induction of natural chromosomal damage in aged seeds have been investigated. Roberts *et al.* (1967) concluded that although visible chromosomal aberrations are a good index of seed age, they are not the cause of seed deterioration. These aberrations reflect more generalized damage to nucleic acids.

In their excellent review on metabolism and spontaneous mutations in plants, D'Amato and Hoffmann-Ostenhof (1956) postulated that spontaneous mutations, at least in higher plants, may be considered as specific responses to special physiological or metabolic conditions in a cell or organism. According to this theory, natural mutagens are formed within a cell and under certain specific conditions they are capable of inducing mutations. Evidence supporting this theory is provided by the ability of extracts from old seeds to induce mutations in seedlings of the same or other species (Schwemmle, 1940; Gisquet *et al.,* 1951; Keck and Hoffmann-Ostenhof, 1952; Jackson, 1959). Old seeds of several species, however, failed to produce any mutagenic effects when applied to germinating seeds (Scarascia, 1955; Abdalla and Roberts, 1968). Nevertheless, when applied to plant tissues in high concentrations, a large number of compounds which are intermediates of vital pathways of living cells have been shown to have mutagenic properties (Woll, 1953). These compounds include purines, amino acids, organic acids, amides, aldehydes, sulfur-containing compounds, alkaloids, phenols, and quinones. These are discussed in detail by D'Amato and Hoffmann-Ostenhof (1956).

Because irradiation treatments of seeds mimic natural aging in that they reduce subsequent seedling growth and increase chromosomal aberrations (Nawashin, 1933a,b; Ehrenberg, 1955; Wolff and Sicard, 1961), it might be suspected that the mechanisms responsible for seed deterioration and for radiation-induced damage are similar. If this were true, more free radicals, such as occur in irradiated seeds (Conger and Randolph, 1959; Conger, 1966), would be expected in old than in new seeds. However, no differences in free radicals were found between old and new seeds of barley, wheat, tomato, onion, and pepper (Conger and Randolph, 1968). On the basis of this evidence Conger and Randolph concluded that although both aging and irradiation induce chromosomal damage in seeds, the mechanisms for inducing damage were different.

## V.  Application of Physiological and Biochemical Changes for Evaluating Deterioration

Many physiological and biochemical changes which take place during seed deterioration are or have been used to assess seed quality. The standard germination test is used to determine the percentage of seeds in a population which can produce normal seedlings under favorable conditions (Association of Official Seed Analysts, 1965). However, because the germination test is not always a reliable measurement for predicting seedling emergence in the field (Munn, 1921, 1926; Whitcomb, 1924; Milton, 1925; Stahl, 1931), other physiological and biochemical methods were employed to help predict field performance. Failure of the germination test to detect certain weaknesses is attributed to the lack of the types and magnitudes of stress which the seeds face in the field during germination (Mark and McKee, 1968). Since field tests usually are impractical because of the time, space, and labor involved, other tests or approaches were needed. To help fill the need for testing for seed quality, first (early) and final germination counts were used. Justification for such measurement is based partly on the observation that deteriorated seeds germinate and grow at a slower rate than do less deteriorated seeds. Seedling growth (shoot, root, or both) is also used as a measure of seed deterioration and this information, in addition to percent germination, allows a better judgment on the quality of a seed lot than that derived from percent germination alone.

Stress tests which could be conducted in the laboratory also were developed for distinguishing healthy vigorous seed lots from deteriorated ones. In the cold test which is designed to simulate poor field conditions, the seed is subjected to cold stress where pathogenic fungi are able to infect seeds of low quality, but not of high quality (Isely, 1957). In the accelerated-aging approach the seed is subjected to high temperature and humidity. This latter approach is based on the hypothesis that the most deteriorated seed lots will lose germinability most rapidly under severe storage conditions (Burns, 1957; Helmer *et al.,* 1962).

Conclusive evidence that abnormally colored seeds of lima beans, snap beans, alfalfa, and clover are less vigorous than normally colored seeds led to the separation of abnormally colored seeds by electronic sorters. Through this routine processing, the quality of a seed sample can be improved significantly by eliminating the less vigorous off-color seeds (Delouche, 1965; Wester, 1970).

Information is accumulating on biochemical changes in deteriorating seeds and some of these changes have been investigated with the hope of using them to measure seed deterioration. The activities of dehydrogenases and glutamic acid decarboxylase enzymes have been found to correlate highly with seed viability. The tetrazolium test is based on reduction of colorless tetrazolium salts and formation of red formazan by the dehydrogenases of the imbibed embryo. The procedure is simple and accurate results can be obtained within a day for most seeds. This test stands at the present time as the most accepted biochemical test that is widely used for assessing seed viability. The literature on its development, modifications, and potential is voluminous. Important contributions have been made by Lakon (1940), Kuhn and Jerchel (1941), Moore (1956, 1963, 1969), and others.

Determination of glutamic acid decarboxylase as an index of seed deterioration did not gain much popularity. Some of the disadvantages of this method lie in the procedures for extracting and assaying the enzyme. Activity of glutamic acid decarboxylase is determined by measuring the $CO_2$ produced in a reaction in which glutamic acid is used for substrate (Linko and Sogn, 1960; Linko, 1961; Grabe, 1964).

Changes in respiration during imbibition of deteriorating seeds are generally reflected in lowered rates of $O_2$ consumption and higher RQ values. These changes, which become pronounced after seed viability declines, have been suggested as a possible index of deterioration. Woodstock (1969) reviewed the potential of respiratory changes for measuring vigor in seeds.

Free fatty acids which build up in seeds infected with storage fungi have been used for several decades as an index of quality. However, this index has severe limitations because the level of free fatty acids is affected by the species of fungus, and the degree and stage of infection (Christensen and Kaufmann, 1969). Its main use is in assessing damage of grains by storage fungi.

Increased leaching of inorganic salts and organic compounds from mechanically damaged and naturally aged seeds has been suggested as an index of seed viability or field performance (Hibbard and Miller, 1928; Matthews and Bradnock, 1967, 1968; Takayanagi and Murakami, 1968). Inorganic salts in the immersion medium are measured by the Wheatstone bridge technique which measures conductivity of the solutions. Soluble sugars are determined colorimetrically by the anthrone procedure (Dent, 1963). Takayanagi and Murakami (1969) suggested the use of urine sugar analysis paper as a rapid test for seed viability by measuring sugar concentration in imbibing media. The practicality of these approaches is attractive. However, they can be useful only where differences in seed deterioration are large.

Rates of carbohydrate and protein synthesis by growing embryos during the early hours of germination have been suggested as an index of seed deterioration (Abdul-Baki, 1969a). The attractiveness of this approach stems from the fact that synthetic reactions are very important components of growing tissues. Deteriorated seeds show major reductions in their capacity to synthesize carbohydrates and proteins. The techniques are somewhat complex and require further simplification before they can be applied in routine testing.

## VI. Concluding Remarks

Our understanding of the sequence of events which takes place as seeds senesce is far from complete. During the 1930's much attention was given to chromosomal damage and disorders, and considerable progress was made. More recently, advances have been made through investigations of changes in individual enzymes, biochemical pathways, metabolites, and membranes of deteriorating seeds.

Although none of the above-mentioned subject areas has been exhaustively investigated, future research should include other areas that have been neglected. For example, the extensive literature on plant hormones presents ample evidence that they regulate a large number of processes in all plant tissues, including seeds (van Overbeek, 1966; Addicott and Lyon, 1969; Galston and Davies, 1969; and others). Yet, with the exception of a few studies (Juel, 1941; Skrabka, 1964), the involvement of hormones in seed deterioration has not been investigated. Studies are also needed on quantitative and qualitative changes in nucleic acids of deteriorated seeds.

Regrettably, progress in seed research has been very slow in the midst of the greatest advances in biochemistry, electron microscopy, and molecular biology. Hopefully, more interest will be devoted in the near future to research on critical changes which are associated with and may account for deterioration of seeds.

## REFERENCES

Abdalla, F. H., and Roberts, E. H. (1968). Effects of temperature, moisture, and oxygen on the induction of chromosome damage in seeds of barley, broad beans, and peas during storage. *Ann. Bot. (London)* [N.S.] **32**, 119.

Abdalla, F. H., and Roberts, E. H. (1969). The effects of temperature and moisture on the induction of genetic changes in seeds of barley, broad beans, and peas during storage. *Ann. Bot. (London)* [N.S.] **33**, 153.

Abdul-Baki, A. A. (1969a). Metabolism of barley seed during early hours of germination. *Plant Physiol.* **44**, 733.

Abdul-Baki, A. A. (1969b). Relationship between seed viability and glucose metabolism in germinating barley and wheat seeds. *Crop Sci.* **9**, 732.

Abdul-Baki, A. A. and Anderson, J. D. (1970). Viability and leaching of sugars from germinating barley. *Crop Sci.* **10**, 31.

Abdul-Baki, A. A., and Baker, J. E. (1969). Some properties of mitochondria from developing, mature, and dry barley (*Hordeum vulgare* L.) seeds. *Abstr. Int. Bot. Congr., 11th, 1969* Vol. 11, p. 1

Abdul-Baki, A. A., and Baker, J. E. (1970). Changes in respiration and cyanide sensitivity of the barley floret during development and maturation. *Plant Physiol.* **45**, 698.

Abu-Shakra, S., and Ching, T. M. (1967). Mitochondrial activity in germinating new and old soybean seeds. *Crop Sci.* **7**, 115.

Addicott, F. T., and Lyon, J. L. (1969). Physiology of abscisic acid and related substances. *Annu. Rev. Plant Physiol.* **20**, 139.

Anderson, J. A., and Alcock, A. W., eds. (1954). "Storage of Cereal Grains and Their Products." Amer. Ass. Cereal Chem., St. Paul, Minnesota.

Anderson, J. D. (1970a). Physiological and biochemical differences in deteriorating barley seed. *Crop Sci.* **10**, 36.

Anderson, J. D. (1970b). Metabolic changes in partially dormant wheat seeds during storage. *Plant Physiol.* **46**, 605.

Association of Official Seed Analysts. (1965). Rules for testing seeds. *Proc. Ass. Off. Seed Anal.* **54**, 27.

Avery, A. G., and Blakeslee, A. F. (1936). Visible mutations from aged seeds. *Amer. Natur.* **70**, 36 (abstr.).

Bain, J. M., and Mercer, F. V. (1966). Subcellular organization of the developing cotyledons of *Pisum sativum* L. *Aust. J. Biol. Sci.* **19**, 49.

Barton, L. V. (1961). "Seed Preservation and Longevity." Wiley (Interscience), New York.

Bautista, G. M., and Linko, P. (1962). Glutamic acid decarboxylase activity as a measure of damage in artificially dried and stored corn. *Cereal Chem.* **39**, 455.

Bautista, G. M., Lugay, J. C., Cruz, L. J., and Juliano, B. O. (1964). Glutamic acid decarboxylase activity as a viability index of artificially dried and stored rice. *Cereal Chem.* **41**, 188.

Borthwick, H. A. (1932). Thresher injury in baby lima bean. *J. Agr. Res.* **44**, 503.

Bradbury, D., Cull, I. M., and MacMasters, M. M. (1956). Structure of the mature wheat kernel. I. Gross anatomy and relationships of parts. *Cereal Chem.* **33**, 329.

Bradnock, W. T., and Matthews, S. (1970). Assessing field emergence potential on wrinkled-seeded peas. *Hort. Res.* **10,** 50.

Brocq-Rosseu, D., and Gain, E. (1908). Sur la durée des peroxy diastases des graines. *C. R. Acad. Sci.* **146,** 545.

Brooks, S. C. (1923). Conductivity as a measure of vitality and death. *J. Gen. Physiol.* **5,** 365.

Burns, R. E. (1957). Effect of age of seed on survival under adverse storage conditions. *Proc. Ass. S. Agr. Workers* **54,** 226.

Cartledge, J. L., and Blakeslee, A. F. (1933). Mutation rate increased by aging as shown by pollen abortion. *Science* **78,** 523.

Cartledge, J. L., and Blakeslee, A. F. (1934). Mutation rate increased by aging seeds as shown by pollen abortion. *Proc. Nat. Acad. Sci. U.S.* **20,** 103.

Ching, T. M., and Schoolcraft, I. (1968). Physiological and chemical differences in aged seeds. *Crop Sci.* **8,** 407.

Chrispeels, M. J., and Varner, J. E. (1967). Gibberellic acid-enhanced synthesis and release of α-amylase and ribonuclease by isolated barley aleurone layers. *Plant Physiol.* **42,** 398.

Christensen, C. M., and Kaufmann, H. H. (1969). "Grain Storage." Univ. of Minnesota Press, Minneapolis.

Clark, B. E., Kaline, D. B., and Worters, E. C., Jr. (1968). Research on seed factors affecting the establishment of vegetable crop stands. *N. Y., Agr. Exp. Sta., Seed Res. Circ.* **3.**

Conger, A. D. (1966). Biological damage and free radicals in irradiated seeds. *Nat. Acad. Sci.—Nat. Res. Counc., Publ.* **43,** 177.

Conger, A. D., and Randolph, M. L. (1959). Magnetic centers (free radicals) produced in cereal embryos by ionizing radiation. *Radiat. Res.* **11,** 54.

Conger, A. D., and Randolph, M. L. (1968). Is age-dependent genetic damage in seeds caused by free radicals? *Radiat. Bot.* **8,** 193.

Crocker, W., and Harrington, G. T. (1918). Catalase and oxidase content of seeds in relation to their dormancy, age, vitality, and respiration. *J. Agr. Res.* **15,** 137.

D'Amato, F. (1954). Di alcuni aspetti fisiologici e genetici dell' invecchiamento dei semi. Contributo al problema della senescenza e della mutabilitè spontanea nei vegetali. *Caryologia* **6,** 217.

D'Amato, F., and Hoffmann-Ostenhof, O. (1956). Metabolism and spontaneous mutations in plants. *Advan. Genet.* **8,** 1.

Darsie, M. L., Elliott, C., and Peirce, G. J. (1914). A study of the germination power of seeds. *Bot. Gaz.* **58,** 101.

Davis, W. C. (1931). Phenolase activity in relation to seed viability. *Plant Physiol.* **6,** 127.

Delouche, J. C. (1965). A preliminary study of methods of separating crimson clover seed on basis of viability. *Proc. Ass. Off. Seed Anal.* **55,** 30.

Delouche, J. C., Rushing, T. T., and Baskin, C. C. (1968). Predicting the relative storability of crop seed lots. *Search* **8,** 1.

Dent, J. W. (1963). Determination of the digestibility and soluble carbohydrate content of fodder crop varieties in trial. *J. Nat. Inst. Agr. Bot.* **9,** 282.

De Vries, H. (1901). "Die Mutationstheorie," Vol. I. Veit. Leipzig.

Dickson, J. G., and Holbert, J. R. (1926). The influence of temperature upon the metabolism and expression of disease resistance in selfed lines of corn. *Agron. J.* **18,** 314.

Ehrenberg, L. (1955). The radiation induced growth inhibition in seedlings. *Bot. Notis.* **108,** 184.

Esbo, H. (1959). Livskraften hos timotejfrö under longtisdslangring. *Vaextodling* **12,** 1.

Filutowics, A., and Bejnar, W. (1954). The influence of the age of sugar beet seeds on their seedling value. *Rocz. Nauk Roln., Ser. A* **69,** 323.

Fleming, J. F., Johnson, J. A., and Miller, B. S. (1960). Effect of malting procedure and wheat storage conditions on alpha-amylase and protease activities. *Cereal Chem.* **37,** 363.

French, R. C. (1959). Formation of embryo starch during germination as an indicator of viability and vigor in heat-damaged barley. *Plant Physiol.* **34,** 500.

Galston, A. W., and Davies, P. J. (1969). Hormonal regulation in higher plants. *Science* **163,** 1288.

George, D. W. (1967). High temperature dormancy in wheat (*Triticum aestivum* L.). *Crop Sci.* **7,** 249.

Gerassimova, H. (1935). The nature and causes of mutations. II. Transmission of mutations arising in aged seeds: Occurrences of "homozygous dislocants" among progeny of plants raised from aged seeds. *Cytologia* **6,** 431.

Gisquet, P., Hitier, H., Izard, C., and Mounot, A. (1951). Mutations naturelles observées chez *N. tabacum* L. et mutations expérimentales provoquées par l'extrait à froid de graines vieilles prématurement. *Ann. Inst. Exp. Tabac Bergerac* **1,** 1.

Gove, P. B., ed. (1965). "Webster's Seventh New Collegiate Dictionary." G. and C. Merriam Co., Springfield, Illinois.

Grabe, D. F. (1964). Glutamic acid decarboxylase activity as a measure of seedling vigor. *Proc. Ass. Off. Seed Anal.* **54,** 100.

Gunthardt, H. M., Smith, L., Haferkamp, M., and Nilan, R. A. (1953). Studies on aged seeds. II. Relation of age of seeds to cytogenetic effects. *Agron. J.* **45,** 438.

Gustafsson, A. (1947). Mutations in agricultural plants. *Hereditas* **33,** 1.

Harrington, J. F. (1967). Seed and pollen storage for conservation of plant gene resources. *FAO Tech. Conf. Explor., Util., Conserv. Plant Gene Resourc.,* Sect. 13 (iii), pp. 1–22.

Harrison, B. J., and McLeish, J. (1954). Abnormalities of stored seed. *Nature (London)* **173,** 593.

Helmer, J. D., Delouche, J. C., and Lienhard, M. (1962). Some indices of vigor and deterioration in seeds of crimson clover. *Proc. Ass. Off. Seed Anal.* **52,** 154.

Hibbard, R. P., and Miller, E. V. (1928). Biochemical studies on seed viability. I. Measurements of conductance. *Plant Physiol.* **3,** 335.

Hutchinson, J. B., Greer, E. N., and Brett, C. C. (1948). Resistance of wheat to sprouting in the ear: Preliminary investigations. *Emp. J. Exp. Agr.* **16,** 23.

Hyde, E. O. (1954). The function of the hilum in some Papilionaceae in relation to the ripening of the seed and the permeability of the testa. *Ann. Bot. (London)* [N.S.] **18,** 241.

Ibanez, M. L. (1964). Role of the cotyledon in sensitivity to cold of cacao seed. *Nature (London)* **201,** 414.

Isely, D. (1957). Vigor tests. *Proc. Ass. Off. Seed Anal.* **47,** 176.

Jachymczyk, W. J., and Cherry, J. H. (1968). Studies on messenger RNA from peanut plants: *In vitro* polysome formation and protein synthesis. *Biochim. Biophys. Acta* **157,** 368.

Jackson, W. D. (1959). The life span of mutagens produced in cells by irradiation. *Radiat. Biol., Proc. Austral. Ass. Conf., 1958* p. 190.

Jacobsen, J. V., and Varner, J. E. (1967). Gibberellic acid-induced synthesis of protease by isolated aleurone layers of barley. *Plant Physiol.* **42,** 1596.

James, E. (1961). An annotated bibliography on seed storage and deterioration. *U.S., Dep. Agr., ARS* **ARS 34-15-1.**

James, E. (1963). An annotated bibliography on seed storage and deterioration. *U.S., Dep. Agr., ARS* **ARS 34-15-2.**

James, E. (1968). Limitations of glutamic acid decarboxylase activity for estimating viability in beans (*Phaseolus vulgaris* L.). *Crop Sci.* **8,** 403.

James, W. O., and James, A. L. (1940). The respiration of barley germinating in the dark. *New Phytol.* **39**, 145.

Jones, D., Divine, J., and Gersdorff, C. (1942). The effect of storage of corn on the chemical properties of its proteins and on its growth-promoting value. *Cereal Chem.* **19**, 819.

Juel, I. (1941). Der Ausingehalt im Samen verschiedenen Alters, sowie einige untersuchungen betreffend die Haltbarkeit der Auxine. *Planta* **32**, 227.

Kearns, V., and Toole, E. H. (1939). Relation of temperature and moisture content to longevity of chewing fescue seed. *U.S. Dep. Agr., Tech. Bull.* **670.**

Keck, K., and Hoffmann-Ostenhof, O. (1952). Chromosome fragmentation in *Allium cepa* induced by seed extracts of *Phaseolus vulgaris. Caryologia* **4**, 289.

Kittock, D. L., and Law, A. G. (1968). Relationship of seedling vigor and tetrazolium chloride reduction by germinating wheat seeds. *Agron. J.* **60**, 286.

Kondo, M., and Okamura, T. (1934). Storage of rice. X. Studies on four lots of unhulled rice stored forty-six to eighty-four years in granaries. *Ber. Ohara Inst. Landwirt. Forsch., Okayama Univ.* **6**, 175.

Koostra, P. T., and Harrington, J. F. (1969). Biochemical effects of age on membranal lipids of *Cucumis sativus* L. seed. *Proc. Int. Seed Test. Ass.* **34**, 329.

Kuhn, R., and Jerchel, D. (1941). Uber Invertseifen. VIII. Mitt. Reduktion von Tetrazoliumsalzen durch Bakterien, garende Hefe und Keimende Samen. *Ber. Deut. Chem. Ges.* **74**, 949.

Lakon, G. (1940). Die topographische Selenmethode, ein neues Verfahren zur Festellung der Keimfahigkeit der Gertreidefruchte ohne Keimversuch. *Proc. Int. Seed Test Ass.* **12**, 1.

Larsen, A. L. (1965). Use of thermogradient plate for studying temperature effects on seed germination. *Proc. Int. Seed Test Ass.* **30**, 861.

Linko, P. (1961). Simple and rapid method for determining GADA (glutamic acid decarboxylase activity) as a quality index in wheat. *J. Agr. Food Chem.* **9**, 310.

Linko, P., and Sogn, L. (1960). Relation of viability and storage to glutamic acid decarboxylase in wheat. *Cereal Chem.* **37**, 489.

Lynch, B., Glass, R., and Geddes, W. F. (1962). Grain storage studies. XXXII. Quantitative changes occurring in the sugars of wheat deteriorating in the presence and absence of molds. *Cereal Chem.* **39**, 256.

McHargue, J. S. (1920). The significance of the peroxidase reaction with reference to the viability of seeds. *J. Amer. Chem. Soc.* **42**, 612.

MacLeod, A. M. (1952). Enzyme activity in relation to barley viability. *Trans. Proc. Bot. Soc. Edinburgh* **36**, 18.

Marcus, A. (1969). Seed germination and the capacity for protein synthesis. *Symp. Soc. Exp. Biol.* **23**, 143.

Marcus, A., and Feeley, J. (1964). Isocitric lyase formation in the dissected peanut cotyledon. *Biochim. Biophys. Acta* **89**, 170.

Marcus, A., Feeley, J., and Volcani, T. (1966). Protein synthesis in imbibed seeds. III. Kinetics of amino acid incorporation, ribosome activation, and polysome formation. *Plant Physiol.* **41**, 1167.

Mark, J. L., and McKee, G. W. (1968). Relationships between five laboratory stress tests, seed vigor, field emergence and seedling establishment in reed canarygrass. *Agron. J.* **60**, 71.

Marotta, D., and Kaminka, R. (1924). Valutazione della vitalita del frumento per via biochimica. *Ann. Chem. Appl.* **14**, 207.

Matthews, S., and Bradnock, W. T. (1967). The detection of seed samples of wrinkle-seeded peas (*Pisum sativum* L.) of potentially low planting value. *Proc. Int. Seed Test. Ass.* **32**, 553.

Matthews, S., and Bradnock, W. T. (1968). Relationship between seed exudation and field emergence in peas and French beans. *Hort. Res.* **8,** 89.

Merry, J., and Goddard, D. R. (1941). A respiratory study of barley grain and seedlings. *Rochester Acad. Sci.* **8,** 28.

Milner, M., and Geddes, W. F. (1946). Grain storage studies. 3. The relation between moisture content, mold growth, and respiration of soybeans. *Cereal Chem.* **23,** 225.

Milner, M., Christensen, C. M., and Geddes, W. F. (1947). Grain storage studies. 6. Wheat respiration in relation to moisture content, mold growth, chemical deterioration, and and heating. *Cereal Chem.* **24,** 182.

Milton, W. E. (1925). An investigation into the soil germination and yield of certain crucifers, clovers, Italian ryegrass and chicory sown at three weekly intervals from May to Nov. *Welsh J. Agr.* **4,** 222.

Moore, R. P. (1956). Slam-bang harvesting is killing our seeds. *Seedsmen's Dig.* **7**(9), 10.

Moore, R. P. (1963). Previous history of seed lots and differential maintenance of seed viability and vigor in storage. *Proc. Int. Seed Test. Ass.* **28,** 691.

Moore, R. P. (1969). History supporting tetrazolium seed testing. *Proc. Int. Seed Test. Ass.* **34,** 233.

Munn, M. T. (1921). Further studies of the fungus associates of germination tests. *Proc. Ass. Off. Seed Anal.* **13,** 57.

Munn, M. T. (1926). Comparing laboratory and field viability tests of seed of garden peas. *Proc. Ass. Off. Seed Anal.* **18,** 55.

Nawashin, M. (1933a). Origin of spontaneous mutations. *Nature (London)* **131,** 436.

Nawashin, M. (1933b). Altern der Samen als Ursache von Chromosomen Mutationen. *Planta* **20,** 233.

Nawashin, M., and Gerassimova, H. (1936a). Natur und Ursachen der Mutationen. I. Das Verhalten und die Zytologie der Pflanzen, die aus infolge Alterns mutierten Keimen stammed. *Cytologia* **7,** 324.

Nawashin, M., and Gerassimova, H. (1936b). Natur und Ursachen der Mutationen. III. Uber die Chromosomenmutationen, die in den Zellen von ruhenden Pflanzenkeimen bei deren Altern auftreten. *Cytologia* **7,** 437.

Nawashin, M., Gerassimova, H., and Belajeva, G. M. (1940). On the course of the process of mutation in the cells of the dormant embryo within the seed. *Dokl. Akad. Nauk SSSR* **26,** 948.

Nemeč, A., and Duchoň, F. (1923). Sur une nouvelle méthode biochemique pour la détermination de la faculté vitale des semences. *Ann. Sci. Agron.* **40,** 121.

Nichols, C., Jr. (1941). Spontaneous chromosome aberrations in *Allium*. *Genetics* **26,** 89.

Nilan, R. A., and Gunthardt, H. M. (1956). Studies on aged seeds. III. Sensitivity of aged wheat seeds to x-radiations. *Caryologia* **8,** 316.

Nilsson, N. H. (1931). Sind die induzierten Mutanten nur selektive Erscheinungen? *Hereditas* **15,** 320.

Nutile, G. E. (1964). Effect of desiccation on viability of seeds. *Crop Sci.* **4,** 325.

Osterhout, W. J. (1918). A method of measuring the electrical conductivity of living tissues. *J. Biol. Chem.* **36,** 357.

Palladin, V. I. (1926). "Plant Physiology" (transl. by B. E. Livingston), 3rd ed. McGraw-Hill (Blakiston), New York.

Parkinson, A. H. (1948). Seed age and storage affect cucumber growth. *Food Res.* **14,** 7.

Peto, F. H. (1933). The effect of aging and heat on the chromosomal mutation rates in maize and barley. *Can. J. Res.* **9,** 261.

Pollock, B. M., and Toole, V. K. (1966). Imbibition period as the critical temperature sensitive stage in germination of lima bean seeds. *Plant Physiol.* **41,** 221.

Pomeranz, Y. (1966). The role of the lipid fraction in growth of cereals, and in their storage and processing. *Wallerstein Lab. Commun.* **29**, 17.

Ranke, J. T. (1865). "Eine physiologische Studie." Engelmann, Leipzig.

Rao, M. N., Viswanatha, T., Mathur, P. B., Swaminathan, M., and Subrahanayan, V. (1954). Effect of storage on the chemical composition of husked, undermilled and milled rice. *J. Sci. Food Agr.* **5**, 405.

Roberts, E. H., Abdalla, F. H., and Owen, R. J. (1967). Nuclear damage and the ageing of seeds. *Symp. Soc. Exp. Biol.* **21**, 65.

Robinson, T. (1964). "The Organic Constituents of Higher Plants." Burgess, Minneapolis, Minnesota.

Sax, K., and Sax, H. J. (1962). Effects of x-rays on the aging of seeds. *Nature (London)* **194**, 459.

Scarascia, G. T. (1955). Action d'extraits de graines vieilles de *Nicotiana tabacum* L. sur la mitose. *Congr. Int. Bot., Rapp. Commun., 1954* Sect. 9, p. 23.

Schwarnikow, P. K. (1937). Über Erhöhung der Mutatuinsrate bei Weizen nach langer Aufbewahrung der Samen. *Genetica* **19**, 188.

Schwemmle, J. (1940). Keimversuche mit alten Samen. *Z. Bot.* **36**, 225.

Skrabka, H. (1964). Interaction of free auxins with ascorbic acid and glutathione in the process of wheat grains germination. Part I. Quantitative investigations of free auxins, ascorbic acid and glutathione in germinating wheat grains with various viability. *Acta Soc. Bot. Pol.* **33**, 689.

Sorger-Domenigg, H., Cuendet, L., Christensen, C. M., and Geddes, W. (1955). Grain storage studies. XVII. Effect of mold growth during temporary exposure of wheat to high moisture contents upon the development of germ damage and other indices of deterioration during subsequent storage. *Cereal Chem.* **32**, 270.

Sreenivasan, A. (1939). Studies on quality in rice. IV. Storage changes in rice after harvest. *Indian J. Agr. Sci.* **9**, 208.

Stahl, C. (1931). Comparative experiments between the laboratory and field germination of seed. *Proc. Int. Seed Test. Ass.* **15–17**, 75.

Sumner, J. B. (1926). The isolation and crystallization of the enzyme urease. *J. Biol. Chem.* **69**, 435.

Täufel, K., and Pohlaudek-Fabini, R. (1955). Keimfahigkeit und Geholt citronensaure bei gelagerten Pflanzensamen. *Biochem. Z.* **326**, 317.

Takayanagi, K., and Murakami, K. (1968). Rapid germinability test with exudate from seed. *Nature (London)* **218**, 493.

Takayanagi, K., and Murakami, K. (1969). Rapid method for testing seed viability by using urine sugar analysis paper. *Jap. Agr. Res. Quart.* **4**, 39.

Throneberry, G. O., and Smith, F. G. (1954). Seed viability in relation to respiration and enzymatic activity. *Proc. Ass. Off. Seed Anal.* **44**, 91.

Throneberry, G. O., and Smith, F. G. (1955). Relation of respiratory and enzymatic activity to corn seed viability. *Plant Physiol.* **30**, 337.

Toole, E. H., and Toole, V. K. (1946). Relation of temperature and seed moisture to the viability of stored soybean seed. *U.S. Dept. Agr., Circ.* **753**.

Toole, E. H., and Toole, V. K. (1953). Relation of storage conditions to germination and to abnormal seedlings of bean. *Proc. Int. Seed Test. Ass.* **18**, 123.

Toole, E. H., and Toole, V. K. (1960). Viability of stored snap bean seed as affected by threshing and processing injury. *U.S., Dep. Agr., Tech. Bull.* **1213**.

Toole, E. H., Toole, V. K., and Gorman, E. A. (1948). Vegetable-seed storage as affected by temperature and relative humidity. *U.S. Dep. Agr., Tech. Bull.* **572**.

Toole, E. H., Toole, V. K., and Borthwick, H. A. (1957). Growth and production of snap beans stored under favorable and unfavorable conditions. *Proc. Int. Seed Test. Ass.* **22,** 418.

van Overbeek, J. (1966). Plant hormones and regulators. *Science* **152,** 721.

Weber, E. (1836). "Questiones physiologicae de phenomenis galvanomagneticis in corpore humano observatis." Breitkopf & Hartel, Leipzig.

West, S. H., and Harris, H. C. (1963). Seed coat color associated with physiological changes in alfalfa and crimson and white clovers. *Crop Sci.* **3,** 190.

Wester, R. E. (1965). Green cotyledon in lima beans—its origin and development. *Seed World* **96,** 30.

Wester, R. E. (1970). Chlorophyll, a genetic marker for quality seed in lima bean. *Seed World* **101,** 24.

Whitcomb, W. O. (1923). Relative value of green and ripe seeds in alfalfa and sweet clover. *Proc. Ass. Off. Seed Anal.* **21,** 37.

Whitcomb, W. O. (1924). Correlation of laboratory and field germination test. *Proc. Ass. Off. Seed Anal.* **16,** 60.

Whitcomb, W. O. (1942). Crop-producing value of discolored alfalfa seed in Montana— 1941 crop correlation of laboratory and field tests. *Proc. Ass. Off. Seed Anal.* **34,** 28.

Whymper, R., and Bradley, A. (1934). Studies on the vitality of wheat. II. Influence of moisture in wheat seeds upon imbibition and speed of germination. *Cereal Chem.* **11,** 546.

Wolff, S., and Sicard, A. M. (1961). Post irradiation storage and the growth of barley seedlings. *Eff. Ioniz. Radiat. Seeds, Proc. Symp. 1960* pp. 171.

Woll, E. (1953). Einwirkung von Nuleinsäuren und ihren Bausteinen auf die Wurzelspitzenmitose. *Chromosoma* **5,** 391.

Woodstock, L. W. (1969). Biochemical tests for seed vigor. *Proc. Int. Seed Test. Ass.* **34,** 253.

Woodstock, L. W., and Combs, M. F. (1965). Effects of gamma-irradiation of corn seed on the respiration and growth of the seedlings. *Amer. J. Bot.* **52,** 563.

Woodstock, L. W., and Feeley, J. (1965). Early seedling growth and initial respiration rates as potential indicators of seed vigor in corn. *Proc. Ass. Off. Seed Anal.* **55,** 131.

Woodstock, L. W., and Grabe, D. F. (1967). Relationships between seed respiration during imbibition and subsequent seedling growth in *Zea mays* L. *Plant Physiol.* **42,** 1071.

Woodstock, L. W., and Justice, O. L. (1967). Radiation-induced changes in respiration of corn, wheat, sorghum, and radish seeds during initial stages of germination in relation to subsequent seedling growth. *Radiat. Bot.* **7,** 129.

Woodstock, L. W., and Pollock, B. M. (1965). Physiological predetermination: Imbibition, respiration, and growth of lima bean seeds. *Science* **150,** 1031.

Woodstock, L. W., Reiss, B., and Combs, M. F. (1967). Inhibition of respiration and seedling growth by chilling treatments in *Cacao theobroma*. *Plant Cell Physiol.* **8,** 339.

Wyttenbach, E. (1955). Der Einfluss verschiedener Lagerungsfaktoren auf die Haltbarkeit von Feldsamereien bei langer dauernder Aufbewahrung. *Landwirt. Jahrb. Schweiz.* **4,** 161.

Zeleney, L. (1954). Chemical, physical, and nutritive changes during storage. *In* "Storage of Cereal Grains and Their Products" (J. A. Anderson and A. W. Alcock, eds.), pp. 46–76. Amer. Ass. Cereal Chem., St. Paul, Minnesota.

Zeleney, L., and Coleman, D. A. (1938). Acidity in cereal and cereal products, its determination and significance. *Cereal Chem.* **15,** 580.

Zeleney, L., and Coleman, D. A. (1939). The chemical determination of soundness in corn. *U.S., Dep. Agr., Tech. Bull.* **644.**

# 5

## SEED PATHOLOGY

### Kenneth F. Baker

## I.   Seed Transmission of Pathogens*

Parasitic fungi and bacteria emerged with land plants from the Paleozoic seas more than 330 million years ago, and evolved in form and physiology as the land flora developed. With the appearance of the seed plants in the Jurassic, and particularly after angiosperms became the dominant flora in the Cretaceous, at least 130 million years ago, seeds became the principal means of plant reproduction. Some fungus pathogens (e.g., *Claviceps* spp.) are thought (Buller, 1950) to have developed mechanisms for seed transmission about this time, and it is probable that many other seed–pathogen relationships also developed very early. Because of the extremely long period during which the highly varied means of seed transmission of plant pathogens have evolved, host–parasite interactions are, as expected, both complex and effective.

However, Atsatt (1965) has cogently pointed out that complex inter-relationships between host and parasite do not necessarily indicate an ancient association. Seeds of the root parasite, *Orthocarpus densiflorus,* are carried in the pappus of its host, *Hypochoeris glabra,* a plant introduced into California less than 100 years ago.

Seed transmission is now recognized as the method par excellence by which plant pathogens (*a*) are introduced into new areas, (*b*) survive periods when the host is lacking, (*c*) are selected and disseminated as host-specific strains, and (*d*) are distributed through the plant population as foci of infection. The subject of seed pathology is, therefore, of concern to man in his unending quest for more and better food.

There are a number of general reference works on seed pathology: Alcock (1931), Baker and Smith (1966), Barton (1967), Bennett (1969), Cain and Groves (1948), Chen (1920), Crocker and Barton (1953), Crowley (1957), Doyer (1938), Dykstra (1961), Groves and Skolko (1944–1946), Ingold (1953), International Seed Testing Association (1958—), LeClerg (1953), Limonard (1968), Malone and Muskett (1964), Marshall (1959–1960), W. C. Moore (1946, 1953), Muskett (1950), Muskett and Colhoun (1947), Neergaard (1940), Noble (1957),

---

*The terms for different types of seed carriage of plant pathogens are explained in the section on terminology (Section I,B). The following abbreviations are used in this chapter:
BCMV = bean common mosaic virus.
Btu = British thermal unit, a measure of quantity of heat.
C = control or prevention procedure.
PP = the plant → plant transfer of a pathogen, with establishment in the field.
PS = the plant → seed transfer of a pathogen.
SP = the seed → plant transfer of a pathogen.
SS = the seed → seed transfer of a pathogen.
TMV = tobacco mosaic virus.

Noble and Richardson (1968), Orton (1931), Porter (1949), Sampson and Western (1954), Skolko and Groves (1948, 1953), and Wallen (1964).

No attempt is made to review all of the voluminous literature of seed pathology in this chapter. What are thought to be the essentials of the field today are briefly discussed, and key references which will lead to other papers are cited. J. C. Walker (1969) and Heald (1933), cited for brevity for many host–parasite associations, will thus provide the principal references. Pathogens used as examples are often those with which the author has had personal experience.

## A. History

A given pathogenic microorganism must have evolved in a localized place on a favorable host plant. Such pathogens generally remained quite restricted in distribution or spread very slowly under natural conditions. The more specialized they were for host and environment, the greater was the likelihood for geographical restriction. In the food-gathering stages of man's development this restriction diminished somewhat, for, in carrying propagules (roots, bulbs, fruit, plants, seeds) about as food, man probably included some infected material which would be discarded in the rubbish pile. The pathogen would thus have been spread over a limited area. After the development of crude cultivation, plant parts were carried about to start the crop in a new area. Taro *(Colocasia esculenta)* was, for example, brought to Hawaii from Tahiti by Polynesians (Whitney *et al.,* 1939). The area of potential spread was thus enormously expanded by man even prior to commencement of permanent agriculture.

With further development of agriculture 8000–9000 years ago in Sumer in the eastern Mediterranean, man began to depend on seeds as a principal means of carrying his main crops, the cereals, from place to place and from season to season. Although plant pathogens were undoubtedly carried with the seeds from the beginning, the earliest confirmed record of this association is from Jarmo, about 8000 years ago (R. B. Stewart and Robertson, 1968).

Early planters must often have noted that certain lots of seed tended to produce a poor crop, and may have speculated how this came about. Fracastoro (1546) came close to the mark when he postulated "germs of contagion" that spread from apple to apple or grape to grape, rotting them, and in "other bodies also that are in contact and putrefy . . . the same thing happens, and by means of the same principle."

Remnant (1637) apparently suspected that smut was carried with wheat seed in some way. He noted that "much of your wheat will . . . become blacke, and spoil all your good wheat in the threshing . . .; therefore pick or lease it out of the sheafe before you thresh it, or else you had need wash it well, . . . before you eate it . . . there is much helpe . . . in adding vigour and helpe to the seed which is sowne, by steeping it in and with a certaine ingredient, before it is sowne, which keepeth it from decay and smut. And of this I purpose hereafter to publish somewhat more . . . ." No further record of this material has been found.

Tull (1733) stated that for wheat smut "there are but two remedies proposed; and these are brining, and the change of seed . . . . But from whatever land the seed be taken, if it were not changed the preceding year, it may possibly be infected, and then there may be danger . . . ."

These observations reflect the growing suspicion that some nebulous deleterious agent was sometimes carried by seed. This idea was given a factual basis through the studies of Hellwig, Micheli, Needham, and Tillet by the middle of the eighteenth century. These studies were the more remarkable because they antedated the demonstration (Prévost, 1807) that fungi were able to infect a plant and cause disease, and the disproof of the idea of spontaneous generation (Pasteur, 1861).

Hellwig (1699) apparently recorded the first case of accompanying transport of a fungus pathogen (*Claviceps purpurea* on rye) with seed, when he advised flotation to separate the ergots from the seeds: ". . . in water the ergot, a long, thick, black tumor, is, considering the comparative proportions, much lighter than true seeds . . . my advice is to wash thoroughly the current season's grain before it goes to the mill, so that the black kernels go to the surface and can be removed from the good seeds . . . ."

Micheli (1723) reported for *Orobanche minor* (a parasitic seed plant) on *Vicia faba* that "it has been proved that if you sow beans in two pots full of uncontaminated soil — namely, only beans in one and beans with Succiamele *(Orobanche)* in the other, you will see that Succiamele will grow only in the one pot where it was sown" (translation by Stephen Wilhelm, personal communication). This is perhaps the first clear demonstration of seed transmission of a plant pathogen and the first instance of accompanying seed transport of a spermatophyte.

Needham (1743, 1745) observed abundant desiccated nematodes *(Anguina tritici)* in wheat seeds, showed that they became active when wetted, and considered them to be animals. He commented that this "enables us to account for the observation of several farmers . . . that blight in wheat, . . . is frequently occasioned by the sowing of seed intermixed with blighted grains." A 30-hour soak in a strong brine, and skimming off the floating grains "will effectively preserve the new crop from any infection of that kind." He thus reported accompanying seed transport. This appears to be the first instance of internal seed transport of a plant pathogen, and positive demonstration of this type of seed transmission came 30 years later (Roffredi, 1775).

Tillet (1755) reported that, for control of wheat bunt, French farmers by 1739 were careful to obtain new seed "and above all to avoid sowing contaminated seed." He reported in 1755 that "the common cause . . . resides in the dust of the bunt balls of diseased wheat; that the clean healthy seed, inoculated with this dust, receives . . . the poison peculiar to it; that it transmits the poison to the kernels . . .; that these kernels, once infested become converted into a black dust and become for others a cause of disease; . . . the treatments I employed have protected the most heavily infested seed against the effects of the contagion . . . ." This is the first known demonstration of external seed transport and transmission of a plant pathogen. Because he showed that the spores were carried in smut balls mixed with seed, he also demonstrated external transport and accompanying transmission of the pathogen.

Roffredi (1775) demonstrated accompanying and internal transmission of the nematode, *Anguina tritici,* on barley and rye. "In early April . . . seeds of rye which I had mixed with the aborted grains, had begun to germinate with disorder in their natural vegetative development; I noted several stools with tortuous weak shoots and yellow color. Microscopic examination showed that these stools were full of eelworms . . . . As for the heads . . . the small grains were filled with eelworms . . . . It is evident to me that the rachitic disease of grains has for immediate cause the little eelworms, which are introduced with the aborted grains mixed in the soil with the healthy grains." Henslow (1841) confirmed this, and Davaine (1855) supplied details of infection of the coleoptile and later invasion of the developing seed.

Bauer (1823) conclusively demonstrated external transmission of this parasite. "I . . . selected some sound grains of wheat, and placed some portions of the mass of worms in the grooves on the posterior sides of the grains, and planted them . . . in October, 1807. Nearly all the seeds came soon up . . . . On the 5*th* of June I found . . . some of the worms . . . within the cavities of the young germens . . . towards the end of June, the germens . . . began to be filled with eggs."

Cysts of the sugar-beet nematode *(Heterodera schachtii)* were undoubtedly carried at an early date in small clods of soil mixed with seed (Shaw, 1915), but this has only recently been clearly shown. Kühn (1858) showed that the stem and bulb nematode, *Ditylenchus dipsaci,* occurred "not only in the seed, also in the pappus and in parts under it, . . . as well as on the receptacle and in the pith of the flowerheads" of *Dipsacus fullonum.* Since this transport was partly in tissues external to the seed coat, it is interpreted as external seed transport. T. Goodey (1943) confirmed that dissemination of plant-parasitic nematodes commonly occurred on the surface of seed.

Frank (1883) demonstrated that the bean anthracnose fungus, *Colle-*

*totrichum lindemuthianum,* was transmitted through seed. "The mycelium often penetrates into the cotyledons of the seed directly beneath a brown spot [on the pod] . . . . We have here, as far as I know, the first case of organic transmission of a parasite from the host plant to its progeny through the embryo . . . . The certainty of a new outbreak of the fungus is assured when the mycelium, which overwinters in the cotyledon of the embryo, sporulates, and this fresh mass of infectious material introduces it in the new planting."

It was not until 1884–1885 that bacteria were conclusively shown (Arthur, 1885) to cause plant disease, and this was still under debate until 1901 (Baker, 1971). It is not surprising, therefore, that conclusive proof of bacterial transmission in seed was fairly recent, although there were early demonstrations of it.

The first evidence of seed transmission of a bacterial pathogen was supplied by Halsted (1893) for *Xanthomonas phaseoli* on bean: ". . . many of the beans . . . were spotted at one or more places by brown, irregular, somewhat sunken pits . . . the germs are carried over from one season to another in the beans themselves . . . . A dozen each of the healthy and of the diseased beans were planted in separate pots of earth and similarly treated. Five feeble plants came up from the diseased seed, and they failed before getting well rooted. The healthy seed sent up good, strong plants . . . ." This almost certainly involved internal transmission, as he suggested. Rolfs (1915) gave somewhat better evidence for internal transmission of *Xanthomonas malvacearum* on cotton by showing that contaminated seed delinted for 20 minutes in concentrated sulfuric acid still carried the viable pathogen. He also demonstrated accompanying transmission of bacteria with the lint. One of the early conclusive demonstrations of internal transmission of bacteria was provided by Clayton (1929) for *Xanthomonas campestris* on cauliflower.

Beach (1892) showed accompanying transport, and possibly external transport, of *Xanthomonas phaseoli* on bean. "This blight affects the foliage and pods . . . and also affects the beans within the pods. Some of the pods . . . produce beans that are discolored by the disease or wrinkled and disfigured with rough spots. It is possible that the blight may be communicated to the crop of the following season in the seed . . . ."

External transport of a bacterial pathogen on seeds apparently was first demonstrated by Harding *et al.* (1904), although its probability had been discussed for over a decade. They gathered seed from cabbage plants with black rot, isolated bacteria washed from them, and showed by inoculation to plants that they were virulent. They "demonstrated the presence of *P. campestris* on the seed of three of the four seed plants examined . . . much of the cabbage seed on the market is contaminated with germs of the black rot disease and . . . some of these germs may survive the winter and become a source of infection to the young cabbage plants." Seed dipped in a bacterial suspension also had virulent bacteria on the surface a year later.

External transmission of a bacterium on seeds apparently was first shown by F. C. Stewart (1897) for *Xanthomonas stewartii* on corn. "The chief method of dissemination of the germ is, probably, by diseased seed . . . seed from diseased plants may contain the germs, usually in great numbers . . . and . . . some, at least . . . retain their vitality long enough to infect the plants of the new crop. Plants of the variety Early Cory were grown in pots of sterilized earth which were watered . . . with sterilized water. Every precaution . . . was taken to prevent contamination and yet several of the plants developed the disease . . . infection must have been brought about by germs which clung to the seed." He may also have shown internal transport. E. F. Smith (1909) repeated these studies, adding a series in which seed, surface treated with 0.1% mercuric chloride, still had 1.5% transmission, as compared with 9.2% in the untreated checks. Rand and Cash (1933) proved internal transmission by repeatedly isolating the bacteria from endosperm of surface-sterilized kernels; glasshouse tests confirmed transmission.

Infectivity of a plant virus was shown in 1886 (Mayer, 1886), and extensive transmission studies were conducted after about 1895. It is therefore to be expected that work on seed transmission of viruses is comparatively recent. Allard (1915) first clearly demonstrated seed transport of a virus. Seeds from tobacco plants infected with tobacco mosaic virus (TMV) were ground and successfully used to inoculate healthy tobacco plants, showing that the virus was on or in the seed. It has since been shown (R. H. Taylor, 1962) that TMV carried on the surface of tobacco seed is responsible for this seed transport. Allard thus showed external transport and, since the contaminated seeds were mixed with those which were virus-free, accompanying transport as well.

Although embryo transport and transmission were implied as early as 1910 for TMV on tomato (Westerdijk, 1910), the first acceptable demonstration of this was for bean common mosaic virus by Reddick and Stewart (1918). The latter workers also postulated transmission of this virus through pollen to the ovule. Reddick (1931) found that pollination of healthy bean plants by pollen from diseased plants gave some infected seed, and Nelson and Down (1933) showed that both parents contributed to seed transmission. Crispin Medina and Grogan (1961) showed that the percentage of transmission through either parent varied greatly with host variety.

Ainsworth (1934) showed that TMV was associated with seed coats of tomato, and Doolittle and Beecher (1937) demonstrated that it was externally transmitted, when they showed that unwashed, freshly harvested tomato seed was more likely than older seed to produce infected seedlings. The explanation for this was developed by R. H. Taylor *et al.* (1961), who showed that the high percentage of infected seedlings from freshly harvested tomato seed largely resulted from transfer of the virus in externally carried or accompanying fruit pulp to the seedling during emergence or transplanting. They thus demonstrated accompanying transport and transmission of the virus. There was little virus transmission if seedlings were not disturbed.

It is clear from this brief historical account, summarized in Table 5.I, that the origins of seed pathology were concurrent with those of plant pathology itself. However, progress in seed pathology has not been commensurate with its importance to agriculture.

## B. *Terminology*

The term seed includes the sexually derived structures of spermatophytes which germinate to produce new plants. Fruits in which the pericarp becomes an integral part of the individual seed (e.g., achenes, caryopses) are included, but vegetative propagules such as seed potatoes are excluded. Seed is thus used in this chapter, not in the precise botanical or broad horticultural senses, but in the long-established usage of the world seed industry.

It is necessary to distinguish between the terms, seed-borne pathogen and seed-borne disease. A seed-borne pathogen must be present in, on, or with the seed, which may or may not show symptoms, that is, be diseased. In a seed-borne disease, on the other hand, the seed must actually be diseased and exhibit symptoms, but may not produce diseased seedlings. Nutritional deficiencies [e.g., marsh spot (manganese deficiency) of peas; Glasscock, 1941; Lewis, 1939; Piper, 1941] may produce diseased seeds which yield healthy seedlings. Infected or infested seeds, diseased or not, may also not produce diseased seedlings, for reasons discussed later. The term, seed-borne disease, should be used only for seeds which exhibit symptoms, despite its frequent use in a more general way (Crocker and Barton, 1953; Doyer, 1938; Dykstra, 1961; W. C. Moore, 1946; Neergaard, 1940; Noble and Richardson, 1968; Wallen, 1964). Seed-borne pathogen is the more correct and inclusive term. Seed-borne parasite (Orton, 1931) is an acceptable alternative, except possibly in cases where the causal entity is a pathogen but not a parasite [e.g., symptoms are produced by toxins formed external to the host, as reported by Durbin (1959), Johann *et al.* (1931), and Koehler and Woodworth (1938)]. Seed borne is an imprecise term commonly used as a general designation,

**TABLE 5.I.**

EARLY DEMONSTRATIONS OF VARIOUS TYPES OF SEED–PARASITE ASSOCIATIONS[a]

| Type of pathogen | Seed Transport | | | Seed Transmission | | |
|---|---|---|---|---|---|---|
| | Accompanying | External | Internal | Accompanying | External | Internal |
| Seed plants | *Orobanche minor* with horse bean (Micheli, 1723) | — | — | *Orobanche minor* with horse bean (Micheli, 1723) | — | — |
| Fungi | *Claviceps purpurea* with rye (Hellwig, 1699) | *Tilletia* sp. on wheat (Tillet, 1755) | *Tilletia* sp. in wheat (Tillet, 1755) | *Tilletia* sp. with wheat (Tillet, 1755) | *Tilletia* sp. on wheat (Tillet, 1755) | *Colletotrichum lindemuthianum* in bean (Frank, 1883) |
| Nematodes | *Anguina tritici* with wheat (Needham, 1745) | *Ditylenchus dipsaci* in pappus of teasel (Kühn, 1858) | *Anguina tritici* in wheat (Needham, 1745) | *Anguina tritici* with rye (Roffredi, 1775) | *Anguina tritici* on wheat (*1807*; Bauer, 1823) | *Anguina tritici* in rye (Roffredi, 1775) |
| Bacteria | *Xanthomonas phaseoli* with bean (Beach, 1892) | *Xanthomonas campestris* on cabbage (Harding et al., 1904) | *Xanthomonas stewartii* in corn (F. C. Stewart, 1897) | *Xanthomonas malvacearum* with cotton (Rolfs, 1915) | *Xanthomonas stewartii* on corn (F. C. Stewart, 1897) | *Xanthomonas phaseoli* in bean (Halsted, 1893) |
| Viruses | Tobacco mosaic virus with tobacco (Allard, 1915) | Tobacco mosaic virus on tobacco (Allard, 1915) | Bean common mosaic virus in bean (Reddick and Stewart, 1918) | Tobacco mosaic virus with tomato (R. H. Taylor et al., 1961) | Tobacco mosaic virus on tomato (Doolittle and Beecher, 1937) | Bean common mosaic virus in bean (Reddick and Stewart, 1918) |

[a] Since transport does not necessarily result in transmission, demonstration of it may precede that of transmission. Year in which study was conducted is given in *italics* in parenthesis followed by reference.

but all too often it simply indicates a lack of information on the exact relationship of the pathogen to the seed.

It is, therefore, necessary to distinguish clearly between three types of association of seeds and pathogens:

1. *Accompanying* — Pathogen independently accompanies the seed of the host, but is not attached to it. This may be as sclerotia (*Sclerotinia sclerotiorum, Claviceps purpurea* with rye), seeds (dodder, broomrape), infected bits of plant tissue (TMV with tobacco, *Puccinia malvacearum* with hollyhock), as nematode galls (*Anguina tritici* with wheat, *Ditylenchus dipsaci* with alfalfa), or as infested soil (*Plasmodiophora brassicae* with turnip, *Heterodera schachtii* with beet).

2. *External* — Pathogen is passively carried on the surface of the seed of the host as sclerotia (*Rhizoctonia solani* on pepper), spores (*Puccinia antirrhini* on snapdragon, *P. carthami* on safflower), vegetative cells (*Corynebacterium fascians* on sweet pea and nasturtium, *R. solani* on pepper), nematodes (*Ditylenchus dipsaci* on alfalfa and onion), or virus (TMV on tomato). A pathogen is external when it is outside of the functional seed or fruit parts essential for production of a new plant.

3. *Internal* — Pathogen is carried internally, imbedded in the tissue, in the seed of the host as fruiting structures (pycnidia of *Septoria apiicola* in celery, *Phoma lingam* in cabbage), spores (oospores of *Phytophthora phaseoli* in lima bean, chlamydospores of *Tilletia caries* in wheat), vegetative structures (*Alternaria zinniae* in zinnia, *Xanthomonas campestris* in cabbage), nematode larvae (*Anguina tritici* in wheat, *Ditylenchus dipsaci* in Compositae), or virus (lettuce mosaic in lettuce, bean common mosaic in bean).

It is important to distinguish between infection of the embryo (in which the seed → progeny (SP) transfer has already occurred) and infection of the endosperm, perisperm, or fruit and seed coats (in which the SP transfer has still to be accomplished). This is especially true for viruses, because embryo invasion usually assures seedling infection, whereas virus in endosperm or seed coats may not effect the SP transfer even when the virus is resistant to inactivation and easily transmitted mechanically.

It is by no means certain that the pathogen in any of the above three groups will produce an infected seedling (see pp. 378–380). It is therefore essential to differentiate between *transport* or carriage of the pathogen with the seed from place to place or season to season, and its successful *transmission* to the progeny. In many published studies it has been erroneously assumed, without evidence, that the presence of the pathogen with, on, or in the seed assures transmission. The presence, for example, of powdery mildew perithecia on pea, parsnip, and zinnia seed has merely implied transmission (Crawford, 1927; Green, 1946; Uppal *et al.,* 1935) or it could not be shown (Baker and Locke, 1946). Only four (Evans, 1968, 1971; Sackston and Martens, 1959; Schippers and Schermer, 1966; Snyder and Wilhelm, 1962) of the many conflicting reports of seed-borne *Verticillium albo-atrum* and *V. dahliae* have actually demonstrated trans-mission. Similarly, the curly-top virus is abundant in the perisperm of sugar-beet seed produced by infected plants, but is not transmitted to the seedlings produced by them (Bennett and Esau, 1936).

Seed which is produced under suitable conditions and is free of a pathogen is often referred to as pathogen-free. This usage is sometimes deplored because it implies that no pathogens are present. It certainly is preferable, when only a single pathogen is involved, to refer to the seed as free of lettuce mosaic virus or halo blight bacteria, for example. Con-venience probably justifies using pathogen-free seed when several pathogens are involved, if it is stated in the paper what the seed is free from. Such statements are, in any case, only as valid as the indexing methods behind them.

## C.   *Approaches to Seed Pathology*

Studies on seed pathology generally have been undertaken because an economically important or scientifically interesting problem has been observed in one or more of the following situations.

### 1.   STUDIES IN THE SEED–PRODUCTION FIELD

Diseased plants, seeds, or both, are observed in the seed fields. Obser-vations of the host–pathogen interactions here provide an excellent opportunity to study the dynamics of the PS transfer of the pathogen, and to devise means of preventing seed infection. Many pathogens are best detected in the growing seed crop (Güssow, 1936), but this is not always possible (Baker, 1950, 1956).

*Seed transport* is shown by demonstrating PS transfer in the seed field.

Parasitic seed plants, such as dodder or broomrape, may produce seed in such a way that it mingles with that of the host during harvest. Micro-organisms may: (*a*) produce structures such as the sclerotia of *Claviceps purpurea* and *Sclerotinia sclerotiorum,* which also mingle with host seed during harvest; (*b*) produce structures which get on the surface of seed during harvest, such as spores of *Tilletia caries* on wheat or bacterial cells of *Corynebacterium fascians* on sweet pea; (*c*) penetrate the seed by growth, as in *Ascochyta pisi* on pea, or by movement, as in *Ditylenchus dipsaci* on alfalfa. Viruses are passively transferred from the host to seed through pollen, as in bean common mosaic, through the vascular system, as in curly top (which is not, however, transmitted to the seedling), or by transfer to the surface of seed, as in TMV on tobacco. Pathogens generally show more than one method of PS transfer.

## 2. STUDIES IN THE SEED-TESTING LABORATORY

Pathological observations of *association of the pathogen with the seed* during routine or specialized laboratory seed testing are important in locating infected or infested seed lots and their sources. When the organism is a well-known pathogen the data may be helpful, but when its disease potential is unknown, the significance of its occurrence is difficult to evaluate without knowledge from both situations 1 and 3. Many entries in published lists of seed-borne pathogens have been obtained in this way, without continuing studies to indicate that a disease may be produced in the given plant.

Examples of this type of report are those of the J. E. Ohlsens Enkes Plantepatologiske Laboratorium (Neergaard, 1936–1951) and the Statens Plantetilsyn Vedrørende Frøpatologisk Kontrol (H. Andersen, 1952–1956, 1961; Neergaard, 1956–1960, 1962–1967) in Copenhagen, and from the New York (Geneva) Agricultural Experiment Station (Clark and Page, 1960–1969).

The Comparative Seed Health Tests conducted annually by the Committee on Plant Diseases of the International Seed Testing Association tabulate disease ratings obtained by pathologists in various parts of the world on uniform seed samples, using uniform methods (International Seed Testing Association, annual). This method is valuable for internationally standardizing methods, and provides satisfactory evidence for *seed transport*. However, the variability of the results limits the value of quantitative information obtained about a given pathogen (Baker and Smith, 1966; Limonard, 1968). Virus indexing (Hollings, 1965) is rarely included in reports of this type.

## 3. STUDIES IN THE CROP—PRODUCTION FIELD

This includes fields whose marketable product is something other than seed used for planting. Following field planting of a given seed lot, the first appearance of a disease in a new area or one appearing with unusual severity, may institute studies on seed transmission, but generally not on the mechanisms involved. This approach has tended to emphasize seed treatment for disease control, rather than production and use of pathogen-free seed. In some cases this is the way the prevalence of a pathogen in a seed crop is discovered. For example, *Alternaria zinniae* on zinnia (Baker and Davis, 1950b; Baker, 1956), which produced no recognizable symptoms in California seed fields, was detected only when the seed was planted in areas of high summer rainfall. On the other hand, appearance of a disease in a crop planted on land for the first time may suggest seed transmission (and is often so reported), but does not prove it. The pathogen may have been unrecognized in weed hosts, or may occur in the rhizosphere of nonhost crop plants. Thus, *Pseudomonas tabaci* and *P. angulata* occur commonly on roots of nonhosts, and will cause wildfire or angular leafspot on tobacco planted there. "There appears to be no reason for assuming that outbreaks of either angular leaf spot or wildfire arise other than from colonies of the respective pathogenic organisms already present in the soil before outbreaks occur" (Valleau *et al.,* 1944). Mishagi and Grogan (1969) found what appeared to be *P. tabaci* of reduced virulence, on nonhost roots near Davis, California, an area where tobacco had not been grown.

*Seed transmission* is demonstrated in studies of this type. Infection may already have occurred in the embryo, as in *Ustilago tritici* in wheat, or seed may be invaded and the fungus spread from it through the soil to infect a neighboring seedling, as in *Rhizoctonia solani* on pepper (Fig. 5.1F). Investigations must be made in the field, laboratory, and greenhouse to reveal details of pathogen transfer from the seed to the seedling produced (SP) and then from plant to plant (PP) in the field.

4. STUDIES IN BOTH THE SEED-TESTING LABORATORY AND IN SEED— AND CROP—PRODUCTION FIELDS

The ideal way to study seed pathology is, of course, to combine all three methods through investigations on specific diagnostic and indexing methods in the seed-testing laboratory, on the PS transfer of the pathogen in seed fields, and on the SP and subsequent PP transfers in crop—production fields. Because this has rarely been possible, the literature largely represents one or another single viewpoint. The investigations of J. C. Walker on seed-borne pathogens of crucifers are outstanding exceptions. Cooperative studies between seed pathologists in the three types of situations would greatly strengthen investigations on seed pathology.

Specialization in seed pathology is, unfortunately, much less common than the importance of the subject merits. The number of members listing seed pathology as a specialty in the membership lists of the American Phytopathological Society are as follows: In 1953, there were 13 such members (1.0%); 1958, 10 (0.6%); 1963, 19 (0.9%) (McCallan, 1953–1963). The classification was not included in 1968. There are, however, many hundreds of published papers demonstrating seed transport or transmission of a wide range of pathogens. These investigations are usually part of a general study of a given pathogen or disease. The authors of such papers are generally pathologists with transient interest in seed pathology as it relates to their particular problem, rather than as a major field of research. There are thus an appreciable number of workers interested in this field, but only a few specialize in it.

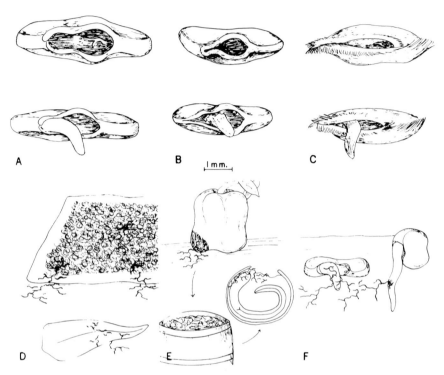

FIG. 5.1. Transmission of *Rhizoctonia solani* in seeds. A–C. Natural large openings in seed coats of pepper (A), eggplant (B), and tomato (C), through which infection occurs. Lower row (A–C) shows beginning germination of seeds. D. Invasion of zinnia flowers piled on canvas on infested soil. *Rhizoctonia* grows through canvas into the flowers piled on it, and through the petals into attached seeds. E. Pepper fruits, infected from contact with infested soil, are chopped up and placed in barrels to ferment prior to separation of seeds; *Rhizoctonia* permeates the mass and penetrates the seeds. F. *Rhizoctonia* grows out from infected seed when planted, and attacks neighboring seeds, as well as infests the soil. [Illustration by Lily H. Davis. A–C from Baker (1947, p. 915). D–F from Baker (1952, p. 38).]

### D. Lists of Seed-Borne Pathogens

The numerous lists of seed-borne pathogens illustrate the restrictiveness of the above approaches. Since many records are based simply on the presence of a pathogen on, in, or accompanying a seed, they largely report seed transport, with actual transmission all too commonly assumed.

The lists have been organized in a number of ways: (*1*) Reports of routine examinations in a seed-testing laboratory (H. Andersen, 1952–1956, 1961; Clark and Page, 1960–1969; numerous reports in Proceedings of the Association of Official Seed Analysts of North America, and Proceedings of the International Seed Testing Association, by W. F. Crosier and C. E. Heit; Neergaard, 1936–1951, 1956–1960, 1962–1967). (2) Reports incidental to general discussions (Chen, 1920; Crocker and Barton, 1953; Doyer, 1938; International Seed Testing Association, 1958—, annual; Neergaard, 1940; Porter, 1949; U. S. Department of Agriculture, 1952). (3) Reports emphasizing types of pathogens (Bennett, 1969; Cain and Groves, 1948; Carter, 1962; Crowley, 1957; Fulton, 1964; Groves and Skolko, 1944–1946; Meehan, 1947; Skolko and Groves, 1948, 1953; Southey, 1965). (4) Reports emphasizing the hosts (LeClerg, 1953; Malone and Muskett, 1964; Muskett and Colhoun, 1947; Noble and Richardson, 1968; Orton, 1931; Sampson and Western, 1954; Conners, 1967).

### E. Significance of Seed Transmission of Pathogens

Transmission of a plant pathogen with any plant propagule is of agricultural importance because an efficient means of its transfer in space (dissemination from place to place) and time (carryover from season to season) is provided. It is, unfortunately, not commonly appreciated that seed transmission has several additional, unique, and significant features. To consider it as just another means of pathogen dispersal or carryover is to miss the biological implications and adaptations involved.

### 1. PROLONGED TRANSMISSIBILITY

Seeds remain viable for a much longer time than do vegetative propagules, thus prolonging the potential transmission period and the possibility of dissemination. Pathogens usually remain viable longer when associated with plant tissue than they do separately or in soil, and this is especially true in dormant seeds. Both the pathogen and generally the seed survive for several years, the disease reappearing in the seedlings produced. However, a few pathogens [*Septoria apiicola* on celery (Gabrielson, 1961; Sheridan, 1966; J. C. Walker, 1969), *Cercospora beticola* on beet (Aebi and Rapin, 1954), *Gloeotinia temulenta* on ryegrass (Sampson and Western, 1954), *Glomerella gossypii* on cotton (Arndt, 1946), cherry (Prunus) necrotic ringspot virus in *Prunus pennsylvanica* (Fulton, 1964), and squash mosaic virus in muskmelon (Rader *et al.,* 1947)] die or are inactivated before seed viability is seriously reduced. This fact is sometimes used in control of these diseases (see pp. 347 and 382).

Survival of spores of *Selenophoma bromigena* on seed of smooth bromegrass dropped rapidly to zero when stored at 70% relative humidity for 16 months, but survived at and below 50%. Seed germination was satisfactory after storage at or below 70% relative humidity, but not above that level (J. D. Smith, 1970). These results may explain some of the contradictory conclusions from studies where humidity was not controlled.

## 2. MAXIMUM INFECTION

Because of the intimate association of host and parasite in seed there is maximum opportunity for progeny infection. When the embryo is infected there is at least as great a chance of successful transmission as with infected vegetative propagules. Because the embryo is already infected, there is decreased opportunity for restriction of primary disease development by unfavorable environment (see p. 378). Seed transmission is more likely to establish a pathogen in a new locality than is aerial dissemination, or importation of research cultures of the pathogen. Despite this, quarantines on cultures are quite strict, but those on seeds are extremely variable and lax (see pp. 342–343).

## 3. DISSEMINATION OVER LONG DISTANCES

Dissemination of a pathogen over long distances is favored by association with seed more than with vegetative propagules for reasons 1 and 2 above. Establishment of pathogens in many isolated areas has undoubtedly resulted from seed transmission. Evans (1971) showed that transmission of *Verticillium dahliae* in fruits of *Xanthium spinosum, X. pungens,* and achenes of *Corthamus lanatus* may have been the means whereby the pathogen was introduced into the Namoi Valley, New South Wales.

Seed transmission is the principal means of dissemination of bean common mosaic virus in beans (Fajardo, 1930), barley stripe mosaic virus in barley and wheat (Slykhuis, 1967), lettuce mosaic virus in lettuce (Grogan *et al.,* 1952), and cucumber mosaic virus in *Marah (Echinocystis)* (Doolittle and Gilbert, 1919), *inter alia,* to new areas.

## 4. PREFERENTIAL SELECTION TOWARD PATHOGENIC STRAINS

There is a strong preferential selection toward strains of a pathogen virulent for a given host or variety in the seed field. Growth of that particular host in the seed-production field will tend to increase the amount of inoculum of pathogen strains able to attack it, and thus increase the probability of infection or contamination of seed by that strain. When the seed is planted in a new area, the pathogen is introduced with it. Strains of *Rhizoctonia solani* virulent to pepper or tomato thus increase in seed fields, infect fruit in contact with the soil, invade the seed, and may be carried to a new area (Figs. 5.1E, F, 5.4; Baker, 1947). Because this pathogenic species is widely distributed (Baker, 1970), and this selective process is not fully appreciated, the significance of seed transmission of *R. solani* is, unfortunately, often minimized. *Fusarium solani* f. sp. *cucurbitae* strain 1 similarly infects squash fruits and becomes seed borne, causing foot rot of the plant (Toussoun and Snyder, 1961). This selection for virulence to fruit and seed may not, however, necessarily select for aggressiveness to stem or roots; thus, although strain 2 of *F. solani* f. sp. *cucurbitae* is seed borne, it does not cause foot rot.

## 5. RANDOM INFECTION FOCI IN CROP–PRODUCTION FIELD

Planting pathogen-carrying seed in a crop–production field introduces the disease entity randomly through the area, providing numerous well-distributed foci of primary infection. Such inoculum is usually more effective in starting a disease outbreak than is that arriving from a distance (e.g., virus-infective insects or wind-blown spores falling in the field) or spreading inward from the margins of the field. This is why seed is the important means of initiating mosaic infection in lettuce fields in California (Grogan *et al.,* 1952). Virus spread by aphids flying in from infected prickly lettuce or other susceptible weeds on the margins of the field is relatively unimportant, whereas if any infected seeds are found in 30,000 indexed, an unacceptable amount of mosaic will result (Greathead, 1964; Zinc *et al.,* 1956).

## 6. PATHOGEN SURVIVAL DEPENDENT ON SEED TRANSMISSION

Some pathogens may be dependent on seed transmission for their survival. According to Slykhuis (1967), barley stripe mosaic virus in barley and wheat essentially is dependent on seed transmission for its perpetuation.

The vectors of nematode-transmitted viruses are obviously far more limited in mobility than are the usual insect vectors of other viruses. Seed transmission, therefore, assumes greater significance for these viruses (Cadman, 1965b; Murant and Lister, 1967), and most of them are transmitted in a high percentage of the seeds of many plant species, often without producing symptoms in the seedlings. Lister and Murant (1967) found that tomato black ring, raspberry ringspot, and arabis mosaic viruses are carried in high percentage in many weed seeds, and Cadman (1965b) considered that "soil-borne plant viruses . . . essentially . . . are pathogens of wild plants." Such seed dissemination probably explains how these viruses, despite having vectors unable to transport them over a distance, have become widespread. Murant and Lister (1967) found tomato black ring and raspberry ringspot viruses in weeds in areas free of nematode vectors, and showed that virus-free populations of vector nematodes acquire these viruses in the field from weed seedlings from infected seeds.

The possibility of wind dispersal of the eggs of the nematode vectors with soil particles seems not to have been explored; the viruses probably would not be spread in this way. Fallowing fields for 9 weeks eliminated infective vectors; when the infected weed seeds germinated, however, nematodes again became infective.

No viruses having fungus vectors which infect plants by root invasion are known to be seed transmitted in any of their hosts (Cadman, 1965b). What means of effective dissemination, independent of human activities, these viruses may have is not known, but wind dispersal of the virus-infected resting spores of the fungi with soil particles seems a likely possibility. The fungi shown to be vectors of plant viruses are *Olpidium brassicae, Synchytrium endobioticum, Polymyxa graminis,* and *Spongospora subterranea* (Calvert and Harrison, 1966; Estes and Brakke, 1965; Grogan and Campbell, 1966). These Phycomycetes (Orders Chytridiales and Plasmodiophorales) form resting spores which could be carried with seed in small lumps of soil [as is *Plasmodiophora brassicae* with turnip (Warne, 1944)] or with dust on the seed. These viruses could thus occasionally be seed transmitted.

An interesting example of the effectiveness of seed transmission of *Ditylenchus dipsaci* on *Hypochoeris radicata* was observed by the late M. B. Linford and me near the summit of Haleakala Crater in Hawaii in 1937. The 10,025-foot summit of this extinct volcano has very deep deposits of ancient fine volcanic ash which is highly permeable to the small amount of rain received; consequently, very few plants grow on this bare site. Isolated small annual *Hypochoeris* were scattered over the area, and a fair percentage of them were infected with the nematode. Because of the isolation of individual plants (often more than a hundred feet from the nearest plant of any kind), the annual nature of the plant in this habitat, and the virtual impossibility of rainfall runoff, infected plants represented individual seed transmission as described by Godfrey (1931). As each plant died, the nematode population around it would also soon perish. Under these conditions survival of the nematode obviously depended on its seed transmission with the host.

7. INCREASED SEED TRANSMISSION ON PLANTS GROWN FROM INFECTED SEED

Plants arising from infected seed may give a higher percentage of seed transmission than from plants infected during the growing season. Fajardo (1930), Nelson (1932), and Harrison (1935) have shown this for bean common mosaic on bean. Tomlinson and Carter (1970) showed that 21–40% of the seed produced by chickweed plants grown from seed naturally infected with cucumber mosaic virus transmitted the virus, whereas only 3–21% of the seed produced by manually-infected plants transmitted it.

8. UNIFORM PATHOGEN INVASION OF PLANT

A plant grown from infected seed may be more uniformly invaded than one infected later (Lister and Murant, 1967), and may be symptomless (see pp. 339, 341, and 387). These plants may thus provide a better source of virus for insect vectors, and are difficult to detect or to rogue from the field.

9. INDUCTION OF DISEASE BY TWO SEED-TRANSMITTED PATHOGENS

Two seed-transmitted pathogens may act together in inducing a disease. Sabet (1954) showed that *Corynebacterium tritici* occurred on the surface of the cockles induced on wheat by *Anguina tritici*. When such seeds are planted in soil, the nematodes carry the bacteria into the coleoptile and produce the yellow slime disease. Inoculations with bacteria, with or without nematodes, were unsuccessful in large plants.

10. POTENTIAL IMPORTANCE OF SEED TRANSMISSION OF DIFFERENT PATHOGENS

Seed transmission of all pathogens is not equally important or dangerous to crop plantings. Arranged in descending order of importance, and neglecting the effect of environment and economic importance of the crop, the types of seed transmission are as follows.

*a.* The pathogen is new to the area and able to survive in soil (*Fusarium oxysporum* f. sp. *callistephi* on China aster) or to become established in persistent hosts (lettuce mosaic virus in weeds) more or less permanently.

*b.* The pathogen is new to the area and able to infest soil for short periods (*Alternaria zinniae* on zinnia) or until host refuse decays (*Septoria apiicola* on celery).

*c.* The pathogen is new to the area and is not soil infesting (e.g., the appearance of *Puccinia antirrhini* on snapdragon in Australia in 1952).

*d.* The pathogen is already present in the area and has infested the soil (*F. oxysporum* f. sp. *callistephi* on China aster, *F. solani* f. sp. *cucurbitae* on squash, *Rhizoctonia solani* on pepper) or the pathogen is established in persistent hosts (lettuce mosaic virus in weeds) more or less permanently.

*e.* The pathogen is already present in the area and is able to infest the soil for short periods (*A. zinniae* on zinnia, *Xanthomonas incanae* on *Mathiola* in California) or until the host refuse has decayed [*S. apiicola* on celery, *Heterosporium* (*Acroconidiella*) *tropaeoli* on nasturtium, *X. campestris* and *Phoma lingam* on cabbage, *Pseudomonas phaseolicola* on bean, TMV on tomato].

*f.* The pathogen is already present in the area and is not soil infesting (*P. antirrhini* on snapdragon in California).

Pathogens able to persist in field soils or in endemic or persistent hosts are obviously the most dangerous to introduce. Once established in this way, it is extremely difficult, or almost impossible in the case of soil-infesting pathogens, to eliminate them. The wider the host range of the pathogen, the more difficult is its eradication.

### F.  Frequency of Seed Transmission of Pathogens

The succession of events, leading from the plant → seed (PS) transfer, through harvesting, cleaning, storage, packaging, shipment, treatment, and planting of the seed, to the seed → plant (SP) transfer is usually very complex. It may, therefore, be successful in only one seed of many thousands planted, and perhaps only in occasional years. This rarity and the difficulties of demonstration have contributed to lack of general appreciation of the significance of seed transmission, but do not diminish its real importance.

Certification schemes often establish tolerance levels for pathogen infection of seeds. de Tempe (1968a) has suggested that, since the level of inoculum necessary for a disease outbreak is affected by environmental conditions after sowing, the "threshold level for pathogenicity" of a given pathogen cannot be a constant. This philosophy seems to be at the heart of the tolerance concept of certification. It emphasizes producing seed only as clean as necessary, not as clean as economically feasible, and thus leads to disease control rather than disease prevention.

Of interest in this connection is the development of an acceptable tolerance for lettuce mosaic in lettuce seed planted in commercial fields in California. The initial figure of 0.1% infection (Zinc et al., 1956) was found to be too high for effective control, and had to be lowered to zero transmission in 30,000 plants. This initial leniency hindered the adoption of effective field control (R. G. Grogan, personal communication).

A lenient viewpoint of tolerance levels implies that it is difficult or impossible to produce pathogen-free seed, or that economic loss from the diseased plants produced will be less than cost of producing seed free of the pathogen or of treating seed to free it of the disease agent. It ignores the importance of introducing a new pathogen or strain of it to a new area, as well as the rapidity with which some pathogens can increase from a small initial population (van der Plank, 1963). It is only safe to adopt such a casual attitude when it is known that the strain being introduced is identical to that already present, and even then infected seeds add to the existing inoculum level.

Tolerance levels of the percentage of infected or infested seed, the number of sclerotia per pound, or some similar designation, vary widely. The bases of these tolerances unfortunately often seem to be more a matter of expediency than of scientific fact. The levels vary with the importance of the crop and prevalence of the pathogen in the given country. They range from 0–35%, for example, in Denmark and the Netherlands. They may vary from year to year, generally rising in years of severe infestation. Canada's tolerance of *Ascochyta* spp. on pea may thus vary from 2–6%, and in Denmark "the tolerances for *Alternaria zinniae* Pape on *Zinnia elegans* and other *Zinnia* species fluctuate somewhat because in some years most seed lots may be considerably infected, and the export would be hampered, if in such years almost all seed lots should be refused health certificates" (Neergaard, 1962). Such an approach to certification of seed would be difficult to defend on legal or scientific grounds. The present attitude on quarantine regulation of seed shipments probably reflects in part the lack of reliable rapid techniques for determining pathogen transport and transmission.

When certification schemes are based on a sound scientific basis, however, they can be extremely valuable in disease prevention. It has been demonstrated (Greathead, 1964; Zinc *et al.,* 1956) that lettuce mosaic, for example, is effectively controlled in the important Salinas Valley, California lettuce-producing area only when the seed planted has less than 0.003% infection (i.e., no mosaic-infected plants found in 30,000 seeds indexed in the greenhouse). This has reduced the disease to minor importance, even though weeds are commonly infected in the vicinity.

*Pseudomonas phaseolicola* was found by J. C. Walker and Patel (1964) to cause severe epidemics in Wisconsin bean fields from as few as twelve infected seeds per acre (about 0.02%).

Lin (1939) found that, if there were ten primary lesions (ten infected seeds) of *Septoria apiicola* in a celery seedbed, there would perhaps be $15 \times 10^5$ spores produced long before the seedlings are transplanted. A few infected seeds can thus give rise to a severe outbreak in the field crop.

The greater the potential rate of spread of a pathogen, the more important seed transmission becomes, and the nearer to zero must the tolerance for seed infection be. A small number of diseased plants may be difficult to detect, and may themselves be economically unimportant in seed fields, but they produce seeds which may cause heavy loss in crop-production fields. This is because of the high rate of increase or spread (generation time and quantity of inoculum produced) (van der Plank, 1963).

The percentage of infected seeds in commercial lots varies enormously. Some pathogens occur only in occasional seeds, as with sun-blotch in some clones of avocado (Wallace and Drake, 1962). Although well-cleaned commercial seed usually has less than 5% of pathogen-infected seed, some lots may run much higher. *Corynebacterium michiganense* may occur in less than 1% of tomato seed (Grogan and Kendrick, 1953), *Ditylenchus dipsaci* in 3% of onion seed (T. Goodey, 1943), *Xanthomonas carotae* in 4.3% of carrot seed (Ark and Gardner, 1944), *Verticillium albo-atrum* in about 9% of spinach seed (Snyder and Wilhelm, 1962), *Septoria apiicola* in up to 40% in celery seed (Sheridan, 1966), *Verticillium dahliae* in 40% of seed of *Xanthium pungens* (Evans, 1968), bean common mosaic well over 50% in commercial seed lots (Schippers, 1963), *Epichloe typhina* in up to 99% of red fescue seed (Sampson and Western, 1954), and *Alternaria zinniae* nearly 100% in zinnia seed (Baker and Smith, 1966). Bennett (1969) listed forty-eight seed-transmitted viruses in about 120 virus–host relationships: of these, sixteen viruses in thirty-one host–virus combinations had 50–100% seed transmission.

The following pathogens have, surprisingly, been reported as having seed transmission: *Plasmodiophora brassicae* on turnip (Warne, 1944); *Puccinia carthami* on safflower, infecting the roots (Zimmer, 1963); and *P. antirrhini* on snapdragon (Baker, 1952; J. Walker, 1954). Seed transport has also been demonstrated for *Erysiphe polygoni* on pea (Crawford, 1927; Uppal *et al.,* 1935) and parsnip (Green, 1946), and *E. cichoracearum* on zinnia (Baker and Locke, 1946), but transmission was merely assumed or was unsuccessful. Because of this situation, seed pathologists have, understandably, become increasingly wary of stating that any pathogen could not be seed borne. The complete roster of seed-transmitted pathogens is undoubtedly much greater than is now known.

### G. Criteria for Demonstrating Seed Transmission

A number of rather obvious, but sometimes overlooked, criteria must be satisfied in order to conclusively demonstrate seed transmission of a pathogen.

1. There must be no alternative source of the pathogen, such as air-borne spores.

2. If the causal agent is insect borne, there must be no insect vectors present. The frequently cited report (Jones, 1944) of up to 96% seed transmission of cineraria streak (spotted wilt virus) is invalid because of admitted uncontrolled thrips in the experiments (Baker and Smith, 1966). The virus occurs in seed coats of cineraria (Crowley, 1957), but this does not lead to seedling infection.

3. If the pathogen is also carried over in soil, it must be demonstrated that it is not present in the soil used.

4. Demonstration of seed transmission of a pathogen consists of more than placing seeds on sterile agar plates or in autoclaved soil. It is clearly

demonstrated (see pp. 375 and 378) that microorganisms on the seed or in the soil have a marked effect on whether infected seed planted in it will produce diseased seedlings (the SP transfer). Seed being studied should be planted in soil treated in the following ways:

*a.* Autoclaved soil free of pathogens and of nearly all saprophytes. Because there will be no inhibitory effect by other microorganisms, this series will indicate the *maximum potential amount* of infection; in this way it resembles planting on agar plates.

*b.* Soil treated with aerated steam at 60°C for 30 minutes. This will eliminate all soil-borne pathogens, but leave much of the remaining flora (Baker, 1962a; Bollen, 1969). This will afford evidence on the *extent to which antagonists are inhibiting seed transmission,* and may provide a clue as to what they are. Selective killing of some of the antagonists may be obtained by treatment of the soil for 30 minutes at different temperatures. The lowest temperature should be the minimum necessary to destroy the pathogen in soil, and the others at 5.6°C increments up to 82.2°–87.8°C.

*c.* Untreated soil which has been demonstrated to be free of the pathogen in question by planting clean seed in it. If possible, this soil should be from fields where the seed is to be planted. The environment should be favorable for development of the disease. This will indicate the *actual potential amount* of infection which may be expected when the seed is sown in the field.

5. Successful transmission of the pathogen in question must be confirmed by suitable tests. The symptoms observed may be due to some other pathogen, and the diseased seedlings must, therefore, be cultured and pathogenicity determined. Symptom expression alone may not be enough to establish transmission. Because of the well-known difficulty of distinguishing bacterial plant pathogens from saprophytic forms and from one another, carefully conducted inoculation tests are often required. Test conditions (e.g., moisture, duration of incubation) must not be too severe. It should be noted that "concentrated bacterial suspensions may produce disease symptoms, but do not necessarily demonstrate ability to function as a natural pathogen" (Mishagi and Grogan, 1969).

Conversely, absence of symptoms may not indicate absence of transmission of a virus, since some viruses may be symptomless in seedlings. This is true, for example, for sun-blotch virus on avocado (Wallace and Drake, 1962), Prunus (necrotic) ringspot virus on peach (Cochran, 1946), elm mosaic virus on elm (Callahan, 1957), raspberry ringspot virus on raspberry and strawberry, and tomato black ring virus on strawberry (Lister and Murant, 1967).

## H. *Importance of Seed Pathology in Seed— and Crop—Production Fields*

The PS transfer of the pathogen is extremely important in the seed production field, and control measures (pp. 380–385) usually aim to prevent it.

A disease which causes severe losses in seed fields and is unimportant or even unknown in crop production because the pathogen is not seed borne is called a production disease. Pea enation mosaic in sweet pea, aster yellows in delphinium, tomato spotted wilt in Compositae, and curly top in petunias, are present in greater or lesser degree in California seed fields each season. Because they are not seed transmitted they present no risk to the seed purchaser. If they become sufficiently damaging to reduce seed yield they may indirectly increase the cost of seed, but otherwise they are unimportant to the consumer. *Pythium* and *Phytophthora* spp. also cause losses from root rot in seed fields, but are extremely rare in lists of seed-borne pathogens. Although *Sclerotinia sclerotiorum* and *S. minor* cause white blight, stem rot, and cottony crown rot in many seed crops, seed transmission has not been demonstrated for most of them, perhaps because sclerotia are eliminated in cleaning operations.

Some diseases are important in seed fields, are seed transmitted, and still are unimportant in crop fields. Fusarium wilt of *Mathiola* is such a case (Baker, 1948a). The plant has a cold requirement for flower induction and is, therefore, grown in the winter months in cut—flower fields, and at low temperatures in glasshouses. *Fusarium oxysporum* f. sp. *mathiolae* does not cause important losses at such temperatures. It requires nearly a year to produce seed of this crop, and the plants flower in winter and mature seed in summer under conditions favorable to fusarium wilt. Losses are thus severe in seed fields, but not elsewhere. *Heterosporium tropaeoli* causes severe disease losses in California nasturtium seed fields, but, for undetermined reasons, it is absent or unimportant elsewhere, though it is commonly seed transmitted (Baker and Davis, 1950a). Control has centered, therefore, on reduction of losses in seed fields rather than on producing *Heterosporium*-free seeds (Baker, 1956).

Some pathogens may not produce detectable symptoms in a seed crop, even though they are present and may infest or infect the seeds produced. *Alternaria zinniae,* for example, rarely produces symptoms in California seed fields. Seedlings from infected seed are killed, and on them the fungus sporulates copiously. Because of the dry climate, only rarely are further above-ground symptoms produced, and the fungas apparently survives in the soil until fall. With maturation of flowers, the dead petals absorb moisture from the heavy dew and frequent fog of fall nights. Flower heads are then attacked by molds such as *Cladosporium, Penicillium,* and *Alternaria.* Among them may be found *A. zinniae,* which grows down the petal and into the fruit coat and the seed (Fig. 5.2A; Baker, 1956). *Corynebacterium fascians* produces no symptoms on sweet pea and nasturtium seed crops in California. Since sweet pea is grown without

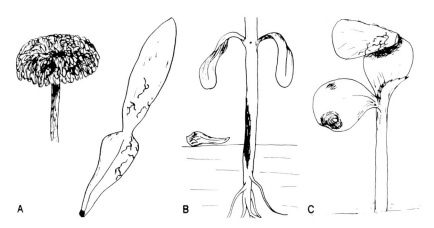

FIG. 5.2. *Alternaria zinniae* transmission in zinnia seeds. A. Senescent flower on the plant absorbs dew and fog at night, remains wet the next day; the heads become moldy from saprophytes and *A. zinniae,* which grows through the petals into attached seeds. B. The seed coat falls onto soil after epigeal germination, and the fungus sporulates on it. Spores spread to the hypocotyl, infect, and produce a girdling lesion. Seedling decays and produces copious spores. C. *Stemphylium callistephi* on China aster seed coat, which adheres to cotyledons during epigeal germination; fungus infects the host. *A. zinniae* does this also. [Illustration by Lily H. Davis. From Baker (1952, p. 38).]

irrigation, conditions are too dry near the crown to favor growth of the bacteria on the plant surface. Although nasturtiums are irrigated by ditches, the surface soil also apparently dries out too rapidly to permit production of symptoms. The bacteria must both increase and persist in these soils, however, because seed from these crops grown in continuously moist soil in the glasshouse produces infected plants with striking proliferation of axial buds. Nasturtium seeds treated with hot water (51.7°C/30 minutes) produce seedlings which grow normally (Baker, 1950).

Some viruses may be symptomless in the host, at least under certain environmental conditions, and still produce infected seed. Sun-blotch virus on avocado (Wallace and Drake, 1962), raspberry ringspot virus on strawberry and *Capsella bursa-pastoris* (Lister and Murant, 1967), and tobacco ringspot virus on tobacco (Valleau, 1932) and soybean (Desjardins *et al.,* 1954) are examples. Point-of-origin inspection and certification in such cases is obviously impossible, and appropriate indexing methods are required.

Seed pathologists are largely concerned with pathogens that are important in both seed— and crop—production fields. *Septoria apiicola* on celery, *Pseudomonas phaseolicola* on bean, and lettuce mosaic on lettuce are examples of this type.

In crop—production fields the SP transfer is the significant one, and control measures (pp. 385–395) aim at its prevention or elimination through planting seed free of or freed from the viable pathogen, or at the reduction of the plant → plant (PP) spread.

### I. Significance of Seed Transmission of Pathogens in Certification and Quarantine

Seed-transmitted plant pathogens probably provide one of the biggest gaps in present-day quarantine protection. It is paradoxical that quarantine regulations generally are less strict for seed than for infected or infested plants, plant parts, or soil, when they are at least as effective in carrying pathogens across political boundaries (W. C. Moore, 1946, 1953, 1954, 1957; Plant Quarantine Division, 1965; Ryan *et al.,* 1969; Sheffield, 1968) ( see pp. 331–338, and 386).

Leppik (1968) has pointed out that importation of seed from the gene centers of a crop plant for breeding purposes may introduce new races of a pathogen, as well as desirable new breeding stock. The gene center may also be the center of origin of the crop's specialized pathogens and pests. He suggested that a postentry quarantine service for all imported seed be established, that it produce and distribute pathogen-free seed derived from imported infected material, and that world distribution of seed-borne pathogens (based on their detection in seed samples) be mapped. The last suggestion could be dangerously inaccurate because the seed company name and location on the seed bag or packet is no assurance that this company produced it. Seed is commonly purchased by companies from other growers in other states or countries, and marketed under their own label. Lack of appreciation of this fact has already led to many erroneous published records.

Quarantine regulations probably represent a realistic recognition of the difficulties of determining whether seeds are carrying pathogens. It is sometimes impossible to determine by point-of-origin inspection of the growing seed crop whether a pathogen is present (pp. 340 and 385). It is also difficult to determine whether a pathogen is present by examination of the harvested seed (pp. 336, 337), and indexing methods may not be effective because embryonically infected seedlings may not exhibit symptoms (pp. 333, 335, 339, 341, 387). There is great need for development of better methods for indexing seed lots for pathogens. However, the variability of results of the Cooperative Seed Health

Tests of the Committee on Plant Diseases of the International Seed Testing Association unfortunately raises questions about the reliability of such tests (Baker and Smith, 1966). For example, uniform lots of pea seed were submitted in 1961 to twelve experienced workers in ten countries, using uniform methods. The percentages of seed found to be infected with *Ascochyta pisi* varied from 0 to 11.0 (see p. 386). Similar samples of carnation seed, examined in 1963 by nineteen laboratories in fourteen countries, were reported to have from 0.3 to 13.8% of the seed infected by *Alternaria dianthicola*. Limonard (1968) comments that "the results of the 'referee tests' for most infections showed a deplorable lack of agreement." He conducted studies at Wageningen, with U.S. Department of Agriculture support, to determine the reasons for such variability of results. "The study . . . led to the view that the factors influencing the results of incubation tests were far more complex than was originally expected." It is unlikely that seed inspection for quarantine purposes would be more accurate.

Properly designed schemes based on sound biological facts provide, however, an effective means of reducing disease losses to unimportance. This is shown by the control of lettuce mosaic in California and Arizona (see p. 337) and of *Ascochyta* spp. on peas in the United States canning industry. The Certified Tobacco Seed program in Rhodesia and Nyasaland (Bates, 1963) seems to be another sound and successful scheme. The essential difference between these programs and quarantine procedures would appear to be their limited scope, permitting greater attention to significant features of host–pathogen relationships.

Indexing of seeds for virus transmission is complicated by the fact (see pp. 333, 335, 339, 341, 387) that there are a number of known examples in which seedlings from virus-infected embryos remain symptomless, and can be detected only by inoculation to healthy seedlings, which will then exhibit symptoms. It is also complicated by the extremely low tolerance level (or even absence) of seed transmission of some viruses (e.g., no infected seedlings in 30,000 for lettuce mosaic) required for successful field control. Because the indexing method must then detect a very rare event, large samples of seeds must be planted under environmental conditions favorable for symptom expression, and the seedlings must be very carefully examined or perhaps inoculated into known healthy seedlings. This procedure obviously is costly and laborious.

The interesting possibility has been suggested verbally by several seed pathologists that some seed-transmitted pathogens have such a tenuous survival that it is well within man's power to eliminate them, at least in restricted areas. This is, of course, already achieved in the production of many glasshouse ornamentals (Baker, 1957; Committee on Seed and Plant Material Certification, 1956; Hollings, 1965) and fruits (Nyland and Goheen, 1969) in nurseries. Such procedures are necessarily based on a zero tolerance for transmission in seeds or other propagules. This is in striking contrast to the existing situation for seeds in many areas today (see p. 336).

### J. Seed Anatomy in Relation to Pathogen Transmission

Because of the intimate relationship of seed ontogeny and structure to infection and transmission of plant pathogens, seed pathology is deeply concerned with seed anatomy. There was intensive study of seed anatomy at the end of the nineteenth century, particularly in Germany, but the subject, unfortunately, has since become unfashionable. A summary of published information in this field, or even a comprehensive annotated bibliography of it, would be a real aid to seed pathologists, and might revive interest in this important and interesting type of anatomical study.

As would be expected, there is great variation in anatomical details among seeds, but numerous generalized features significant in seed transmission of pathogens (Eames and MacDaniels, 1947; Esau, 1938, 1948, 1965; Hayward, 1938; Maheshwari, 1950, 1963; Robbins *et al.*, 1965; U.S. Department of Agriculture, 1952) are discussed here.

A dome-shaped mass of cells, which arises from the placenta of the ovary, differentiates into the ovule, which is connected to the placenta by the funiculus or stalk. The outermost layers of the ovule develop into integuments, which enclose the nucellus or central meristematic tissue in which the megaspore is formed. The ovule may be loosely enclosed in a fruit (e.g., a pod or capsule), or have closely adherent fruit coats (e.g., an achene or caryopsis). All of these structures arise, then, from the sporophytic mother tissue and, with the exception of the megaspores, remain in the diploid condition, and probably are cytoplasmically connected to the mother plant. It would be expected, therefore, that viruses would move into the seed and fruit coats, nucellus, and sometimes the embryo sac formed from the megaspore (Crispin Medina and Grogan, 1961; Crowley, 1959; Gold *et al.,* 1954; Schippers, 1963).

Stamens differentiate from the diploid floral primordium, and vascularization extends through the filaments into the anthers. A compact mass of diploid pollen mother cells differentiates from the parenchyma in the center of the anther primoridum. Each pollen mother cell divides by meiosis to give a tetrad of haploid microspores, which round up to form pollen grains. Cytoplasmic connections occur between the tapetum and the pollen mother cells, and these may persist through the tetrad stage in the pollen mother cells, perhaps until the walls of the pollen grain are laid down (Heslop-Harrison, 1966). One may speculate that virus can move into the tetrad and microspores until the walls of pollen grains are formed. The time of virus invasion would thus be very important, and if maturation of microspores is not uniform throughout the anther, the percentage of seed transmission attributable to pollen would be expected to vary. Furthermore, if this maturation is progressive, outward from the point of attachment of the anther to the filament or linearly in the anther, a method of determining the relationship of the stage of pollen maturation to invasion by the virus is available. There appears to be a greater probability of cytoplasmic connections between the plant and the developing pollen grains than between it and the developing egg cells. This is in accord with the fact that there is greater transmission of bean common mosaic through pollen than through egg cells (Crispin Medina and Grogan, 1961), but this is not always true (Bennett, 1969). There are examples [TMV on tomato (Broadbent, 1965), sowbane mosaic virus on *Atriplex coulteri* (Bennett and Costa, 1961), and apple chlorotic leafspot virus on two *Chenopodium* spp. (Cadman, 1965a)] in which the pollen is infected, but embryo infection apparently does not occur. The above phenomena may be related to the insusceptibility of meristematic tissue, or immunity of the pollen to invasion (Bennett, 1969; Hollings, 1965; Kassanis, 1957; R. H. Taylor *et al.,* 1961).

Pollen sterility is produced by a number of seed-transmitted viruses, and may play a part in preventing production of infected embryos. Examples are tobacco ringspot virus in tobacco (Valleau, 1932), lettuce mosaic virus in lettuce (Ryder, 1964), barley stripe mosaic virus in barley (Inouye, 1962), and Datura quercina virus on *Datura stramonium* (Blakeslee, 1921).

The nucleus of the megaspore mother cell undergoes meiosis, and one of the resulting haploid megaspores becomes the embryo sac. In this structure, mitosis endogenously produces a uninucleate egg cell and two polar nuclei. It is unlikely that the haploid status of the egg cell is responsible for its virus-free condition, since pollen grains may be infected. Furthermore, haploid, diploid, triploid, or tetraploid sugar beets of comparable lines do not differ in resistance to curly top, mosaic, or yellows viruses (Bennett, 1969).

If cytoplasmic connections occur between the nucellus and embryo sac, they would permit virus entry, but one may speculate that the endogenously formed egg cell lacks such connections with the nucellus, thus sealing off the egg from virus invasion. Perhaps, however, if a virus entered the embryo sac before the membrane of the egg cell is formed, the pathogen could be included in it, and the embryo would be infected. After cell walls form around the embryo, there is usually no virus movement into it (Bennett, 1969).

The presence of virus in the embryo does not always result in seed transmission (Cheo, 1955; Gold *et al.,* 1954; Zaumeyer and Harter, 1943). Lack of seed transmission may result from virus inactivation as the seed dries or matures, or it may be related to the insusceptibility of meristematic tissue alluded to above. Since time from differentiation of egg cell and pollen tetrad to initiation of the embryo may be as little as a day (Maheshwari, 1950), delayed movement of the virus into meristematic tissue may be crucial in determining whether embryo infection occurs. The earlier the virus infection of the mother plant, the better the chance of transmission in the seed produced.

Powell and Schlegel (1970a) suggested, on the basis of studies with labeled antibodies, that virus (squash mosaic in cantaloupe) may be located in only small groups of cells in the embryo. They found that 22% of the seeds contained detectable amounts of virus in their embryos, and that 12% of seedlings grown from them contained virus.

Although a virus must have the potential for seed transmission, the host and the environment determine its expression. It has been found that certain species or selected lines of some hosts may have little or no seed transmission of a given virus: lettuce mosaic virus in lettuce (Broadbent *et al.,* 1951; Grogan *et al.,* 1952; Kassanis, 1947); bean common mosaic virus in bean (F. L. Smith and Hewitt, 1938; Crispin Medina and Grogan, 1961); sun-blotch virus in avocado (Wallace and Drake, 1962); citrus psorosis virus in *Citrus* spp. (Childs and Johnson, 1966); coffee ringspot virus on *Coffea* spp. (Reyes, 1961); squash mosaic virus in squash (Grogan *et al.,* 1959); soybean mosaic virus in soybean (Ross, 1963); barley stripe mosaic virus (Eslick and Afanasiev, 1955; Inouye, 1962; McKinney and Greeley, 1965); and tobacco ringspot virus in lettuce (Grogan and Schnathorst, 1955). See also Table 5.II. Some viruses are seed transmitted in one host but not in others [dodder latent mosaic virus transmitted in *Cuscuta* spp. but not in systemically infected cantaloupe, buckwheat, or pokeweed (Bennett, 1944); cucumber mosaic virus in *Marah (Echinocystis) lobata,* but not in cucumber, squash, muskmelon, or pumpkin (Doolittle and Walker, 1925)]. (See also p. 384). As Bennett (1969) stated, "Perhaps one of the most characteristic features associated with seed transmission of viruses is the ability of embryos, infected in the very early stages of formation, to survive, grow, and mature into plants that usually are productive and often almost normal."

Seed transmission is also affected by the environment. Bean common mosaic virus had 16–25% seed transmission in seeds produced at 20°C, but none in those produced at 16.5°–18.5°C (Crowley, 1957). On the other hand, bean southern mosaic virus in kidney bean gave 95% embryo infection in seeds produced at 16°–20°C, but only 55% in those produced at 28°–30°C (Crowley, 1959).

When the pollen grain germinates on a stigma, it develops a pollen tube which grows through the style into the embryo sac through the opening (micropyle) between the integuments, and releases therein two male gametes. One of these gametes unites with the egg cell and, if it carries a virus, the resulting diploid embryo may be infected. The other gamete unites with the polar nuclei and, if it carries a virus, the haploid, diploid, or triploid endosperm which is formed may be infected.

Virus-infected pollen may, in some instances, infect another plant as well as the embryos produced by it. Examples are cherry yellows (Prunus dwarf) virus in sour and sweet cherries (George and Davidson, 1963; Gilmer and Way, 1963), cherry (Prunus) necrotic ringspot virus in cherry (George and Davidson, 1963), raspberry bushy dwarf (apple chlorotic leafspot) virus in raspberry (Cadman, 1965a), and Prunus (necrotic) ringspot virus in squash (Das *et al.,* 1961). It is probable that such transmission occurs by the escape of virus from the infected ovule into the mother plant. As expected, this is rare because rapid movement of viruses is correlated with carbohydrate transport in the phloem, which is more or less unidirectionally toward the fruit. Most seed-transmitted viruses may be mechanically transmitted and are able to invade and move through parenchyma. However, movement through parenchyma cells of the pedicel would be too slow for the virus to invade the mother plant before the fruit was removed in harvesting (Bennett, 1969).

A few viruses appear to be able to invade the embryo directly. Eslick and Afanasiev (1955) found seed transmission of barley stripe mosaic virus in barley even when inoculated in the hard-dough stage of maturity. Crowley (1959) confirmed this direct infection of the embryo by this virus, and by bean southern mosaic virus in bean. The mechanism of such invasion is not understood.

Apparently viruses that are primarily concentrated in the phloem, that are transmitted by leaf hoppers, or are persistent type and aphid transmitted, are not seed transmitted. On the other hand, most nematode-transmitted viruses appear to be seed transmitted (Bennett, 1969). The reasons for these relationships are unknown.

Viruses may retain infectivity for only a short time in embryos [bean southern mosaic virus in bean (Cheo, 1955)] or perisperm [curly-top virus in beet (Bennett, 1942)], or they may persist for long periods. Examples are bean common mosaic in bean for 30 years (Pierce and Hungerford, 1929), sowbane mosaic virus in *Chenopodium murale* for 14 years and Lychnis ringspot virus in *Lychnis divaricata* for 9 years (Bennett, 1969),

**TABLE 5.II.**

Transmission and Nontransmission of Viruses in Seeds of Some Crop Plants

| Virus and host | Variety and % seed transmission | | Varieties showing no seed transmission | Reference |
|---|---|---|---|---|
| Barley stripe mosaic | | | | |
| In barley | Chevalier | 52.1 | | Inouye, 1962 |
| | Harbin | 15.7 | | Inouye, 1962 |
| | Ko-ran | 1.9 | | Inouye, 1962 |
| | Imperial | 0.2 | | Inouye, 1962 |
| | Compana | 63.7 | | Eslick and Afanasiev, 1955 |
| | Titan | 4.4 | | Eslick and Afanasiev, 1955 |
| In oats (several strains) | Cherokee | 8.5 | Ajax, Brandon, Manchuria, Olli, Stateville | McKinney and Greeley, 1965 |
| | Letoria | 9.5 | | McKinney and Greeley, 1965 |
| Bean common mosaic | New York Marrow | 66.1 | Great Northern, Michigan Robust | F. L. Smith and Hewitt, 1938 |
| | Small White | 56.2 | | F. L. Smith and Hewitt, 1938 |
| | Tender Green | 40.0 | | F. L. Smith and Hewitt, 1938 |
| | White Navy | 7.5 | | F. L. Smith and Hewitt, 1938 |
| | Kentucky Wonder | 3.9 | | F. L. Smith and Hewitt, 1938 |
| | Scotia | 2.2 | | F. L. Smith and Hewitt, 1938 |
| | Great Northern × Pinto | 86.0 | Red Mexican × Pinto; Red Mexican × Small White; Idaho Refugee × Pinto | Crispin Medina and Grogan, 1961 |
| | Sutter Pink × Great Northern | 25.0 | | Crispin Medina and Grogan, 1961 |

| Citrus psorosis | Carizzo citrange | 15.0–31.0 | Common citrus varieties | Childs and Johnson, 1966 |
|---|---|---|---|---|
| Coffee ringspot | Coffea excelsa | 10.6 | Coffea arabica | Reyes, 1961 |
| Lettuce mosaic in lettuce | Bibb | 8.0 | | Grogan et al., 1952 |
| | Varieties common in California | 1.0–3.0 | | Grogan et al., 1952 |
| | | | Cheshunt Early Giant | Kassanis, 1947 |
| Soybean mosaic in soybean | Harosay | 24.4 | Merit | Kennedy and Cooper, 1967 |
| | Acme | 9.0 | | Kennedy and Cooper, 1967 |
| | Lee | 11.1 | | Ross, 1963 |
| | Hill | < 1 | | Ross, 1963 |
| Squash mosaic in squash | Zucchini | 5.1 | Early Summer Golden Crookneck | Grogan et al., 1959 |
| | Buttercup | 1.3 | Butternut | Grogan et al., 1959 |
| Tobacco ringspot in lettuce | Paris Island Cos | 3.0 | Imperial 615 | Grogan and Schnathorst, 1955 |

and tobacco ringspot virus in tobacco for 5.5 years (Valleau, 1939). The interesting possibility is suggested by Bennett (1969) that some stable viruses may remain active after the seed is dead; this point has not been investigated. TMV would probably remain infectious in seed coats for very long periods, since it survives in dried stem and leaf tissues for at least 52 years (Johnson and Valleau, 1935); however, this would not likely result in seed transmission.

Virus invasion of endosperm may possibly occur through endosperm haustoria (Maheshwari, 1950). If these structures, which ramify through the nucellus, are able to establish virus-passing cytoplasmic connections with it in a manner similar to haustoria of *Cuscuta* spp., virus invasion could result. The greater surface of the endosperm than the embryo also possibly provides more opportunities for cytoplasmic connections and for virus invasion.

Some of the nucellus may be retained in the mature seed (e.g., in beet) as perisperm storage tissue. As would be expected from its origin, this tissue contains virus (Bennett and Esau, 1936).

Adventitious embryos which may arise from nucellus or inner integuments, and apogametic embryos (those which arise from synergids, antipodal cells, or endosperm) are quite common in many genera (Johansen, 1950; Maheshwari, 1950, 1963). There may be as many as thirteen nucellar embryos in one citrus seed. This condition has been used commercially as a form of vegetative propagation which provides virus-free planting material, although vegetative cuttings do not (Weathers and Calavan, 1959). These virus-free embryos are perhaps formed before the virus moves into this meristematic tissue. Since an occasional embryo is apparently virus-infected, movement into them may be similar to that into apical meristems.

Seed coats contain vascular strands, which may be extensive in large seeds. For example, in lettuce, bean, flax, cotton, castor bean, celery, cucurbits, and tomato, this vascular tissue may extend for some distance, fused with the seed coats. This tissue, the raphe, is a continuation of the funiculus. In some cases there may be vascularization of the nucellus as well (Maheshwari, 1950). Bits of funiculus containing vascular elements often remain attached to the seed at the hilum, or point of abscission. Seeds formed from anatropous or hemitropous ovules (Hayward, 1938), and having such vascularization, usually have particularly favorable sites for internal transmission of vascular pathogens. Microorganisms may spread through the vascular system, and be retained in the raphe following seed abscission. Seeds formed from campylotropous and atropous ovules, on the other hand, contain less internal vascular tissue because most of it is in the funiculus.

Fruit coats have more extensive vascularization than do those of seeds. Vascular pathogens are, therefore, more likely to be transmitted in seeds with adherent fruit parts. *Verticillium* spp. are transmitted in achenes of sunflower, *Senecio,* spinach, and *Corthamus,* and burs of *Xanthium. Cephalosporium gramineum* is commonly transmitted in the lemma surrounding the caryopsis of barley, but is rarely transmitted in wheat seed (G. W. Bruehl, personal communication).

Viruses that are restricted to vascular tissues have not been shown to be seed transmitted, although many [curly top in sugar beet (Bennett and Esau, 1936)] may be seed transported.

There are several openings in seeds through which microorganisms may gain entry. The micropyle is often closed during tissue growth, and rarely is a good infection court. The large natural opening through seed coats of peppers, eggplant, and tomato into the space between the seed coats and endosperm cuticle is easily penetrated by microorganisms (Figs. 5.1A–C, 5.4B; Baker, 1947). The hilum, which is highly absorptive of water, provides a favorable point of entry for bacteria, or at least a protected site for transmission of them (Grogan and Kimble, 1967; Kendrick and Baker, 1942). Cracks in seed coats from threshing injuries or from wetting the seed also provide protected sites for bacteria (Grogan and Kimble, 1967). Details of the legume seed coat are well illustrated by McKee (1970).

Seeds, possibly excepting achenes, develop in a moisture-saturated atmosphere inside the fruit until time of ripening and drying. Any microorganism which gains access to such a natural moist chamber is likely to spread rapidly, infect, and perhaps decay the contained seeds. For example, *Heterosporium eschscholtziae* infects through a small inconspicuous lesion on the capsule and spreads through the interior, producing a moldy mass (Fig. 5.3A–F; Davis, 1952). *Fusarium oxysporum* f. sp. *mathiolae* and *Xanthomonas incanae* enter the silique of *Mathiola* through the vascular system and then spread through the interior, infecting the enclosed seeds (Baker, 1948a; Kendrick and Baker, 1942). Bean pods are directly invaded by *Colletotrichum lindemuthianum,* and pea pods by *Ascochyta pisi,* but only seeds immediately beneath the external lesion are usually infected (Heald, 1933; J. C. Walker, 1969). Only slightly greater spread is exhibited by *Pseudomonas phaseolicola* (Grogan and Kimble, 1967). Whether the phytoalexins, pisatin and phaseolin (Cruickshank, 1965, 1966), are involved in this phenomenon has not been determined.

Fruits with broken exteriors may be invaded by many microorganisms, ranging from saprophytes to obligate parasites [e.g., powdery mildew on pea seeds in opened pods (Crawford, 1927; Uppal *et al.,* 1935)]. Fruits

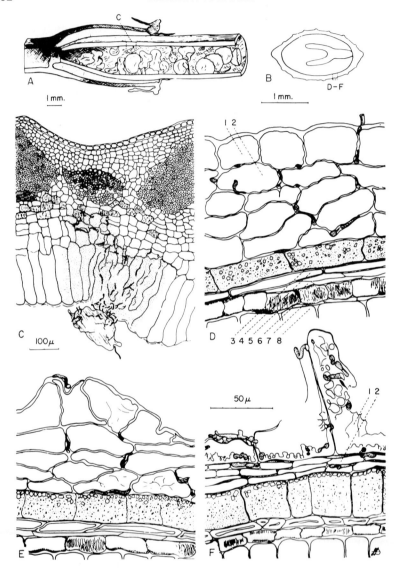

FIG. 5.3. Anatomical aspects of seed infection of California poppy by *Heterosporium eschscholtziae*. A. Diagrammatic section of capsule showing shriveled seeds and darkened areas representing infection. Mycelium passes from the receptacle into the lumen of the capsule through the point of attachment. Location of enlargement C is shown. B. Section of seed showing location of enlargements D–F. C. Section of capsule wall, showing mycelium in wall and in rotted seed, and necrotic area produced. D–F. Sections of very young, nearly mature, and dry mature seeds, respectively, showing mycelium in the integuments. Explanation of numbers: 1, epidermis; 2, parenchyma; 3, crystal layer; 4, fiber layer; 5, cross cells; 6, pigment layer; 7, inner cells; 8, endosperm. [Illustration from Davis (1952, p. 372).]

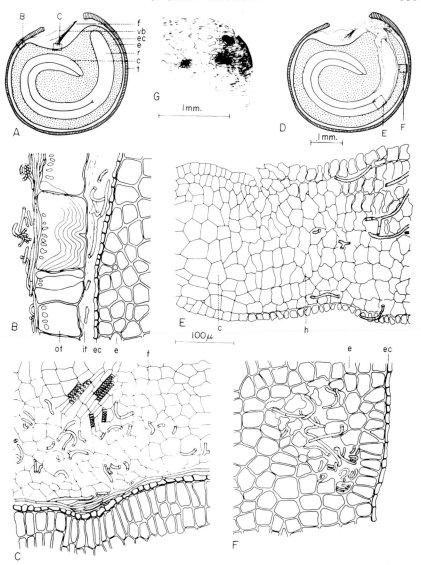

Fig. 5.4. Anatomical aspects of seed infection of bell pepper by *Rhizoctonia solani*. A. Section of seed showing location of enlargements B–C. B. Mycelium on the exterior and in inner layer of the seed coat. C. Mycelium in an attached remnant of funiculus. D. Section of seed showing location of enlargements E–F, with infection of embryo and endosperm. E. Mycelium in hypocotyl. F. Mycelium in endosperm. G. Surface of seed, showing mycelium and small attached sclerotia. Abbreviations: c, cotyledon; e, endosperm; ec, endosperm cuticle; f, remnant of funiculus; h, hypocotyl; it, inner layer of seed coat; ot, outer layer of seed coat; r, radicle; t, seed coat; vb, vascular bundle of funiculus. [Illustration by Katharine C. Baker. From Baker (1947, p. 914).]

Chart 5.1. Schematic Spectrum of Dynamics of Seed Transmission and Transport, Correlated with Importance
of Methods of Preventive Seed Treatment or Manipulation.

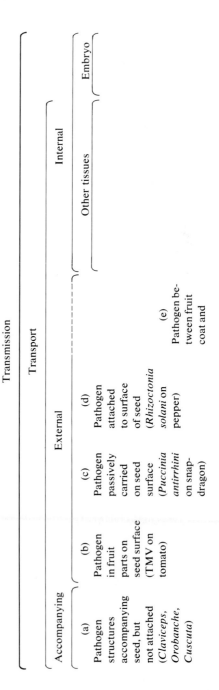

Transmission

| Accompanying | | External | | Internal | |
| --- | --- | --- | --- | --- | --- |
| | | | | Other tissues | Embryo |
| (a) | (b) | (c) | (d) | (e) | |
| Pathogen structures accompanying seed, but not attached (*Claviceps, Orobanche, Cuscuta*) | Pathogen in fruit parts on seed surface (TMV on tomato) | Pathogen passively carried on seed surface (*Puccinia antirrhini* on snapdragon) | Pathogen attached to surface of seed (*Rhizoctonia solani* on pepper) | Pathogen between fruit coat and | |

Transport

attached flower parts (*Ustilago nigra* on barley)

Pathogen in fruit or seed coats or in attached flower parts (*Septoria apiicola* in celery)

(f)

Pathogen in endosperm or perisperm (*Xanthomonas stewartii* in corn)

(g)

Pathogen in embryo (*Ustilago tritici* in wheat)

(h)

Separation

Chemical treatment

Thermotherapy

Selection for non-transmission

Management practices; selection of growing areas; seed-field inspection; indexing; nuclear stock resting on the ground may also be infected by soil microorganisms such as *Rhizoctonia solani,* which thus penetrates into bean pods (Hedgecock, 1904) or tomato, pepper, and eggplant fruits (Figs. 5.1E, 5.4A–G; Baker, 1947), infecting the seeds. The exposed seeds (fruits) of Compositae may be invaded directly by pathogens [*Alternaria zinniae* on zinnia (Baker and Davis, 1950b), *Stemphylium callistephi* on China aster (Fig. 2C; Baker and Davis, 1950c)] which grow down the senescent petal into the fruit coat. In any of the above examples, decayed seeds are usually removed in cleaning operations, but those only slightly invaded are not eliminated, and are an important means of selective spread of strains virulent to that host.

### K. Dynamics of Transmission of Pathogens by Seed

The sites of seed transport and transmission form a spectrum from accompanying pathogen structures to embryo infection (a to h, in Chart 5.I), with roughly increasing probability of successful transmission with this progression.

Types of seed-transmitted pathogens include fungi, bacteria, viruses, seed plants, and nematodes. They range from obligate parasites such as rusts, downy mildews, and viruses, through endophytic fungi of uncertain role (e.g., those in *Lolium temulentum*), to facultative types (such as *Fusarium* and bacteria) and saprophytes common on seed *(Botrytis).* In the last category occur *Coprinus lagopus* on *Beta* and *Tetragonia* (Crosier *et al.,* 1949), and *Aspergillus flavus* which, acting as a pathogen but not as a parasite, may induce albinism in citrus seedlings (Durbin, 1959).

Several mycorrhizal fungi have been thought to be cyclic in nature (carried by seed, infecting the plant systemically, and then invading the coat of the new seeds). Among these are: (*1*) A series of endophytes of *Lolium* and *Festuca* spp. (Neill, 1942; Sampson and Western, 1954), which cause no symptoms on the host and are carried in the seed (fruit) coat. This mixed group of fungi apparently includes nonpathogenic infections by *Gloeotinia temulenta* and *Epichloe typhina,* as well as other unidentified organisms. (*2*) An endophyte of *Helianthemum chamaecistus* and other Cistaceae (Boursnell, 1950). An unidentified fungus is said to invade the seedling from the seed coats, and to become systemic without producing symptoms. (*3*) An endophyte on *Calluna vulgaris,* considered to be seed borne and to infect systemically (Rayner and Levisohn, 1940), without producing symptoms. The fungus, *Phoma radicis,* was said to infect the young seedlings. (*4*) An endophyte on *Casuarina equisetifolia* (Bose, 1947), found to be seed borne in the integuments and throughout the plant. The fungus, *Phomopsis casuarinae,* was thought to begin as a symbiont and to later become parasitic, eventually killing part or all of the plant.

These examples apparently require confirmation before acceptance, and the concept of cyclic mycorrhizal infection needs further study (Harley, 1969). However, the establishment of *Rhizobium* spp. in roots of legumes when inoculated on the surface of the seed (Nutman, 1965) is well established. Invasion only occurs in the roots in this case, and is localized in the nodules induced; it is not cyclic in character.

Many kinds of structures may enable pathogens to survive during seed transmission: seeds *(Cuscuta, Orobanche)*; modified seeds *(Tilletia caries, Claviceps purpurea)*; fragments of plants (many foliage pathogens, *Melampsora lini*, TMV on tomato); galls *(Anguina tritici)*; lumps of soil, or dust *(Heterodera schachtii, Corynebacterium fascians, Plasmodiophora brassicae)*; sclerotia *(Sclerotinia sclerotiorum, Sclerotium rolfsii)*; fruiting bodies *(Phoma lingam, Septoria apiicola)*; spores *(Fusarium oxysporum* f. sp. *callistephi, Puccinia carthami, P. antirrhini, Phytophthora phaseoli)*; vegetative cells *(Rhizoctonia solani*, bacterial pathogens); nematodes *(Ditylenchus dipsaci)*; cysts *(Heterodera rostochiensis)*; viruses (lettuce mosaic and bean common mosaic).

Specific examples of the successive phases in seed transmission are presented below. These phases are: the plant → seed (PS) transfer; the seed → seed (SS) transfer; the seed → plant (SP) transfer; establishment in the field, often with plant → plant (PP) transfer. As would be expected, some of these forty-five pathogens exhibit several mechanisms of PS and SP transfer. Each example is cross indexed (PS1, SP1) for the methods of transfer and for the appropriate control procedure (C1) for the PS or SP phase of the given pathogen and host. Further studies on seed pathogens will, undoubtedly, reveal more types of transfer. The types listed below are simply those apparent at this time.

1. INFECTION OR INFESTATION OF SEEDS BY PATHOGENS FROM THE PLANT — THE PS TRANSFER

**PS1** — PATHOGEN ACCOMPANIES SEED, BUT IS INDEPENDENT OF IT. Most, if not all, pathogens may be transported with plant or other debris mixed with seed, but do not necessarily infect the seedling. Transmission may be by any of the structures listed above.

*Cuscuta* spp. (dodder on clover; SP1; C1a, c, f, h, C2, 5–7). Dodder seeds are mixed with the crop seed during harvest (Dawson *et al.,* 1969; Heald, 1933; Kuijt, 1969). *Orobanche* spp. (broomrape of tomato; SP1; C1a, c, C2, 3, 5–7, 9) (Wilhelm *et al.,* 1958), and *Orthocarpus densiflorus* on wild *Hypochoeris glabra* (SP1) (Atsatt, 1965) are variants of the type. *Orthocarpus* is unique in that its seeds are caught in the pappus of its host and thus attain coordinated dispersal with it.

*Claviceps purpurea* (ergot of grains; PS5; SP2; C1a, c, d, h, C2, 5, 7). Infected seeds develop into sclerotia and are mixed with crop seed during threshing (J. C. Walker, 1969).

*Sclerotinia sclerotiorum* (cottony rot and white blight of many crops; SP1, 2; C1a, c, h, C2, 3, 5–7, 8). Sclerotia develop in or on plant parts and become mixed with seed in threshing; their shape and size may be similar to the seed, increasing the difficulty of removing them in cleaning operations (Fig. 5.5A; Baker and Davis, 1951).

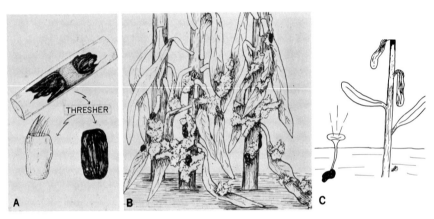

Fɪɢ. 5.5. Sclerotia of *Sclerotinia sclerotiorum* accompanying seed of *Centaurea cyanus*. A. Sclerotia formed in small stems are the same size and shape as the seeds, and are mixed with them in threshing. B. Sclerotia sown with the seed may germinate to mycelium, which invades and decays the basal stems and leaves. C. Sclerotia sown with seed may produce apothecia, from which spores are discharged and air borne to the aerial parts, which they infect, producing white-blight lesions. [Illustration by Lily H. Davis. A and C from Baker (1952, p. 38).]

*Tilletia caries* (wheat bunt; PS2; SP4; C1a–c, C2, 7, 8). Infected seeds are converted into smut balls; these are mixed with seed in threshing; they may break and release spores onto the seed (Heald, 1933).

*Puccinia malvacearum* (hollyhock rust; SP2; C1a, c, h, C5, 7). Teliospores on bits of involucral bracts are mixed with seed in threshing (J. C. Walker, 1969). *Melampsora lini* (flax rust; SP2; C1a, c, h, C2, 4, 5–7) is a variant of the type (Muskett and Colhoun, 1947).

*Plasmodiophora brassicae* (club root of turnips; SP1; C1a, c, e, h, C2–4, 7, 8). Resting spores in bits of soil are mixed with seed during threshing (Warne, 1944).

*Corynebacterium michiganense* (bacterial canker of tomato; PS2, 3;

SP1, 5; C1a, b, e, f, C2, 5, 6, 8, 9). Bacteria invade fruit through the vascular elements; infected pulp mingles with seed during the seed extraction process (J. C. Walker, 1969).

*Anguina tritici* (wheat cockles; SP1; C1a, c, f, h, C2, 5–7, 9). Nematode larvae remain near the growing point as the seedling elongates; they enter flower parts and produce galls in which eggs are laid. The hatched larvae become hypobiotic as seeds mature, and galls are mixed with seeds in threshing (Thorne, 1961). *Ditylenchus dipsaci* (stem and bulb nematode of alfalfa; PS2, 3; SP1, 5; C1a, c, f, h, C2, 5, 6, 9) is carried in plant parts mixed with seeds (Edwards, 1932), and *Heterodera schachtii* (sugar-beet nematode; SP1; C1a, c, f, h, C2, 5, 6, 9) as cysts in clods of soil mixed with seed (Shaw, 1915).

Tobacco mosaic virus on tobacco (transport only, transmission not demonstrated). TMV may occur in trash accompanying tobacco seed, but it rarely if ever infects the resulting seedlings (R. H. Taylor, 1962).

**PS2**—PATHOGEN IS A PASSIVE CONTAMINANT ON THE EXTERIOR OF THE SEED. The pathogen may be present as sclerotia, spores, vegetative cells, nematodes, or virus particles.

*Puccinia antirrhini* (snapdragon rust; SP3; C1a, c, h, C2, 4, 5–7, 9). The abundant urediospores from infected plants contaminate the seed in threshing; these spores formed in California seed fields are long lived (Fig. 5.8A, p. 369; Baker, 1952; J. Walker, 1954). *Puccinia carthami* (safflower rust; SP2; C1a, f, C2, 4–6, 8). Teliospores from infected plants get on seed in threshing (Klisiewicz, 1965; Zimmer, 1963).

*Tilletia caries* (wheat bunt; PS1; SP4; C1a–c, C2, 7, 8). Spores contaminate the surface of the seed in threshing (Heald, 1933).

*Fusarium oxysporum* f. sp. *callistephi* (fusarium wilt of China aster; SP5; C1a, e, C2–6, 8). Macrospores produced at the base of infected stems get on seeds during the threshing of plants cut off at soil level (Fig. 5.6; Baker, 1953).

*Rhizoctonia solani* (damping-off of pepper, eggplant, and tomato; PS3; SP5, 6; C1a, c–e, h, C2, 3, 5–7, 9). Mycelium in the soil infects fruits in contact with it, decays the pulp, and may form sclerotia or mycelium on the surface of seeds (Figs. 5.1E, 5.4G; Baker, 1947).

*Ascochyta pinodella* (ascochyta foot rot of pea; PS3; SP4; C1a, d, h, C2, 5, 6, 8). Seed from the Pacific Northwest may carry chlamydospores with the dust getting on seed in threshing (L. O. Lawyer, see p. 386).

*Pseudomonas phaseolicola* (halo blight of bean; PS3; SP5; C1a, d–f, C2, 5–7). Bacteria in dust get on the surface and into natural openings and cracks in the seed coat during threshing (Grogan and Kimble, 1967).

*Corynebacterium fascians* (fasciation of sweet pea and nasturtium; SP4; C1a, d, e, C2, 3, 6, 8, 9). Bacteria produce no symptoms in plants

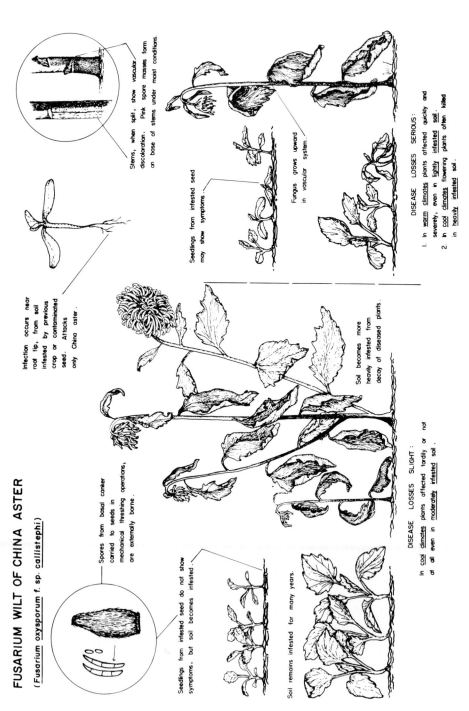

FIG. 5.6. Life-history chart of *Fusarium oxysporum* f. sp. *callistephi* on China aster, showing seed transmission. (Illustration by Margaret S. Parr, Department of Plant Pathology, University of California, Los Angeles.)

in California seed fields because of dry conditions, but they multiply sufficiently in soil to get on seed with dust during threshing operations (sweet pea), or on fruit coats in contact with soil when growing in the field (nasturtium) (Baker, 1950).

*Corynebacterium michiganense* (bacterial canker of tomato; PS1, 3; SP1, 5; C1a, b, e, f, C2, 5, 6, 8, 9). Bacteria from infected fruit pulp contaminate the surface of seed during the seed extraction process (J. C. Walker, 1969).

*Ditylenchus dipsaci* (stem and bulb nematode on alfalfa; PS1, 3; SP1, 5; C1a, c, f, h, C2, 5, 6, 9). Larvae move onto the surface of the seed as it matures (Edwards, 1932).

Tobacco mosaic virus on tomato (PS3; SP5; C1a, c, e, f, h, C6–8). TMV is carried into fruit pulp, and remains on the seed surface after extraction of the seed (R. H. Taylor *et al.,* 1961).

PS3—PATHOGEN SPREADS INTO THE SEED FROM THE FRUIT. Some of the most important seed-borne pathogens are of this type; they are usually internally borne and difficult to control. Transmission is by vegetative cells, pycnidia, spores, nematodes, or virus particles.

*Fusarium solani* f. sp. *cucurbitae* (fusarium foot rot of squash; SP1, 5; C1a, e, C2, 5–7, 9). Mycelium in soil infects the fruit, decays the pulp, and may enter the seeds (Toussoun and Snyder, 1961).

*Rhizoctonia solani* (damping-off of pepper, eggplant, tomato, and zinnia; PS2; SP5, 6; C1a, c, e, h, C2, 3, 5–7, 9). Fruits are invaded by mycelium when in contact with the soil; seeds may be wholly or partially decayed. The mycelium may spread through the mass of minced fruit pulp and seeds during the fermentation process prior to separation of the seeds (Figs. 5.1E, 5.4). Zinnia flower heads are piled on canvas on the ground as they are picked throughout the season; moisture from the soil permeates the pile, and *R. solani* grows from the soil through the canvas into flowers and seeds (Fig. 5.1D; Baker, 1947).

*Alternaria zinniae* (alternaria disease of zinnia; SP5, 6; C1a, d, e, C2, 3, 5–9). Although growing plants only rarely show symptoms in California seed fields, the senescent flower heads, which are soaked with dew, support a mixed microflora, including *A. zinniae*. The fungus probably survives from the time of the dead seedlings to flower maturation in the soil, in an occasional spot on basal leaves, and on root lesions. Seeds are invaded from the petals, and are partially to wholly decayed (Fig. 5.2A; Baker, 1956; Baker and Davis, 1950b).

*Heterosporium eschscholtziae* (heterosporium disease of California poppy; SP5; C1a, c, d, C2, 5–7, 9). Conidia from infected stems and leaves produce tiny lesions on the capsules; the mycelium permeates inside the capsule, invading and sometimes decaying the seeds (Fig. 5.3A–F; Davis, 1952).

*Ascochyta* spp. (ascochyta blight of pea; SP4; C1a, d, h, C2, 5, 6, 8). Conidia from diseased foliage infect the pods and grow through them into the seeds (J. C. Walker, 1969). *Colletotrichum lindemuthianum* (bean anthracnose; SP5; C1a, d, C2, 4–7). Infection similar to that of *Ascochyta* above (J. C. Walker, 1969). *Phoma lingam* (black leg of cabbage; SP5; C1a, d, C2, 5, 6, 9). Spores from plant lesions infect the siliques and penetrate the seed, on which the fungus may produce pycnidia (J. C. Walker, 1969). *Phytophthora phaseoli* (downy mildew of lima bean; SP5; C1a, d, C2, 5, 6, 8). Conidia from infected foliage infect pods, and grow through them into the seeds, where they form oospores (Wester *et al.,* 1966).

*Botrytis cinerea* (gray mold of flowers of Compositae; SP4, 6; C1a, d, e, h, C2, 6–8). Spores of this low-grade pathogen are common in the air; under cool moist conditions the petals are infected, and the mycelium spreads down them into the seeds (Baker, 1946).

*Pseudomonas phaseolicola* (halo blight of bean; PS2; SP5; C1a, d–f, C2, 5–7). The bacteria may systemically invade the plant; they may spread from leaf lesions to infect pods and invade the seed directly, or bacteria may, during threshing, get into injuries of the seeds (Grogan and Kimble, 1967).

*Corynebacterium michiganense* (bacterial canker of tomato; PS1, 2; SP1, 5; C1a, b, e, f, C2, 5, 6, 8, 9). Fruit infections are produced under humid conditions, may spread and invade seed (J. C. Walker, 1969).

*Ditylenchus dipsaci* (stem and bulb nematode on alfalfa; PS1, 2; SP1, 5; C1a, c, f, h, C2, 5, 6, 9). Larvae penetrate stems and buds, are carried upward by the elongating axis, and enter flower buds and seeds (Edwards, 1932).

Tobacco mosaic virus in tomato (PS2; SP5; C1a, c, e, f, h, C6–8). TMV moves from the fruit into seed coats and endosperm (Broadbent, 1965) in early stages of development (p. 344).

PS4 – PATHOGEN PENETRATES SEED THROUGH THE VASCULAR SYS-TEM. Relatively few pathogens are known to infect seeds through the vascular elements; most of the pathogens are in the vascular elements of the seed coat (p. 350).

*Verticillium albo-atrum* (verticillium wilt of spinach and sunflower; SP5; C1a, C2, 3, 6). The fungus grows into vascular elements of sunflower fruits; it grows into spinach fruits by infection from spores formed on the plant (Sackston and Martens, 1959; Snyder and Wilhelm, 1962). Apparently environmental conditions for seed infection are critical; this results in numerous conflicting reports.

*Fusarium oxysporum* f. sp. *mathiolae* (fusarium wilt of stock; SP5; C1a, C2, 3, 5, 6). The pathogen grows through xylem into the seed, where it is internally transmitted, a situation apparently rare in vascular fusaria. This disease is important in seed fields, but not in flower fields (Baker, 1948a).

*Xanthomonas campestris* (black rot of cabbage; SP5; C1a, d, f, C2, 5–7, 9). Bacteria are systemic in xylem, and move into the siliques and seeds (J. C. Walker, 1969). A suggestion that the pathogen may pass successively from seed, through the plant to the seed produced, without inducing symptoms, is not yet supported by data. *Xanthomonas incanae* (bacterial blight of stock; SP5; C1a, d, C2, 5–7, 9). Infection is similar to that of *X. campestris* (Fig. 5.10, p. 373; Baker, 1956; Kendrick and Baker, 1942).

Bean common mosaic virus of bean (PS5; SP7; C1a, c, C2, 4–6). BCMV is not limited to the vascular system, but may enter seeds through it, and be seed transmitted (Schippers, 1963).

Curly-top virus of beet (transport only; does not transmit through seed). This virus is phloem restricted and accumulates in the perisperm of seeds during food storage (Bennett and Esau, 1936).

PS5—PATHOGEN PENETRATES THE EMBRYO THROUGH THE PISTIL OR OVARY. Published data often merely report internal transmission of a pathogen without indicating whether the embryo or other tissue is invaded. The examples here are confined to positive reports.

*Ustilago tritici* (loose smut of wheat; SP7; C1a, d, C2, 5, 6, 8, 9). Chlamydospores from diseased flowers are wind borne to other flowers; they germinate on the stigmas, where conjugation occurs, and the ovule is infected. The mycelium develops in the embryo and then becomes dormant in a normal appearing seed (Heald, 1933).

*Claviceps purpurea* (ergot of grains; PS1; SP2; C1a, c, d, h, C2, 5, 7). Ascospores infect at the base of the ovary during flowering, and produce a mycelial mass on which the abundant sphacelial stage develops. These conidia are spread by insects to other flowers. The mycelial mass develops into an ergot, replacing the seed and is mixed with seeds during harvest (Heald, 1933).

*Gloeotinia temulenta* (blind seed disease of ryegrass; SP2; C1a–c, g, C2, 5–9). Infected ovules are destroyed; abundant conidia are produced and provide secondary inoculum. Infections after the embryo and endosperm have developed give "blind" (nonviable) seed; later infections may not prevent germination (Sampson and Western, 1954).

*Botrytis anthophila* (anther mold of red clover; SP7 [2?]; C1a, h, C2, 5, 7). Spores are spread by bees from diseased anthers to healthy flowers; they germinate on the stigma and reach the ovule, where they grow slightly and then remain dormant. The disease causes reduced seed germination (Sampson and Western, 1954).

Bean common mosaic virus in bean (PS4; SP7; C1a, c, C2, 4–6). If virus infection of the mother plant occurs early, the virus may spread into the ovule. Embryo infection commonly occurs from fertilization of the egg cell by pollen from an infected plant (Schippers, 1963).

Lettuce mosaic virus in lettuce (SP7; C1a, f, C2, 4–6). The ability of this virus to penetrate the embryo (as indicated by transmission to seedlings) varies with lettuce varieties (Broadbent *et al.*, 1951; Couch, 1955; Grogan *et al.*, 1952; Kassanis, 1947).

PS6—PATHOGEN ACTIVELY AND DIRECTLY PENETRATES THE SEED. The "seed" (in this type usually a fruit in the botanical sense) is exposed, and direct invasion by the pathogen occurs.

*Heterosporium (Acroconidiella) tropaeoli* (heterosporium disease of nasturtium; SP4; C1a, d, C2, 5, 9). Spores from leaf and stem lesions infect the seed while it is green (Fig. 5.7D), but the fungus develops only after the seed ripens; it then remains dormant in the mature fruit coat and in the seed (Figs. 5.7A–C, E, F, 5.9A). Largely a seed-production disease; apparently unimportant in gardens (Baker and Davis, 1950a; Baker, 1956).

*Septoria apiicola* (late blight of celery; SP5; C1a, c, d, g, C2, 5, 6, 8, 9). Pycnidia form on infected seedling plants, and the spores are spread by water as the plant grows. Seeds are infected directly, and may develop pycnidia in the fruit coats (Gabrielson, 1961; Maude, 1964; Sheridan, 1966).

*Erysiphe cichoracearum* (powdery mildew of zinnia) and *E. polygoni* (powdery mildew of pea and parsnip) (transport only; transmission not demonstrated). Mycelium grows over zinnia and parsnip flowers and seeds, and forms perithecia (Baker and Locke, 1946; Green, 1946). *Erysiphe polygoni* grows into ruptured pea pods and forms perithecia on seeds (Crawford, 1927; Uppal *et al.*, 1935).

*Xanthomonas carotae* (bacterial blight of carrot; SP5; C1a, d, C2, 5, 6, 9). Bacteria are spread by water to flowers from leaves and stems, and produce lesions on seeds (Ark and Gardner, 1944). *Xanthomonas translucens* (black chaff of wheat; SP4; C1a, C2, 5–8). Bacteria infect the seed through breaks in the seed coat, probably in threshing operations (Wallin, 1946).

2. TRANSFER OF THE PATHOGEN FROM SEED TO SEED— THE SS TRANSFER

Seed undergoes many different handling operations between threshing and planting, and in all of these there is opportunity for transfer of the

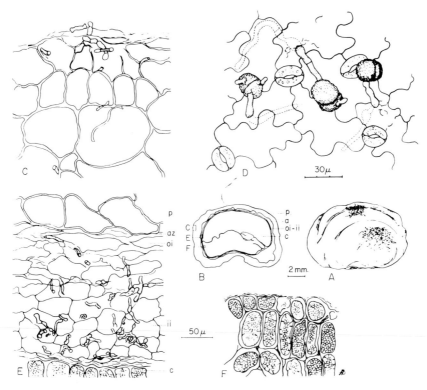

Fig. 5.7. Anatomical aspects of seed infection of nasturtium by *Heterosporium (Acroconidiella) tropaeoli*. A. External view of fruit, showing dark mycelial masses in mature pericarp. B. Section of fruit showing location of enlargements C, E, and F. Abbreviations: a, air space; c, cotyledon; ii, inner layer of seed coat; oi, outer layer of seed coat; p, pericarp. C. Section of pericarp showing mycelium. D. Conidia germinating on lower leaf surface, showing direct penetration after 17 hours. E. Section showing infected tissues between inner pericarp and cotyledon. Abbreviations: az, zone in which air space develops between pericarp and outer layer of seed coat. F. Section of infected cotyledon. [Illustration by Lily H. Davis. From Baker and Davis (1950a, p. 556).]

pathogen from one seed to another. This occurs commonly with bacterial pathogens *(Pseudomonas phaseolicola, Xanthomonas incanae)* which form scales or flakes of dried exudate, and with fungi which produce dry air-borne spores *(Alternaria zinniae, Puccinia antirrhini, Tilletia caries)*. Both of these types of structures blow about in handling, and probably spread rather uniformly over all the seeds. Fungi *(Colletotrichum lindemuthianum, Septoria apiicola, Fusarium oxysporum* f. sp. *callistephi)* and bacterial pathogens with spores or vegetative cells which are sticky when moist, tend to be smeared and to stick to other seeds they may contact.

During threshing in the field a great deal of dust and spores usually blow around. Wheat seed may become contaminated with bunt or smut spores (Heald, 1933), sweet pea seed with *Corynebacterium fascians* (Baker, 1950), and turnip seed with *Plasmodiophora brassicae* (Warne, 1944) during this operation. In order to reduce mechanical injury to the brittle seed, threshing of flower and vegetable seed is usually done in the early morning while the plants have a high moisture content from dew. Pathogens with sticky spores or vegetative cells are then hydrated and tend to adhere to other seeds; macrospores of *F. oxysporum* f. sp. *callistephi* on the basal stems of China aster are thus spread to the seed during threshing.

Field-threshed seed must be routinely cleaned, often several times, to eliminate plant parts, weed seed, immature or dead seed, and dirt. Great ingenuity has been shown in devising many special types of screens, variable-pitch vibrating tables with controlled vacuum or air flow, specific gravity and spiral separators, and washing, scouring, and polishing equipment for routine use in various combinations (Harmond *et al.*, 1968). Even more elaborate equipment is available for special situations: the powdered iron-magnetic method (Foy, 1924; Harmond *et al.*, 1968) and the velvet-lined Dosser machine (Heald, 1933; Harmond *et al.*, 1968) for separating dodder seeds from legumes; specific-gravity flotation to separate ergot from grains (Heald, 1933; Hellwig, 1699). Washing equipment may accumulate large quantities of spores if the water is reused from a tank; constant-waste flow equipment should be used. These routine cleaning treatments are very important in reducing the inoculum of pathogens by eliminating infected shriveled seed, plant parts, and dirt, but they are not sufficient to eliminate the pathogens. The beneficial effect is offset to some extent by spreading the pathogens more uniformly over all the seeds in a given lot. Since it is very difficult to free equipment of contaminant pathogens, spread between seed lots also occurs from this source. The cleaned seed is stored in bags, and is run through weighing and packaging equipment before sale, with further opportunity for SS spread.

Different seed lots may be mingled to (*a*) complete a large order exceeding the available supplies of one lot, (*b*) raise the germination percentage in one lot by mixing with a high-germination lot, and (*c*) produce a "formula mix," for example, to obtain a specifically proportioned color range in flower seed. If one lot is contaminated and the other is not, direct transfer and intermingling of accompanying pathogens inevitably result.

Seed-treatment equipment may also spread pathogens. If the treatment material (e.g., an insecticide application) does not kill the pathogen, and particularly if it is a wet treatment, undoubtedly a good deal of SS transfer results.

Finally, in the planting operation itself, the SS transfer may occur. A pathogen from one seed lot may contaminate the planter or other handling equipment, and thus spread to clean seed lots.

3. SEEDLING INFECTION BY SEED-BORNE PATHOGENS — THE SP TRANSFER

SP1—ACCOMPANYING STRUCTURES DEVELOP AND DIRECTLY INFECT THE HOST SEEDLING. The accompanying pathogen structures germinate and grow *(Fusarium solani* f. sp. *cucurbitae)* or hatch and migrate *(Anguina tritici)* through the soil for a short distance to infect adjacent seedlings.

*Cuscuta* spp. (dodder on clover; PS1; C1a, c, f, h, C2, 5–7). Dodder seed is planted with clover seed, and germinates to form a twining stem which must attach parasitically to a suitable host or it dies (Dawson *et al.,* 1969; Heald, 1933; Kuijt, 1969). *Orobanche* spp. (broomrape of tomato; PS1; C1a, c, C2, 3, 5–7, 9). Seeds germinate and attach parasitically to host roots, then produce erect stems which bear abundant seeds (Wilhelm *et al.,* 1958). *Orthocarpus densiflorus* (PS1) similarly infects *Hypochoeris* (Atsatt, 1965).

*Sclerotinia sclerotiorum* (cottony rot of many crops; PS1; SP2; C1a, c, h, C2, 3, 5–7, 8). Sclerotia planted with the seed may produce mycelium which infects the crown of the host (Fig. 5.5B; J. C. Walker, 1969).

*Plasmodiophora brassicae* (club root of turnip; PS1; C1a, c, e, h, C2–4, 7, 8). Resting spores germinate to zoospores which infect the roots. Spores develop in "clubs" formed in the roots, and are released into soil when roots decay (Warne, 1944; J. C. Walker, 1969).

*Fusarium solani* f. sp. *cucurbitae* (fusarium foot rot of squash; PS3; SP5; C1a, e, C2, 5–7, 9). Mycelium in seed infests soil; races 1 and 2 grow through soil and infect fruit resting on it. Race 1 may also infect seedlings, and may cause foot rot of mature plants (Toussoun and Snyder, 1961).

*Corynebacterium michiganense* (bacterial canker of tomato; PS1–3; SP5; C1a, b, e, f, C2, 5, 6, 8, 9). Bacteria infect seedlings through transplanting wounds; they spread through the plant in the xylem, and enter the fruits (J. C. Walker, 1969).

*Anguina tritici* (wheat cockles; PS1; C1a, c, f, h, C2, 5–7, 9). Nematode larvae in cockles planted with the seed revive and migrate to seedlings, penetrate the leaf sheaths, and feed in growing tissues (Thorne, 1961). *Ditylenchus dipsaci* (stem and bulb nematode of alfalfa; PS1–3;

SP5; C1a, c, f, h, C2, 5, 6, 9). Larvae in plant fragments and on the seed surface migrate through soil and enter seedling buds and stems. They migrate upward as the plant elongates (Edwards, 1932). *Heterodera schachtii* (sugar – beet nematode; PS1; C1a, c, f, h, C2, 5, 6, 9). Eggs in the cysts hatch, and the larvae invade roots, producing galls, from which cysts are later released into the soil by root decay (Shaw, 1915; Thorne, 1961).

SP2 – PATHOGEN STRUCTURES ACCOMPANYING SEEDS PRODUCE FRUIT-ING BODIES WHICH ACTIVELY DISCHARGE INFECTIVE SPORES. Seeds with this type of fungus usually infect a greater number of plants than those of types SP1 and SP3–6, because of the forcibly discharged spores.

*Sclerotinia sclerotiorum* (white blight of many crops; PS1; SP1; C1a, c, h, C2, 3, 5–7, 8). Sclerotia planted with seed may produce apothecia which discharge vast numbers of ascospores which infect aerial parts of host plants (Fig. 5.5C; J. C. Walker, 1969).

*Claviceps purpurea* (ergot of grains; PS1, 5; C1a, c, d, h, C2, 5, 7). Sclerotia planted with the seed produce perithecia-containing stromata which discharge ascospores which infect the ovule through the flower. Sticky conidia formed there are carried by insects to other flowers (J. C. Walker, 1969).

*Puccinia malvacearum* (hollyhock rust; PS1; C1a, c, h, C5, 7). Bits of bracts carrying teliospores may fall on the soil surface when seeds are planted; these spores germinate to produce air-borne sporidia which in-fect seedling leaves and stems; telial sori are produced which provide the secondary inoculum (J. C. Walker, 1969). *Puccinia carthami* (safflower rust; PS2; C1a, f, C2, 4–6, 8). Teliospores on planted seeds produce sporidia which infect underground and aerial parts through the epidermis. The stem is girdled below ground, and the seedling may die. Aerial in-fections result in spermogonia, aecia, uredia, and telia; aeciospores and urediospores produce secondary aerial spread (Klisiewicz, 1965; Shus-ter, 1956; Zimmer, 1963). *Melampsora lini* (flax rust; PS1; C1a, c, h, C2, 4, 5–7). Teliospores on plant fragments fall on the ground during seeding, germinate to produce sporidia which infect flax seedlings. Pycnia and aecia are formed, and aeciospores provide initial inoculum; urediospores then form copiously. Teliospores form as the plants mature, but cannot germinate until spring (Muskett and Colhoun, 1947).

*Gloeotinia temulenta* (blind seed disease of ryegrass; PS5; C1a–c, g, C2, 5–9). Shallowly planted infected seeds give rise to apothecia; dis-charged ascospores infect the ovules at flowering time (Sampson and Western, 1954).

SP3 – PATHOGEN AIR BORNE FROM PLANT FRAGMENTS OR SEED SUR-FACE TO INFECT ADJACENT SEEDLINGS. *Puccinia antirrhini* (snapdragon rust; PS2; C1a, c, h, C2, 4–7, 9). Snapdragon seed is sown successively in nurseries to provide a constant supply of seedlings; urediospores may be scattered and air borne during hand sowing of the seed flats, and fall on leaves of nearby seedlings, infecting them through stomata. Contaminated seed covered with soil produces healthy seedlings, since the spores germinate before the seeds, and subterranean stomata are lacking (Fig. 5.8B; Baker, 1952; J. Walker, 1954).

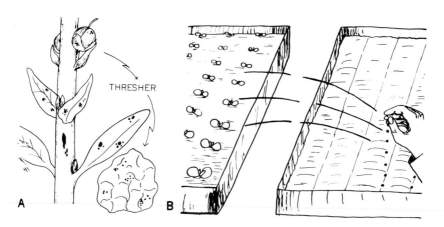

FIG. 5.8. *Puccinia antirrhini* transmission on seeds of snapdragon. A. Copious powdery urediospores from leaves and stems get on the surface of seeds in threshing. Spores produced under California seed—field conditions are long-lived. B. During planting of seed flats, spores may be air borne to nearby seedlings, infecting them through stomata. The spores on seeds underground do not infect, as stomata are lacking there. [Illustration by Lily H. Davis. From Baker (1952, p. 38).]

**SP4 — PATHOGEN GROWS FROM HYPOGEALLY GERMINATING SEED INTO THE STEM OR ROOTS.** The SP transfer here is by direct growth from the subterranean seed to the seedling it produces. The old germinated seed in some cases (*Ascochyta* spp. on pea, *Heterosporium tropaeoli* on nasturtium, *Botrytis cinerea* on Compositae) provides the food base for invasion of the seedling under the favorable subterranean conditions.

*Ascochyta* spp. (ascochyta blight and foot rot of pea; PS2, 3; C1a, d, h, C2, 5, 6, 8). Spores of *A. pinodella* germinate and infect roots. Mycelium of *A. pinodes* and *A. pisi* in seed resumes growth after planting, invades the stem, and kills the emerged seedling. Pycnidia form on the dead seedling and extrude spores, which are splashed by rain onto foliage and stems of adjacent seedlings, infecting them (J. C. Walker, 1969). *Heterosporium tropaeoli* (heterosporium disease of nasturtium; PS6, C1a, d, C2, 5, 9) is similar to *Ascochyta* spp. above (Fig. 5.9B; Baker and Davis, 1950a; Baker, 1956).

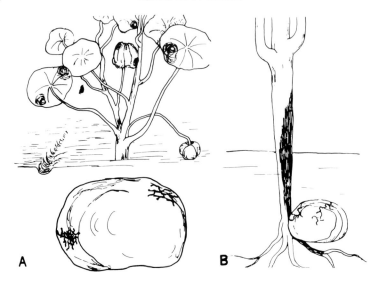

Fig. 5.9. *Heterosporium tropaeoli* transmission in seeds of nasturtium. A. Green seed under a canopy of diseased leaves is infected by spores under the prevailing moist conditions. The fungus is latent until the seed yellows and matures. B. Infected seed remains attached underground to the stem in this hypogeal germination. The fungus advances into the stem and produces a lesion, on which spores are produced; the spores are air borne and infect the leaves. [Illustration by Lily H. Davis. From Baker (1952, p. 38).]

*Tilletia caries* (wheat bunt; PS1, 2; C1a–c, C2, 7, 8). Spores on seeds germinate and form sporidia, which fuse in pairs and infect the seedling shoot. Mycelium grows as plant elongates, enters and fills the ovaries, and is then converted to smut spores (Heald, 1933).

*Botrytis cinerea* (gray mold of flowers of Compositae; PS3; SP6; C1a, d, e, h, C2, 6–8). Mycelium in seed may decay the seed after planting, or the seedling during or after germination, particularly in cool soil. The mycelium may spread to closely planted adjacent seedlings (Baker, 1946).

*Corynebacterium fascians* (fasciation of sweet pea and nasturtium; PS2; C1a, d, e, C2, 3, 6, 8, 9). Bacteria on seed spread to the seedling, grow on its surface, and may, under continued moist conditions, stimulate buds to break dormancy. This produces a cauliflower-like appearance or a witches' broom effect (Baker, 1950).

*Xanthomonas translucens* (black chaff of wheat; PS6; C1a, C2, 5–8). Bacteria spread from the seed to the coleoptile, infect through stomata, and are carried to the soil surface by elongation during germination. Bacterial exudate is spread by rain and insects to leaves and spikes, and to adjacent plants (Wallin, 1946).

**SP5**—Pathogen spreads from epigeally germinating seed into

COTYLEDONS, STEM, OR ROOTS. The many pathogens with this type of SP transfer fall into two groups: (*a*) those which establish and survive in the soil for long periods (*Rhizoctonia solani, Fusarium solani* f. sp. *cucurbitae, Verticillium albo-atrum*), and (*b*) those which survive for only short periods in the soil (*Alternaria zinniae*), or disappear when host refuse is decayed (*Heterosporium eschscholtziae, Phoma lingam, Pseudomonas phaseolicola, Xanthomonas campestris*). Those which establish and persist in field soils are obviously the most dangerous to introduce because, once established, it is virtually impossible to eliminate them.

*Fusarium solani* f. sp. *cucurbitae* (fusarium foot rot of squash; PS3; SP1; C1a, e, C2, 5-7, 9). Mycelium in seed infests soil; that of race 1 may also kill the seedling produced and may cause foot rot of mature plants (Toussoun and Snyder, 1961).

*Rhizoctonia solani* (damping-off of pepper, eggplant, tomato, and zinnia; PS2, 3; SP6; C1a, c, e, h, C2, 3, 5-7, 9). Infected seed may give rise to diseased seedlings in soil, and may spread to adjacent plants from this food base (Fig. 5.1F; Baker, 1947).

*Alternaria zinniae* (alternaria disease of zinnia; PS3; SP6; C1a, d, e, C2, 3, 5-9). Infected seeds may give rise to diseased seedlings, on which spores are copiously produced. Air-borne spores infect leaves, stems, and flowers in moist areas and, very rarely, in semiarid climates (Fig. 5.2B; Baker, 1956; Baker and Davis, 1950b). *Heterosporium eschscholtziae* (heterosporium disease of California poppy; PS3; C1a, c, d, C2, 5-7, 9). Infection similar to *A. zinniae* above (Davis, 1952).

*Colletotrichum lindemuthianum* (bean anthracnose; PS3; C1a, d, C2, 4-7). Cotyledons are invaded from infected seed; fungus produces acervuli from which spores are spread by water to leaves and stems of adjacent plants (J. C. Walker, 1969).

*Phoma lingam* (black leg of cabbage; PS3; C1a, d, C2, 5, 6, 9). Cotyledons are invaded from the adhering infected seed coat; pycnidia are produced from which spores may be spread by water to leaves and stems (J. C. Walker, 1969). *Septoria apiicola* (late blight of celery; PS6; C1a, c, d, g, C2, 5, 6, 8, 9). Infection similar to *P. lingam* above (Gabrielson, 1961; Maude, 1964; Sheridan, 1966).

*Fusarium oxysporum* f. sp. *callistephi* (fusarium wilt of China aster; PS2; C1a, e, C2-6, 8). Spores on seed coat germinate and infect the seedling roots. Mycelium invades xylem from the cortex and may kill the plant in warm soil. The fungus spreads into the cortex at the stem base, and there produces, under moist conditions, copious macrospores (Fig. 5.6; Baker, 1953). *Fusarium oxysporum* f. sp. *mathiolae* (fusarium wilt of stock; PS4; C1a, C2, 3, 5, 6). The seedling is invaded from the internally infected seed. Mycelium invades the xylem and may kill the plant, or may move through vascular strands into siliques and seeds (Baker, 1948a). *Verticillium albo-atrum* (verticillium wilt of spinach and sunflower; PS4; C1a, C2, 3, 6). The seedling is invaded from infected seed, and mycelium spreads through the xylem. The fungus moves into the cortex under moist conditions, and sporulates on the surface of stems (Sackston and Martens, 1959; Snyder and Wilhelm, 1962).

*Phytophthora phaseoli* (downy mildew of lima bean; PS3; C1a, d, C2, 5, 6, 8). Following epigeal germination of seed, the oospores in the seed coat infect the seedlings (Wester *et al.,* 1966).

*Pseudomonas phaseolicola* (halo blight of bean; PS2, 3; C1a, d–f, C2, 5–7). Bacteria spread from the seed to the surface of cotyledons and penetrate through injuries, spreading through intercellular spaces of the cortex. When the xylem is reached, infection becomes systemic, and may give rise to leaf lesions and stem cankers. Bacteria in the exudate from lesions may be spread by water to other plants (Grogan and Kimble, 1967; Zaumeyer, 1932). *Xanthomonas campestris* (black rot of cabbage; PS4; C1a, d, f, C2, 5–7, 9). Cotyledons are invaded through stomata by bacteria from the adhering seed coat. They reach the xylem, become systemic in it, and may develop leaf lesions from which exudate appears. Bacteria spread from ooze by rain, infecting leaves through the hydathodes (J. C. Walker, 1969). *Xanthomonas incanae* (bacterial blight of stock; PS4; C1a, d, C2, 5–7, 9). Similar to *X. campestris* above (Fig. 5.10; Baker, 1956; Kendrick and Baker, 1942). *Xanthomonas carotae* (bacterial blight of carrot; PS6; C1a, d, C2, 5, 6, 9). Similar to *X. campestris* above (Ark and Gardner, 1944).

*Ditylenchus dipsaci* (stem and bulb nematode of alfalfa; PS1–3; SP1; C1a, c, f, h, C2, 5, 6, 9). Larvae on the seed surface enter the seedling buds and stems. They migrate upward as the plant elongates (Edwards, 1932).

*Corynebacterium michiganense* (bacterial canker of tomato; PS1–3; SP1; C1a, b, e, f, C2, 5, 6, 8, 9). Bacteria spread from the seed coat to wounds on roots, stems, leaves, and also penetrate through stomata. They then spread through the plant in the xylem and enter the fruits (J. C. Walker, 1969).

Tobacco mosaic virus on tomato (PS2, 3; C1a, c, e, f, h, C6–8). The virus is apparently in all parts of the seed except the embryo, which may rarely be invaded. Seedlings may be infected in soil from the plant refuse containing TMV; infection from such refuse accompanying seed (PS1) might be expected, but has not been clearly demonstrated. Seedling infection from the seed only occurs through wounding during transplanting or growth (Broadbent, 1965; R. H. Taylor *et al.,* 1961).

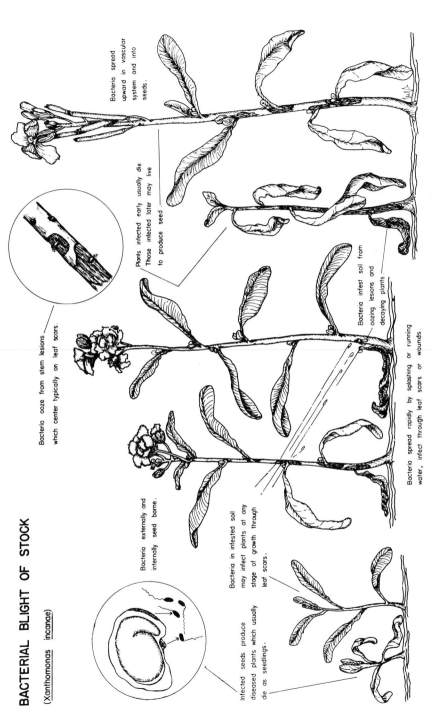

# BACTERIAL BLIGHT OF STOCK

(*Xanthomonas incanae*)

Bacteria ooze from stem lesions which center typically on leaf scars.

Bacteria externally and internally seed borne.

Bacteria in infested soil may infect plants at any stage of growth through leaf scars.

Infected seeds produce diseased plants which usually die as seedlings.

Bacteria spread upward in vascular system and into seeds.

Plants infected early usually die. Those infected later may live to produce seed.

Bacteria infest soil from oozing lesions and decaying plants.

Bacteria spread rapidly by splashing or running water, infect through leaf scars or wounds.

FIG. 5.10. Life-history chart of *Xanthomonas incanae* on garden stock, showing seed transmission. (Illustration by Margaret S. Parr, Department of Plant Pathology, University of California, Los Angeles.)

SP6—PATHOGEN SPREADS FROM DECAYING SEED TO SURROUNDING SEEDLINGS. Seed decayed during its formation and not removed by cleaning, as well as that which decays after planting, is usually an infection hazard in crop fields.

*Rhizoctonia solani* (damping-off of pepper, eggplant, tomato, and zinnia; PS2, 3; SP5; C1a, c, e, h, C2, 3, 5–7, 9). Seeds wholly or partially decayed, but not removed in cleaning, serve as foci of infection for surrounding seedlings when they are planted (Fig. 5.1F; Baker, 1947).

*Botrytis cinerea* (gray mold of flowers of Compositae; PS3; SP4; C1a, d, e, h, C2, 6–8). Mycelium may spread from this food base to contacting or adjacent seed in closely planted seedbeds (Baker, 1946).

*Alternaria zinniae* (alternaria disease of zinnia; PS3; SP5; C1a, d, e, C2, 3, 5–9). Decaying seed in soil provides a food base for spread to adjacent seedlings and to roots of older plants which may contact it (Baker, 1956; Baker and Davis, 1950b).

SP7—INFECTED EMBRYOS GIVE RISE TO DISEASED SEEDLINGS; TRANSMISSION IN OTHER SEED TISSUES DOES NOT OCCUR OR IS UNIMPORTANT. An infected embryo is usually the biological equivalent of an infected seedling. Although a diseased plant usually results, a particularly unfavorable environment may prevent further development of the pathogen [e.g., *Ustilago tritici* (Heald, 1933)]. Some viruses may become inactivated in the embryo (p. 346).

*Ustilago tritici* (loose smut of wheat; PS5; C1a, d, C2, 5, 6, 8, 9). Mycelium grows upward with the plant from the infected seed, usually without producing conspicuous symptoms until time of flowering. The flower parts are replaced by copious spore masses at the normal time of flowering, and the air-borne spores infect healthy flowers (Heald, 1933).

*Botrytis anthophila* (anther mold of red clover; PS5; [SP2?]; C1a, h, C2, 5, 7). Mycelium from infected seed grows as the plant elongates, but produces no symptoms until time of flowering. Anthers are invaded and spores are produced instead of pollen. This disease is important only in seed fields, not in crop production (Sampson and Western, 1954).

Bean common mosaic virus (PS4, 5; C1a, c, C2, 4–6). Virus persists in the seedling which develops from the infected embryo (Schippers, 1963). Lettuce mosaic virus (PS5; C1a, f, C2, 4–6) is similar to BCMV above (Grogan *et al.,* 1952).

4.   ESTABLISHMENT AND DISSEMINATION OF PATHOGENS IN THE FIELD—THE PP TRANSFER

Transport of a pathogen does not necessarily lead to its transmission, and transmission does not insure establishment of it in the planting.

Among factors which determine the success of such establishment, the following may be mentioned:

1. The environment may limit infection and expression of symptoms (Baker and Snyder, 1950). Infection of the seedling may be prevented [*Puccinia carthami* of safflower at a soil temperature of 30°C (Schuster, 1956)]. Environment may also limit pathogen spread from the initial infected seedling, and thus suppress the disease in the field so effectively that seed free of the pathogen may be produced. *Ascochyta*-infected peas and *Alternaria*-infected zinnia seed planted in the dry interior valleys of California may thus produce seed free of these pathogens (Baker, 1956; Mackie *et al.,* 1945; Middleton and Snyder, 1947). Much of the United States seed industry is located in the semiarid western states, in large part because of this effect on seed transmission of pathogens (Kreitlow *et al.,* 1961). It should be noted, however, that bean seed infected with *Pseudomonas phaseolicola* may occasionally produce diseased plants under semiarid conditions, and the bacteria may be spread to the surface of the seed in threshing operations (Grogan and Kimble, 1967). *P. lachrymans* causes angular leafspot in California on cucumber plants repeatedly picked for processing while plants are wet with dew, being spread in that way. It does not occur, however, in seed fields which are harvested only once, and is not seed transmitted from such plants (Grogan *et al.,* 1971). The effect of temperature on seed transmission of viruses has already been discussed on pages 346–347.

2. The absence or scarcity of a suitable vector may limit spread of the disease. The paucity of *Diabrotica* beetles in the Imperial Valley of California thus greatly restricts spread of squash mosaic on cantaloupe there (Grogan *et al.,* 1959). Similarly, the near-absence of four aphid vectors during the summer in the Swan Hill region of Australia's Murray River Valley prevents spread of lettuce mosaic (Stubbs and O'Loughlin, 1962). Some lettuce seed for commercial planting in California has, for this reason, been produced there. Scarcity of the thrips vectors of spotted wilt virus in the San Joaquin Valley, and of *Circulifer tenellus,* the vector of beet curly-top virus, in the coastal valleys of California, also limit the occurrence of these viruses in these areas.

3. The pathogen in or on the seed may die before the seed loses satisfactory germinability (see pp. 331, 347, and 382).

4. When pathogen-infected seed is planted in field soil, the microflora there may prove inhibitory or antagonistic to the pathogen, and thus suppress the disease. It has been repeatedly shown that seed infected or infested by a pathogen will produce a larger number of diseased seedlings in nearly sterile than in untreated field soil. Examples are *Polyspora lini* and *Colletotrichum linicolum* on flax seed (Henry and Campbell, 1938),

*Helminthosporium sativum* on wheat (Ledingham *et al.,* 1949; Machacek and Wallace, 1942; Simmonds, 1947), *H. teres* on barley (Machacek and Wallace, 1942), *H. victoriae* on oats (Tveit and Moore, 1954), *Fusarium nivale* on oats (Tveit and Wood, 1955), and *Verticillium albo-atrum* on *Senecio vulgaris* (Schippers and Schermer, 1966). The effect on the amount of seed transmission is shown in Table 5.III. This effect is ob-

TABLE 5.III.

THE EFFECT OF THE SOIL MICROFLORA ON SEED TRANSMISSION OF PLANT PATHOGENS

| | | % Infected Seedlings | | |
| | | Untreated | Treated | |
| Pathogen on seed | Host | soil | soil | Reference |
|---|---|---|---|---|
| *Polyspora lini* (natural) | Flax | 1.7[a] | 15.0[a] | Henry and Campbell, 1938 |
| *Colletotrichum linicolum* (inoc.) | Flax | 17.0[a] | 66.3[a] | Henry and Campbell, 1938 |
| *Helminthosporium sativum* (natural) | Wheat | 14.8 | 38.6 | Machacek and Wallace, 1942 |
| *Helminthosporium victoriae* (nat.) | Oats | 68.3 | 97.9[b] | Tveit and Moore, 1954 |
| *Helminthosporium victoriae* (nat. +*Chaetomium* sp.) | Oats | 20.8 | 3.8[b] | Tveit and Moore, 1954 |
| *Verticillium albo-atrum* (natural) | *Senecio vulgaris* | 0 | 39.2 | Schippers and Schermer, 1966 |
| | | 0 | 7.5[c] | Schippers and Schermer, 1966 |

[a] Average of three trials.
[b] *Helminthosporium victoriae* and *Chaetomium* sp. in soil, not on seeds.
[c] Autoclaved soil contaminated by exposure to air for 24–72 hours.

viously produced by antagonists in the soil, but these microorganisms have not been studied. Such antagonists would be expected to be most effective against externally borne pathogens and those which spread from plant to plant through the soil. *Verticillium albo-atrum,* which has only a limited ability to grow through soil, was transmitted by *Senecio* seed in sterile, but not in untreated soil (Schippers and Schermer, 1966). Antagonists on the seed also inhibit infection by seed-borne pathogens (Chang and Kommedahl, 1968; Ledingham *et al.,* 1949; Tveit and Moore, 1954; Tveit and Wood, 1955). Limonard (1967) found that soaking wheat seeds in water at 20°C for 24 hours before testing them on blotters reduced the average percentage which developed *Fusarium* spp. from 28.1 to 13.1. When 50 ppm of the antibiotic terramycin was added to the water, the average percentage of *Fusarium* spp. was 23.8, due to inhibition of the bacteria antagonistic to *Fusarium*. The comparable percentages for *Phoma betae* on beet seeds were 52.8, 25.1, and 47.9. Bacteria on the seed surface undoubtedly affect the seed transmission of pathogens in a similar way in soil.

Saprophytic soil microorganisms may, on the other hand, aggravate plant injury from a seed-borne pathogen. Damage to crucifers from *Plasmodiophora brassicae* may result in part from decay of the clubbed roots by secondary invaders. Damage to cauliflower and broccoli crops has been greatly reduced by fungicidal application in the hole at time of transplanting, although the severity of root clubbing was not markedly affected, because the secondary invaders were controlled and the clubs were not decayed (Lear *et al.,* 1966). The question may be raised whether the decay of safflower roots infected by *Puccinia carthami* (Klisiewicz, 1965; Schuster, 1956) may not be partly a result of secondary invaders.

5. The type of germination may impose a natural restriction on infection by a seed-borne pathogen. Hypogeal germination thus limits transmission of pathogens which can only infect aerial parts (most rusts, downy mildews, and powdery mildews). Transmission of powdery mildews on seed is contrary to what would be anticipated. W. C. Snyder (personal communication) has been unable to confirm reports (Crawford, 1927; Uppal *et al.,* 1935) of transmission of *Erysiphe polygoni* by pea seed, and Baker and Locke (1946) obtained no transmission of *E. cichoracearum* on zinnia seed. On the other hand, hypogeal germination favors infection by pathogens which invade stems and roots (*Ascochyta* spp. on pea, *Tilletia caries* on wheat) because of the more uniformly favorable conditions in the soil. Epigeal germination favors transmission of pathogens which infect aerial parts (*Colletotrichum lindemuthianum* on bean, *Xanthomonas campestris* on cabbage), but may also be quite effective for subterranean pathogens (*Fusarium oxysporum* f. sp. *callistephi* on China aster, *F. solani* f. sp. *cucurbitae* on squash). The pathogen may in such cases infest the soil from the seed coats as they are pushed upward in germination, or are occasionally detached prior to elevation above ground.

Some examples of seeds with hypogeal germination are: all Gramineae, nasturtium, palms, pea, sweet pea, and *Prunus*. Seeds with epigeal germination are alfalfa, bean, beet, cabbage, California poppy, carrot, celery, China aster, cotton, flax, garden stock, lettuce, onion, pepper, pines, *Ricinus,* snapdragon, spinach, squash, sunflower, tomato, and zinnia.

The possibility of a seed-borne pathogen initiating a disease in a crop increases progressively from seed transport of a pathogen, through seed transmission, to successful production of a disease in seedlings, to an outbreak in the field.

## L. *Factors Limiting Effectiveness of Seed Transfer*

The presence of a virulent viable pathogen in, on, or with a given lot of seed fortunately does not mean that a disease will necessarily result. For example, over the years, hundreds of tons of pea seed have been brought into California and sown for canning, freezing, and seed production. Unquestionably, a good deal of this seed was carrying spores of *Fusarium oxysporum* f. sp. *pisi,* but the fusarium wilt it causes has been found only in two small coastal areas, even though susceptible varieties have been grown (W. C. Snyder, personal communication). Apparently there is a somewhat similar situation with this disease in Illinois (L. O. Lawyer, personal communication) and in New South Wales, Australia (L. R. Fraser, personal communication). These areas are in striking contrast to those in Washington (Kadow and Jones, 1932) and Wisconsin (J. C. Walker and Hare, 1943), where the disease became quickly established in certain soils, and necessitated the use of resistant varieties.

The following factors have been shown to affect the degree of success of seed transmission of plant pathogens under commerical conditions.

### 1. THE PHYSICAL ENVIRONMENT

If China aster seed contaminated with *F. oxysporum* f. sp. *callistephi* is sown in steamed soil and held at a temperature of 15.6°C, no evident disease will develop in that planting, even though some plants may actually be infected. Some of the same seed sown in steamed soil and held at 25°–26.7°C, however, will develop conspicuously diseased plants (Fig. 5.6; Baker, 1953). *Ascochyta*-infected pea seed (Middleton and Snyder, 1947) and *Alternaria*-infected zinnia seed (Baker, 1956) sown in the semiarid interior valleys of California produce seed crops free of these pathogens because the dry conditions are unfavorable to infection of aerial parts. The effect of temperature on invasion of the embryo by viruses has already been mentioned on pages 346–347.

### 2. LENGTH OF SURVIVAL OF THE PATHOGEN ON OR IN SEED

Most pathogens survive as long as the seed remains viable. Some, however, may die or become inactivated before seed viability is reduced (pp. 331, 347, 382). Some pathogens may perhaps survive longer than the seed.

### 3. MICROFLORA OF THE SEED AND SOIL

The antagonistic microflora of the seed, soil, or both, may play a very important role in determining the efficacy of seed transmission. Oat varieties from Brazil were found to be highly resistant to infection by *Helminthosporium victoriae* applied to the seeds or to the soil in which they were planted. Hot-water treatment of the seeds destroyed this resistance, however, and led to the discovery that *Chaetomium cochlioides* and *C. globosum* on the surface of the seed were responsible for the resistance (Tveit and Moore, 1954). Later studies (Tveit and Wood, 1955) showed a similar protective effect by these fungi against *Fusarium nivale* on oats. Chang and Kommedahl (1968) obtained striking control of seedling blight by inoculating corn seed with antagonists prior to planting. In the field there was a 43% seedling stand without seed inoculation, 66% when inoculated with *Chaetomium globosum,* and 73% when inoculated with *Bacillus subtilis.* The figures for untreated soil inoculated with *Fusarium roseum* in a glasshouse test were 23%, 87%, and 57%; in a series with uninoculated soil and seed, the stand was 83%. For autoclaved soil, the figures were 40%, 99%, 53%, and 40%. Brewer *et al.* (1970) showed that *Chaetomium cochlioides* and *C. globosum* produced a metabolite, cochliodinol, antibiotic to *Fusarium* spp. and a number of other fungi and bacteria. The influence of the antagonistic soil microflora on effective seed transmission of pathogens has already been discussed on pages 339 and 375. The restrictive effect of some soils on *F. oxysproum* f. sp. *pisi* mentioned above is an outstanding example of the effectiveness of this factor.

4. TYPE OF SEED GERMINATION

The relation of epigeal and hypogeal germination to seed transmission of pathogens was discussed on page 377.

5. LEVEL OF TRANSMISSION OF THE PATHOGEN IN A GIVEN HOST SPECIES OR VARIETY

It has been found that some varieties have little or no seed transmission of a virus, whereas others may have very high transmission (p. 346). There is little clear evidence of similar differences in seed transmission of fungi, bacteria, or nematodes, perhaps because the subject has been so little studied.

6. LEVEL OF POLLEN OR OVULE STERILITY INDUCED BY THE PATHOGEN

Some viruses are known to cause a high level of sterility in pollen (p. 345), but there is still a significant amount of seed transmission in these cases. However, some viruses (aster yellows; curly top in petunia) do induce quite complete sterility; other factors are also involved in their nontransmission by seeds. Fungi and nematodes which induce seed sterility (*Gloeotinia temulenta* of rye grass, *Botrytis anthophila* on red clover, *Anguina tritici* on wheat) do not sustain significant reduced transmission because of this, since the aborted structures themselves serve to transport the pathogen.

### 7. Cultural Practices

Depth of planting, direct seeding, density of seeding, time of planting, avoidance of overhead watering, time of harvesting, selection of growing area, and use of fungicidally treated ground canvases for harvesting operations, *inter alia,* will influence the degree and success of seed transmission (pp. 381–382).

### 8. Separation of Damaged from Healthy Seeds

The use of special seed-cleaning equipment to eliminate or reduce the percentage of seed transmission has been discussed (see p. 366). Some seeds may be so reduced in size that careful cleaning will reduce the level of seed transmission [squash mosaic in squash (Middleton, 1944); *Anguina tritici* in wheat (Noble and Richardson, 1968)].

### 9. Ability of Meristematic Tissue to Resist Penetration by a Virus, or to Inactivate it if Penetration does Occur

These topics have been discussed already (pp. 345, 347). Meristematic tissue seems to have the capacity to inactivate virus if the concentration has not reached a high level (Bennett, 1969). Very young tomato seedlings are thus resistant or immune to infection by TMV (R. H. Taylor *et al.,* 1961).

### M. Methods for Preventing Seed Transmission of Pathogens

A brief outline of some of the common methods for preventing seed transmission of pathogens is presented, using the code designations (C1, 2) referred to in the preceding sections. Practices C1 through C5 pertain to the seed — production fields and are preventive in nature, C6 through C9 pertain to crop — production fields and are curative measures for seed already contaminated. Practices C5 and C6 comprise certification procedures and are intermediate between preventive and curative measures.

### 1. Preventive Measures in Seed Fields

Procedures C1 through C3 include the so-called cultural practices of control.

C1 — Management practices.
*a. Seed source.* Seed used for planting the commerical seed-production fields should be free of the pathogens; this may be obtained from a carefully maintained nuclear or mother block (Baker, 1956) or from an area known to be free of the pathogen.

*b. Sowing methods.* Deep planting may supress some pathogens (*Gloeotinia temulenta* on ryegrass), but favor others (*Tilletia caries* of wheat). Direct seeding in the field, as compared with transplanting from seed beds, has practically controlled some diseases (*Corynebacterium michiganense* and TMV on tomato). If seedbeds are used, careful sanitation should be practiced to prevent introduction and spread of the pathogen. *Orobanche* on tomato has been spread in California from infested seedbeds. Increased density of seeding may reduce losses from some diseases (curly-top virus of tomato), but may increase others (the gray mold, *Botrytis cinerea,* on Compositae, *Rhizoctonia solani* in seedbeds, *Corynebacterium michiganense* on tomato). Late planting may decrease some diseases (loose smut of wheat, *Ustilago tritici*), but increase others, (wheat bunt, *Tilletia caries*). It is clear that specific localized knowledge concerning the pathogen involved and the epidemiology of the disease is required for successful application of this control practice.

*c. Pathogen control in the seed field.* Control of weed hosts in and around the seed field will reduce the source of virus hosts and vectors (lettuce mosaic virus). It is desirable to isolate seed fields from crop–production fields to avoid spread by air-borne spores (see p. 384) or insect vectors. Spraying or dusting with fungicides to control leaf diseases which spread to and infect the seed (*Septoria apiicola* on celery), contaminate it during threshing (*Puccinia antirrhini* on snapdragon), or merely to increase seed yield (*Erysiphe polygoni* on pea) is sometimes practiced. Patches of dodder in alfalfa seed fields may be destroyed by flaming or cutting prior to harvest.

Spraying or dusting seed fields with insecticides to control insect vectors of viruses is less satisfactory. Nonpersistent viruses, which may be seed transmitted, are not effectively controlled in this way because they are acquired and transmitted by aphids in brief feeding periods, and the insects immediately (but not permanently) become infective. Insecticide applications have been ineffective in preventing spread of such viruses, or have increased it by disturbing the aphids and causing them to feed on more plants (Bawden, 1964). Little benefit has been realized, therefore, in reduction of these viruses either in crop production or in seed transmission by application of insecticides. The spread of leafhopper-borne and persistent aphid-borne viruses, which are not seed transmitted, may, however, be reduced by insecticide application.

*d. Avoidance of overhead watering.* Ditch irrigation, rather than overhead sprinkling, of seed corps is necessary. This avoids dissemination of water-borne pathogens, and avoids making the environment favorable for infection by a wide range of pathogens (*Pseudomonas phaseolicola* and *Colletotrichum lindemuthianum* on bean, *Septoria apiicola* on celery,

*Xanthomonas campestris* and *Phoma lingam* on cabbage, and *Ascochyta* spp. on pea). This is particularly true in semiarid areas where foliage would otherwise remain uninfected. However, in many such areas real-estate developments are today taking the level irrigable land, and forcing agriculture to use hill slopes which can only be watered by overhead sprinkling. This is particularly serious for seed production, and may result in a wider geographical scattering of the industry, and a rise in disease incidence.

*e. Harvesting methods.* Delay in harvesting the seed crop may extend it into the rainy or foggy season, favoring some pathogens such as *Botrytis cinerea* on flowers of Compositae, and *Alternaria zinniae* on zinnia. Careful selection of pepper fruits free of soil rot caused by *Rhizoctonia solani* will eliminate seed transmission and spread of the fungus through the pulped fruit and seed during the fermentation process preliminary to cleaning (Baker, 1947). Use of canvas sheets chemically treated to prevent fabric decay, on which to pile harvested plants or flowers in the field prior to threshing, will prevent pathogens such as *R. solani* growing through them from the soil into flowers and seeds (Baker, 1956). Use of plastic sheets to prevent the movement of moisture from the soil into the plant piles may be advisable.

*f. Eradication of host plants.* A newly introduced pathogen may sometimes be eradicated in limited areas before it has spread more widely, as for example, a small area of *Cuscuta* in a seed field. Crop rotation and fallowing are forms of temporary eradication of host plants to break the pathogen cycle. Weed hosts in the area surrounding the seed crop may prove a disease hazard to it, and should be removed or killed with chemicals (see also p. 333). The wider the host range of a pathogen, the more difficult are these various types of eradication. Soil-borne pathogens generally are much more difficult to eradicate than those not residing in soil.

*g. Aging of seed.* Although the pathogen may die or be inactivated in some seeds in 1–3 years (see pp. 331 and 347), this phenomenon is so variable from the effect of environmental conditions as to be an unreliable and even dangerous method of control (Arndt, 1946; Gabrielson, 1961).

*h. Cleaning and treating seed.* Seed used in planting seed-production fields should be free of pathogens. Cleaning equipment may be used to eliminate accompanying pathogens and infected seed which is smaller or lighter than normal, but this is usually insufficient to eliminate the pathogen. Eradication, rather than reduction, of the pathogen is the objective, and thermotherapy may be effectively used for this purpose (Baker, 1956). For example, hot-water treatment of nasturtium seed infected with *Heterosporium tropaeoli* has been found to make possible the production of a healthy crop in isolated localities in California. Because the fungus only causes a production disease, however, the treatment has been used to keep the disease at an unimportant level by seed treatment every second or third year. *Xanthomonas incanae,* on the other hand, may produce severe losses in garden stock in both seed and cut-flower fields. It has been the practice in California for more than two decades to hot-water treat seed planted in seed fields to insure producing propagules free of the bacteria. Because of the mucilaginous character of the seed coat, such treatment is troublesome and unlikely to be undertaken by growers of cut flowers. Control of this disease in the seed — production field has proved to be the best procedure and, because the pathogen survives only briefly in soil, has been very successful. A third pathogen, *Alternaria zinniae,* may similarly be controlled by thermotherapy. However, this disease is unimportant in California seed fields, even though the pathogen infects seed produced there, and the fungus may survive in soil from season to season. It is possible to produce a seed crop free of the pathogen by planting in dry interior California valleys in land not previously used for zinnias. This disease may, therefore, be controlled in the seed field by site selection, and in the home yard by thermotherapy if the soil is not infested. If it is, soil treatment also is necessary. Because of the inconvenience of the seed treatment, and soil carryover of the pathogen, the disease has not been effectively controlled, even though methods are available to do so.

C2—SELECTION OF THE SEED—PRODUCTION AREA. Seed should be grown in areas where, for climatic or other reasons, the pathogen is unable to establish or maintain itself, or the vector is absent or infrequent. Pea seed free of *Ascochyta* spp. may thus be produced in south-central California because of the semiarid conditions, and pea seed free of *Fusarium oxysporum* f. sp. *pisi* also may be produced apparently because of soil antagonists. The beet leafhopper, *Circulifer tenellus,* may infrequently carry the curly-top virus to petunia seed fields in coastal California from the interior valleys, where they are abundant (see also p. 375). Some pathogens may persist in the soil without producing either symptoms on above-ground parts or seed transmission, as in *Phoma lingam* of brussels sprouts in coastal California (Snyder and Baker, 1950); this is, therefore, a subterranean disease because of the dry climate. Similarly, bean seed free of *Colletotrichum lindemuthianum,* pea seed free of *Ascochyta* spp., and cucurbit seed free of *C. lagenarium* and *Pseudomonas lachrymans* (Baker and Snyder, 1950) are produced in rain-free growing areas.

It is possible also to grow seed crops in isolated areas where the pathogen, vector, or both, are not present. It is imperative in such cases to plant only seed free of the pathogen, or treated to be freed of it, since there may be no operating natural restriction once the pathogen is introduced. It is good practice to isolate seed fields from any other plantings likely to produce inoculum and introduce the pathogen or to contaminate the seed. Nasturtium seed fields are thus grown as far as possible from volunteer or perennial nasturtiums in home yards or old seed fields, in order to reduce introduction of air-borne spores of *Heterosporium tropaeoli* (Baker, 1956). Such isolation is also very important in the production of virus-free seed because of the mobility of insect vectors. For example, lettuce seed production in the San Joaquin Valley, California, might have to be abandoned if commercial lettuce production was to be started in the area, because it would provide a possible source of lettuce mosaic (R. G. Grogan, personal communication).

C3—TREATMENT OF FIELD SOIL. Freedom from soil-borne plant pathogens on, in, or with seed produced in fields would, in almost all cases, be enhanced by treating the soil. When production of seed free of a soil-borne pathogen is to be achieved, treatment of field soil may be a necessity. Eradicatory soil treatment in seed fields is certainly economic with valuable crops (*Fusarium oxysporum* f. sp. *mathiolae* on stock), in soil where the severity of disease loss makes it imperative if the crop is to be grown, or in situations where failure to treat may lead to greater expense through infesting a larger area of land [*Orobanche* spp. on tomato (Wilhelm *et al.,* 1958)]. Reduction of plant infection by *Sclerotinia sclerotiorum* has been achieved through discing 800–1000 pounds of calcium cyanamide per acre into the moist soil about 1 month before planting (Brooks *et al.,* 1945), and reduction of soil infestation is achieved by flooding for 23–45 days (W. D. Moore, 1949).

Soil treatment of the crop — production field may augment disease control from use of pathogen-free seed.

C4—SELECTION AND BREEDING FOR RESISTANCE OR FOR NONTRANSMISSION OF THE PATHOGEN. Selection and breeding for resistance to attack by a pathogen will also reduce incidence of seed transmission and its importance in the perpetuation of the microorganism. Because of host resistance, the amount of inoculum to get in, on, or to accompany the seed is reduced, and distribution of the pathogen through the vascular elements into the seed is restricted. Resistant varieties thus limit incidence of disease in the production crop by reducing the amount of inoculum, as well as the susceptibility of the host to infection.

There is another type of resistance available against seed-transmitted virus diseases. It has been found that some species and cultivars have little or no seed transmission of a given virus, whereas others may have a high percentage of it (Table 5.II and p. 346). Crispin Medina and Grogan (1961) showed that dominant factors for resistance to bean common mosaic virus in bean may preclude its seed transmission. That some viruses are inactivated in the embryo [bean southern mosaic virus in bean (Zaumeyer and Harter, 1943; Cheo, 1955)] may also be useful in this regard. The level and type of resistance, as measured by symptom expression on the host, may in some cases be related to the amount of seed transmission (Bennett, 1969), but in others the greatest amount of transmission occurs from symptomless plants [sun-blotch virus on avocado (Wallace and Drake, 1962)]. However, the possibility of developing crop varieties that do not transmit viruses through seed is sufficiently promising that it should be fully studied and utilized. The fact that some viruses are seed transmitted in one species but not another [sowbane mosaic virus in *Atriplex pacifica* had 21.4% such transmission, but in five other species there was none (Bennett and Costa, 1961)] may also be useful in a breeding program.

C5—SEED-FIELD INSPECTION. This is a useful general technique for determining whether a disease is present in a seed field or growing area so as to be able to predict pathogen transmission in the seed produced (Güssow, 1936; McCubbin, 1954). However, seed which is clean in the field may subsequently be contaminated during threshing or cleaning (W. C. Moore, 1946). Furthermore, some diseases may not be detected in the seed field, even though seed produced may transmit the pathogen [*Corynebacterium fascians* on sweet pea and nasturtium (Baker, 1950); *Alternaria zinniae* in zinnia seed (Baker, 1956)]. This is, then, a useful but not infallible method for certifying the health of seed; it must be used in conjunction with other laboratory techniques when negative findings are obtained in the seed field. Halo blight *(Pseudomonas phaseolicola)* of bean spreads very little under dry California conditions, and is very difficult to detect by field inspection; a vermiculite pot test of the harvested and cleaned seed is a much more effective means of appraisal (Grogan and Kimble, 1967).

## 2. CURATIVE MEASURES FOR SEED ALREADY CONTAMINATED

The SP transfer is the important one in crop—production fields, and control practices aim at its prevention or elimination through use of seed free of or freed from the pathogen.

C6—SEED INDEXING. This is the customary procedure of seed-testing laboratories incidental to purity and germination analyses. Special direct techniques which have been devised for specific pathogens (Doyer, 1938; Grogan and Kimble, 1967; International Seed Testing Association,

1958 ——, annual; Limonard, 1968; Malone and Muskett, 1964) may be used in laboratories giving particular attention to seed-borne pathogens.

Various indirect methods may also be used. Among these are the use of (a) microscopic examination of the embryo for detecting *Ustilago nuda* in barley seed (Marshall, 1959–1960) and *U. tritici* in wheat (Morton *et al.,* 1960; Popp, 1958); (b) bacteriophage for detection of *Xanthomonas phaseoli* in bean (Katznelson and Sutton, 1951; Katznelson *et al.,* 1954); (c) serological methods for detecting *Pseudomonas phaseolicola* in bean seeds (Guthrie *et al.,* 1965); (d) various selective media (Tsao, 1970); (e) virus-specific radioautographic labeling for detecting squash mosaic in cantaloupe (Powell and Schlegel, 1970b); (f) inoculation of seedlings with ground seeds, used for detecting *Corynebacterium michiganense* on tomato (Thyr, 1969); (g) immunofluorescence techniques (Paton, 1964); (h) selective incubation temperature (Malone, 1962); (i) use of selective fungicides, bactericides, etc. (Limonard, 1968); and (j) isolation from ground seeds on special media, used for *P. phaseolicola* in bean seeds (J. D. Taylor, 1970).

Much present-day seed testing is done with various modifications of the blotter or agar techniques because of their convenience and ease of use. These may, however, give data at variance with results of field sowing. For example, de Tempe (1968b) cited unpublished data of K. T. J. Inia and J. Miedema of an instance where *Ascochyta* spp. caused severe loss in canning pea crops in the Netherlands in 1965 although the seed had showed scarcely any infection in blotter trials. Del Monte Corporation found that pea seed produced in the Pacific Northwest was contaminated with dust containing chlamydospores of *Ascochyta pinodella*. When such seed was treated with calcium hypochlorite, as was customary prior to laboratory testing, *Ascochyta* failed to develop on blotters. Without this pretreatment, *Ascochyta* developed there, providing good correlation with foot rot in field plantings. Seed treatment with Captan gave effective field control for these seed lots (L. O. Lawyer, personal communication). One wonders whether some of the reported successful controls of ascochyta blight of peas by chemical treatments (Maude *et al.,* 1969) may not have been due to such surface contamination, rather than internal infection.

In the final analysis, the best method of indexing is probably to plant the seed in soil (untreated, and steamed at 60° and 100°C/30 minutes; see p. 339) under optimal environmental conditions for infection and disease development, and with adequate safeguards to prevent contamination from other sources.

The question of acceptable tolerances for numbers of contaminated seeds arises in certification (see pp. 336 and 342). Such tolerances seem to be more indicative of commercial demands and inadequate testing methods than of pathological requirements (Güssow, 1936). It is instructive that some of the most effective cases of disease control through pathogen-free seed have had essentially zero tolerances and have been supported by commercial companies (e.g., lettuce — mosaic control in California and Arizona lettuce crops, ascochyta — blight control in the cannery pea crop in the United States). There is opportunity for much wider adoption of such standards.

Indexing of seeds for virus transmission raises special problems because of the necessity of using the seeds or their seedlings to inoculate known healthy plants. The assumption that seedlings grown from virus-infected embryos of seeds will, under favorable environmental conditions, be detectable because of visible symptoms, has now been questioned (see pp. 333, 334, 339, and 343). A number of viruses have been studied in which seedlings from infected embryos can only be detected by transmission to healthy seedlings, which will then exhibit symptoms. As Bennett (1969) stated, "if the mother plant recovers to a high degree, seedlings tend to be symptomless. It appears . . . that the recovery stage of a virus disease is usually carried through the seed."

Broadbent (1965) has commercially combined heat treatment and indexing of tomato seed to free it of TMV. The seed lots are heated in a dry oven for 3–4 days at 70°C and then indexed by planting samples of the seed; those lots which still contain active TMV are discarded, those which do not, qualify as Virus-Tested Seed.

C7 — SEPARATORY PROCEDURES. Commercial procedures of cleaning seed to free it of debris, weed seeds, and nongerminating and light seeds may reduce total pathogen transmission of types SP1–SP6 (e.g., *Anguina tritici* on wheat), but may also spread such pathogens as bacteria more uniformly over the remaining seeds. Although squash seeds infected with squash mosaic may be lighter in weight than healthy seeds, and therefore separable from them (Middleton, 1944), this would not eliminate a high percentage of infected seeds. Washing seeds with flowing water to free them of surface spores was used by Tillet (1755) as an effective control of bunt of wheat. Although often overlooked in this era of fungicides, it is still a useful method. See pages 366 and 380 for types of equipment used for cleaning seeds.

C8 — CHEMICAL SEED TREATMENTS. Protective treatments aimed at preventing attack of the seed by soil microorganisms are of little or no value in preventing seed transmission of pathogens (J. C. Walker, 1948). Eradicative chemical treatments are generally used to kill pathogens on the surface of seed, between the glumes and the seed, or in shallow infections of the fruit or seed coats. General discussions of seed treatments are given by Leukel (1948) and J. C. Walker (1969).

A prolonged water-soak treatment has been used against the loose smuts of cereals (Chinn and Russell, 1958; Tyner, 1957) and peanuts (Ivanoff, 1958). The treatment (21°–29°C for 48–72 hours) controlled the pathogens. Seed injury was reduced by addition of 0.02% mercuric chloride or 1.0% sodium chloride to the water (Ivanoff, 1958; Russell and Chinn, 1958). Various other chemicals have also been used in the water to combine fungicidal effects with that of the water soak.

The thiram-soak treatment (0.2% aqueous suspension of tetramethyl-thiuram disulfide at 30°C for 24 hours) has recently been recommended (Maude, 1970; Maude *et al.,* 1969) for control of internally borne: *Ascochyta* spp. on pea; *Septoria apiicola* on celery; *Helminthosporium avenae* on oats; *Alternaria dauci* and *Stemphylium radicinum* on carrot; *Phoma lingam, Alternaria brassicae,* and *A. brassicicola* on cabbage, kale, and brussels sprouts; *Phoma betae, Colletotrichum spinaceae,* and *Fusarium* sp. on beet; *Ascochyta imperfecta* on alfalfa; *Botrytis cinerea* on flax; *Septoria nodorum* and *Tilletia caries* on wheat; *Ustilago nuda* and *Helminthosporium gramineum* on barley; *Ascochyta fabae* on horse bean; *Alternaria* sp. on *Lobelia.* Injury was said to be negligible, except for lettuce. The thiram suspension contained 10 ppm in solution, and this was taken up by the seeds and apparently acted fungicidally. The treatment is ineffective against bacteria. The seed must be dried following treatment.

The mechanisms of action of these soak treatments have not been clarified, and the method, although apparently effective, remains empirical (Limonard, 1968). I have found a great deal of swelling, splitting, and embryo exposure of pea, bean, and zinnia seeds soaked in water for 24 hours in these treatments. Cabbage seed withstood the treatment rather well. Since the Red Kidney bean, and Alaska and Progress peas used have tough seed coats, it is unlikely that the method can safely be used, on these crops at least. The effect of these soak treatments on pathogen-infected hard seed of legumes, which do not absorb water, apparently has not been considered.

Allen (1963) has reported injury from thiram dust on *Nemesia, Phlox drummondii, Petunia, Tagetes erecta, T. patula, Celosia plumosa, Salvia, Dianthus,* and *Lobelia. Cheiranthus, Zinnia, Callistephus,* and *Aster* were not injured.

Fumigation of seeds with methyl bromide has been used to free them of nematodes. Onion, teasel, red clover, and alfalfa seeds infected by *Ditylenchus dipsaci* were freed of the nematodes by treatment for 24 hours at 25 mg/liter at 24°C (J. B. Goodey, 1949; T. Goodey, 1945).

Maude and Kyle (1970) reported good control of ascochyta blight of pea with a benomyl dust seed treatment. Jacobsen and Williams (1971) obtained control of internally borne *Phoma lingam* in broccoli seed with a 24-hour soak in 0.2% aqueous suspensions of benomyl.

Fermentation (96 hours at 21°C or below) and the acetic acid treatment (0.6% solution for 24 hours at 21°C or below), commonly recommended to free tomato seed of *Corynebacterium michiganense* (Blood, 1933, 1937, 1942; J. C. Walker, 1969), apparently are useful for lots with a high percentage of infected seeds. The lots reported by Blood showed the following percentages of diseased plants when sown in the field: 14.9%, reduced to 0.9% and 0% by 24— and 48—hour fermentation, respectively; 81.3%, reduced to 0.2% by 96-hour fermentation, and to 0.08% by the acetic acid treatment. This level of control is insufficient for seed sown in seedbeds for field transplanting. Seed lots tested by Grogan and Kendrick (1953) had less than 1% infected seed, and the above treatments failed to reduce this small amount of infection. Direct field seeding has provided effective control (R. G. Grogan, personal communication). Treatment of tomato seed with concentrated hydrochloric acid for 30 minutes frees it of surface-borne TMV (Broadbent, 1965); this treatment is required for tomato seed to be imported into New Zealand or Australia.

Resistance in pathogens to certain proprietary seed-treatment fungicides has been appearing in various parts of the world from natural selection (Georgopoulos, 1969). Kuiper (1965) demonstrated that hexachlorobenzene failed to control *Tilletia caries* and *T. foetida* on wheat in Victoria, Australia after 1964. Organomercurials have similarly become ineffective against *Helminthosporium (Pyrenophora) avenae* on oats in Scotland (Noble *et al.,* 1966), Ireland (Malone, 1968), Netherlands (Limonard, 1968), New Zealand (Sheridan *et al.,* 1968), New York, and perhaps Austria (Crosier *et al.,* 1970). Greenaway (1971) showed that mercury resistance in this fungus is associated with production of red anthroquinone pigments which apparently bind the phenyl-mercury ions.

**C9**—THERMOTHERAPY OF SEED. Heat treatment of seed may be used to destroy many types of pathogens.

*i. Pathogens which accompany seed. Orobanche* spp. was shown by Garman (1903) to be killed by hot-water treatment at 60°C for 10 minutes without injury to hemp seed. *Heterodera schachtii* may be killed in small lumps of soil with beet seed by dry heat at 65°–70°C for 5–10 minutes (Shaw, 1915).

*ii. Pathogens on the surface of seed. Ditylenchus dipsaci* on alfalfa seed may be killed in hot water at 48°–49°C for 15 minutes (Edwards, 1932); *Corynebacterium fascians* may be eliminated from nasturtium seeds by a presoak in water for 1 hour, followed by treatment in hot water at 51.7°C for 30 minutes (Baker, 1950).

*iii. Pathogen in shallow infections in seed. Septoria apiicola* may be destroyed in celery seed by hot-water treatment at 48°–49°C for 30 minutes (J. C. Walker, 1969).

*iv. Pathogen internal in seed. Fusarium solani* f. sp. *cucurbitae* may be killed in squash by hot-water treatment at 55°C for 15 minutes (Gries, 1946); *Heterosporium tropaeoli* in nasturtium seed at 51.7°C for 30 minutes following a soak of 1 hour in water (Baker and Davis, 1950a; Baker, 1956); *Phoma lingam* in cabbage seed at 50°C for 30 minutes (J. C. Walker, 1969); *Xanthomonas campestris* in cabbage seed at 50°C for 30 minutes (J. C. Walker, 1969); *X. incanae* in stock seed at 54°–55°C for 10 minutes (Baker, 1956); *Ustilago tritici* in wheat at 52°C for 11 minutes following a 5-hour presoak at 21°C (J. C. Walker, 1969).

Thermotherapy is usually used against pathogens so situated as to be protected from chemicals. This is, however, an unnecessarily restricted usage.

There are three general methods for application of thermotherapy to seeds (Baker, 1962a,b, 1969). In decreasing order of efficiency of heat transfer to seeds, these are:

1. Hot water. This method was the first one used in seed thermotherapy; J. L. Jensen used it in Denmark in 1888 against the loose smuts of oats and barley *(Ustilago avenae* and *U. nuda)*. Water is five times as efficient a thermal exchange medium as dry air, and twice as efficient as saturated water vapor (aerated steam). It is, therefore, to be expected that temperatures or time of treatment increase from water (usually in the range, 49°–57°C for 30 minutes), through aerated steam (54°–60°C for 30 minutes), to dry air (from 54°C for 5 hours to 95°–100°C for 12 hours). Temperature and time are to some extent mutually compensating, but the relationship is not direct and cannot be plotted by a general formula (Baker, 1962b). It is, therefore, desirable to standardize on one time interval in order to bring more rationality into treatment schedules. It should be noted that some increase in time may be necessary for treatment of very large seeds (e.g., avocado) to permit heat penetration. The great majority of seeds may be treated for a 30-minute period. A presoak in cool water is desirable for some seeds to displace the air (a good insulator) between the fruit and seed coat [nasturtium (Baker and Davis, 1950a)], or between the glumes and seed [oats (Jensen, 1888)]. Details of the various hot-water methods are presented by Baker (1962b).

Treatment injury may be evidenced as (*a*) a 1–15 day delay in germination, (*b*) slight stunting or somewhat retarded development of seedlings, and slow weak growth of seedlings (which may later become normal), (c) development of only the radicle, or (d) death of the seed. When seed is planted in soil, these injuries are seen as delayed and decreased seedling emergence, since the weakened plants lack sufficient vitality to push through the soil. Such injuries occur particularly in old, low-vitality, or mechanically injured seeds, or those of especially susceptible kinds.

Following any thermotherapy, cooling, and drying, the seed should be treated with a mild protectant fungicide. Since seed is inevitably weakened to some degree and germination retarded by heat treatment, it should be thus protected from attack by soil microorganisms until germination.

2. Aerated steam. Treatment of seed with aerated steam came into use about 1960 (Baker, 1962a,b, 1969) because of inherent disadvantages in the use of hot water:

a. Seeds treated in hot water absorb much water, undergo severe leaching of soluble materials, and require drying before storage or planting. Bean seed treated with aerated steam and with hot water was found (Baker, 1969) to have 18–21% water absorption (6.5–8.0% after evaporative cooling) and 32.2–50.5%, respectively. The figures for pea seed were 19.2–20.0% (8.0–9.0% after evaporative cooling) and 42.1%, respectively; for barley they were 14.0% (6.0% after evaporative cooling) and 49.0%.

b. Equipment for hot-water treatment is expensive and cumbersome. Because of the increased heat sensitivity of hydrated seed, temperature regulation must be precise or germination may be reduced. Large drying areas are required; these are avoided with aerated-steam treatment.

c. Some seed coats (beans, peas) may rupture during hot-water treatment, and others (stock, alyssum, flax) exude mucilaginous material which sticks the seeds together and makes handling and drying extremely difficult. These conditions are avoided by use of aerated steam. The margin of safety between effective pathogen control and excessive seed injury is generally greater for aerated-steam than hot-water or dry-air treatments. Dry air has been found not to be effective against most seed-borne pathogens, and requires very long exposures. Aerated-steam treatment is an effective compromise between hot water and dry air.

Mixing air with steam dilutes it and thus lowers its temperature to any desired level. The simplest way to produce aerated steam is to join together lines of flowing steam and air; a needle valve in each line provides satisfactory temperature control. The principles behind the use of aerated-steam treatment have been elucidated (Baker, 1962a, 1969), and proper methods of application to seeds have been described and illustrated

(Baker, 1969). The equipment there described required a flow rate of about 20 cu ft/min of oil-free compressed air. An improved unit has been developed in Australia which does not require a source of compressed air, is extremely accurate and stable, and can treat up to seventeen lots of seed simultaneously; this unit will soon be available in limited quantities (Fig. 5.11). Whatever the type of equipment used, there are certain minimal requirements that must be observed:

a. The aerated steam must move by pressure flow *downward through* the seed mass being treated in a double-walled container. The flow rate of aerated steam must be great enough that prompt heating of the seed occurs. Treatment temperatures must be reached and the seed cooled *promptly and uniformly;* this is not possible in steam chambers or in trays or cloth or plastic mesh bags, because of poor heat penetration. Some of the deficiencies of the several kinds of equipment devised and used in various parts of the world have been described (Baker, 1969). It is preferable to pass the vapor down through the seed mass, rather than up through it, because any condensate will then be blown out the bottom, and the seed will be drier at the end of the treatment.

b. The temperatures must be carefully adjusted and maintained, and timing must be precise. For these reasons, only a quantity of seed should be treated which the unit will heat in about 2 minutes. The flow rate of aerated steam must be sufficient to heat the whole mass nearly simultaneously, rather than in progressive strata, and the temperature at the output during this initial heating should not be more than 1.5°C below that at the input. Timing should begin when the two temperatures are approximately equal (i.e., when the seed is no longer absorbing Btu). Some seeds, such as those of celery, are highly hygroscopic, and may attain temperatures 2°C higher than the vapor flowing past them, owing to heat of wetting. Once sorption declines, the temperature will return to ambient. During this wetting phase such seeds may be maintained at the proper temperature by slightly increasing the air flow.

c. At the end of the treatment period the steam is shut off and temperature is rapidly lowered by evaporative cooling. This should be continued until the temperature falls to about 32°C, and then until the seed reaches the desired level of dryness. If the area in which seed treatment is performed is dusty or laden with spores of pathogens, it will be necessary to filter the air blown through the seed, to prevent recontamination. Many inexpensive effective air filters are available.

d. Temperatures must be measured with thermometers or thermocouples whose accuracy *in the treatment range* has been checked against one of known accuracy. Expanded-range thermometers, individually calibrated, are excellent for the purpose. Thermocouples placed at several points in the seed mass (and actually inside large seeds) provide the most accurate temperature measurements.

FIG. 5.11. Aerated-steam equipment for simultaneously treating multiple lots of seeds. The unit is self-contained except for a steam source, and is fully adjustable for temperature and vapor flow rate. The air is supplied by a blower with a filter on the intake, and the steam-air mixture is free of entrained water drops. Thermocouples are placed with the seed in the treatment tubes to accurately measure the attained temperature. (Courtesy of A. J. Newport and Son Pty. Ltd., Dundas, New South Wales, Australia.)

Germination is usually not reduced as much with aerated steam as with hot water, but reduction may still be severe with old, low-vitality, or injured seed, or seed of particularly susceptible kinds. Seed used for treatment should be fresh, of high vigor and germination, and free from mechanical injury.

Aerated-steam treatments may sometimes be made more effective by holding seed in moisture-saturated air at room temperature for 1–3 days prior to thermotherapy. Apparently bacteria and fungi are made more thermally sensitive by this hydration. In some cases this effect is greater for the pathogen than for the seed, and thus increases the margin of safety. Since this is not always so, it is necessary to try the method in each instance. Hard seeds of legumes are not hydrated by this procedure and, if infected with a pathogen, may provide a nucleus of infection when planted. It may be necessary to scarify lots having hard seeds before hydration and treatment.

Some examples of aerated-steam treatments for various pathogen-host combinations are:

*Alternaria zinniae* on zinnia seed. 57°C for 30 minutes on non-humidified seed.

*Phoma betae* on table beet seed. 56°C for 30 minutes following humidification for 3 days.

*Phoma lingam* on cabbage seed. 56°C for 30 minutes following humidification for 3 days.

*Alternaria brassicae* on cabbage seed. 56°C for 30 minutes following humidification for 3 days.

*Fusarium oxysporum* f. sp. *mathiolae* on stock seed. 54°C for 15 minutes on nonhumidified seed. The seed is not rendered mucilaginous, as it is in hot-water treatments.

*Botrytis cinerea* and *Alternaria tenuis* on China aster seed. 53°C for 30 minutes on nonhumidified seed (Bertus, 1967).

*Botrytis cinerea* on table beet seed. 52°C for 10 minutes on non-humidified seed (Miller and McWhorter, 1948).

*Itersonilia pastinaceae* on parsnip seed. 45.5°C for 30 minutes on nonhumidified seed (P. R. Smith, 1966).

3. Hot dry air. Because microorganisms are less sensitive to dry than to moist heat, hot dry air for seed treatment has been little used, despite its attractive physical attributes.

Cotton seed with anthracnose (*Glomerella gossypii*) is treated in two steps: predried at 60°–65°C for 20–24 hours to lower moisture to 3.62% or less (dry weight basis), and then treated at 95°–100°C for 12 hours (Lehman, 1925).

Snapdragon rust (*Puccinia antirrhini*) urediospores are killed in 1–2 hours at 34°–39°C (Aronescu-Săvulescu, 1938), or in a range from 34°C for 45 hours, 37°C for 16 hours, 38°C for 10 hours, 43°C for 2 hours, 40 minutes, to 46°C for 1 hour (Yarwood, 1948). It is of interest that the greater resistance of California-produced urediospores, as compared with those in Romania is consistent with their longevity (Baker, 1952). Hot-air treatments of seed at 46°C for 2–3 hours should be tried to free it of viable urediospores, which appear to be the means of intercontinental spread of the pathogen (Baker, 1952; J. Walker, 1954).

The sugar—beet nematode *(Heterodera schachtii)* may be killed in sugar—beet seed by hot-air treatment at 65°–70°C for 5–10 minutes (Shaw, 1915).

Treatment of seed in carbon tetrachloride has been suggested (Watson *et al.,* 1951) to avoid the disadvantages of hot-water soaks. It has been little used, because results have been quite variable and frequently ineffective. This method is regarded (Baker, 1962b) as a type of hot-air treatment, since the temperatures required fall in the same range, and the specific heat of carbon tetrachloride (0.2) is the same as that of air. The toxicity of carbon tetrachloride to man requires that it be used with adequate ventilation.

Several other treatments, which perhaps involve heating the seeds, have not been markedly successful: exposure to high-frequency fields (Grainger and Simpson, 1950; Seaman and Wallen, 1967); exposure to cathode rays (Lambou *et al.,* 1952); and exposure to ultrasonic waves (A. M. Andersen *et al.,* 1964; Gordon, 1964).

## N. Relation of Seed Vitality and Vigor to Seedling Disease

The ability of a seed to produce a viable seedling when planted in soil in the usual manner may be referred to as its vitality, and its ability to produce a rapidly growing normal seedling is considered to be its vigor. This topic is discussed more fully in Volume I, Chapter 6, and is considered here from the standpoint of its relation to the SP transfer of pathogens and to disease control.

It is well known that a number of factors contribute to a loss of vigorous germination of seeds:

1. As seed ages, both the percentage and the vigor of germination decline. It is an interesting but still unexplained fact that the ability of seed to tolerate thermotherapy declines much more rapidly than does germination. Zinnia seed thus retains satisfactory germinability for 6–7 years, but reduction of germination from hot-water treatment increases after the first year. One-year-old seed averaged 15.6% germination reduction below the checks, 2-year-old averaged 38.9%, and 3-year-old seed 78.8% reduction (Baker and Davis, 1950b).

2. Heredity factors are important in seed survival. Since China aster seed has a much shorter life than that of zinnia, 2-year-old aster seed should be more susceptible to mold fungi than zinnia. China aster seed is, in fact, one of the principal sources of trouble with molds among flower seeds (Crosier and Heit, 1948).

3. Unfavorable environment of seed—production fields is important. Plants grown under humid conditions and frequent rains produce seed which is likely to be moldy, short-lived, and low in vitality and vigor. Furthermore, the chance of its transmitting pathogens is greatly increased.

The seed industries of the world are, therefore, largely found in areas of low rainfall and humidity during the growing seasons (Kreitlow *et al.,* 1961).

4. Storage under unfavorable conditions shortens seed vitality and life. The modern sealed can or packet of dehydrated seeds has been developed to protect against this effect, and has been particularly valuable in moist subtropical areas.

5. Injuries to seed and fruit coats decrease longevity and vitality of seed, as well as provide infection sites for molds and pathogens. In addition, such injuries increase damage from fungicidal treatments or thermotherapy. Thus, wheat seed with unbroken coats gave 98% germination, both when untreated or treated in hot water; that with broken coats gave 98% germination when untreated, but only 4% when treated in hot water (Tapke, 1924).

6. Saprophytic microorganisms on the seed weaken it (see p. 401). These develop in the field under humid conditions, and may wholly or partly decay the seeds. This reduces the volume of storage tissue and the amount of stored material available to the developing seedling, thus increasing its susceptibility to disease. Metabolic products of these microorganisms may also affect seedling growth, sometimes strikingly. Finally, when seed is planted, the saprophytes will resume growth in the favorable conditions of the soil, and may then decay the seed. The routine aerated-steam treatment (54.4°C for 30 minutes) of seeds practiced by several Australian bedding-plant growers undoubtedly is effective in part through elimination of these molds.

Any condition which retards the germination process will increase damage from seedling diseases. Seed of low vitality and vigor, excessively deep seeding, and conditions unfavorable to seed germination, all favor seedling diseases (Fig. 5.12; Baker, 1970). In general, the more unfavorable a given condition is to the plant, without drastically reducing growth of the pathogen (or the saprophyte in some cases), the worse the disease will be (Baker, 1957).

The use of a 2, 4-D treatment or freezing of seed prior to running a blotter test for the presence of pathogens (Limonard, 1968) is essentially an application of this principle, since it aims at destroying host resistance and vitality, and thus favoring the pathogen.

## II. Nontransmissible Pathological Conditions

### A. *Mechanical Injuries*

Crop seeds are today largely harvested or threshed by machinery, and are exposed to an additional sequence of mechanical operations during cleaning, polishing, and packaging. If the seeds are subjected to an impact during these operations there may be fracturing of parts of the embryo, or cracking, chipping, or flaking of the seed or fruit coat. These injuries

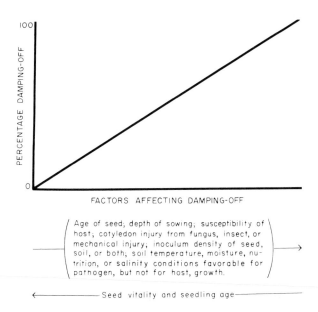

FIG. 5.12. Schematic diagram of the effect of several factors on the percentage of damping-off of seedlings. The influence of these conditions on seed transmission of most pathogens is similar. (Diagram by W. H. Fuller, Department of Plant Pathology, University of California, Berkeley.)

may occur in 80% of some kinds of seeds and not uncommonly reach 20%. There are usually marked differences in varietal susceptibility to such injuries.

In general, the lower the moisture content of the seeds, the more brittle they become, and the more severe the injury. However, corn is apt to exhibit increasing injury to the seed (fruit) coat as the moisture content reaches much above or below 12%. Injury is commercially reduced in most crops by threshing in the early morning while the seeds have a high moisture content, or seasonally before the seeds have become excessively dry. Seeds should not be reduced to a very low moisture level before cleaning operations are completed.

Low cylinder speeds should be used in threshing machinery, and sharp impacts should be avoided in all stages of handling. The injuries can largely be avoided by hand threshing of the seed when this is economically feasible.

### 1. INJURIES TO THE EMBRYO

Embryo injuries often are not evident until germination, when various seedling abnormalities are produced. An injury to bean seed, called bald-head or snakehead, is caused by impact in threshing. Seedlings are produced which may have no growing point, one or no cotyledons, no radicle, or a cracked hypocotyl. Varieties with hard thick seed coats sustain least injury (Borthwick, 1932; Zaumeyer and Thomas, 1957).

### 2. INJURIES TO THE SEED OR FRUIT COAT

Parts of the seed coat of beans may be flaked away, exposing the cotyledons (Zaumeyer and Thomas, 1957). The coats of flax seed may be cracked or fractured, and sometimes partly lost, exposing the cotyledons; the appearance of seedlings produced suggests that embryo damage also occurs (Kommedahl et al., 1955). The seed (fruit) coat of wheat commonly may be cracked or partly broken away over the embryo during threshing. The tip of the embryo may be exposed (Hurd, 1921; Mead et al., 1942). Corn seed may develop cracks in the seed (fruit) coat over the embryo or at various points over the endosperm, or bits of the coats may be flaked out (Koehler, 1957).

These injuries are particularly damaging because they provide entry for pathogenic and saprophytic microorganisms. The latter may decay the seed after planting, or deprive the seedling of nutrients from cotyledons. Protective fungicidal seed treatments may reduce the amount of damage resulting.

### 3. RELATION TO ENTRY OF PATHOGENS AND FUNGICIDES

It is a common observation that seeds with injuries and natural openings through the outer coats are prone to invasion by microorganisms, particularly under moist conditions (Machacek and Greaney, 1936; Koehler, 1957; Kommedahl et al., 1955). The moldy-seed condition then commonly results. Seeds which are infected but not decayed at time of planting may decay in the soil. Protective fungicides applied to the seed before sowing have given beneficial results (Koehler, 1957; Kommedahl et al., 1955). However, fungicides must be used which are not injurious to the exposed tissues (Foster, 1947). Bertus (1967) controlled *Botrytis cinerea* and *Alternaria tenuis* on China aster seed by treatment with aerated steam at 53°C for 30 minutes. Similar treatment of flower seeds at 54.4°C for 30 minutes, commonly practiced by Australian bedding-plant nurserymen, provides good control of such seed decay.

### B. *Congenital Defects*

Split seed coats in flax apparently result from a genetic weakness, a defect aggravated by environmental conditions. The two halves of the coat separate at the small end of the seed, exposing the cotyledons. Splitting of seed coats is worst in yellow-seeded varieties, and is not associated with harvesting methods (Kommedahl *et al.,* 1955). It is also reported on some bean varieties (Zaumeyer and Thomas, 1957).

The seeds (fruits) of Compositae are firmly attached to the receptable, and when removed in threshing, an opening results at the basal end. This is particularly true of *Callistephus chinensis,* in which the embryo may thus be exposed. It is, therefore, not surprising that these seeds are short-lived (1–2 years), very prone to attack by molds, and very sensitive to fungicide injury.

Seeds of pepper, eggplant, and tomato (Figs. 5.1A–C, 5.4B) normally have a conspicuous break in the seed coat which leads into a chamber between the seed coat and endosperm cuticle (Baker, 1947). Except as an entry point for pathogens, this apparently causes no damage. Pepper seeds also frequently have a coat-bound condition in germination in which the thick seed coat continues to confine the cotyledons and growing point after the seedling emerges. Under dry conditions, these seedlings often are never released because the inflexible seed coat is not softened by water absorption (Baker, 1948b). The situation may be controlled by deeper planting, keeping the soil surface moist during seedling emergence, and by avoiding both application of a sand cover on the seedbed and use of subirrigation. This condition can occur only with seeds having epigeal germination.

Embryoless seeds in wheat (Lyon, 1928) and barley (Harlan and Pope, 1925) were thought to be due to defective fertilization.

## C. Injuries Due to Environment

Seeds which become wet while still in the field or piled on canvases are likely to sustain injuries to their seed coats. Beans may develop small fissures or cracks over the surface, seeds of garden stock may adhere together in clumps, and many kinds of seeds become moldy.

A different situation commonly occurs with seed of Compositae. The petals senesce but remain attached, forming a water absorptive and retentive mass. Night dew or fog wets these flower heads, whether on the plant or piled on canvases on the ground, and they are likely to remain wet all through the following day. The flower heads then become permeated with saprophytic fungi which grow on or into the seed coats, decaying some of the seeds in place, and infecting them all. Zinnia seed (fruit) coats often have a retted appearance from such infection.

The effect of excessive dryness of seed in mechanical injury during threshing is discussed elsewhere (see p. 397).

### D. Injuries Due to Insects

1. SEED PITTING

While it is probable that a number of sucking insects may eventually be found that cause injuries to seeds, there have been few cases so far studied. *Lygus hesperus* and *L. elisus* have been shown to cause a destructive pitting of several kinds of beans in California and elsewhere, and to cause shriveling of alfalfa and beet seed. Young bean pods may be shed as a result of their feeding; on older pods no external symptoms may be produced. If the insect punctures seed through the pod, the injury on the mature bean varies from a tiny sunken pin point to large, irregular, crater-like yellow or brown lesions in the cotyledons, over which the testa is destroyed. The cavity is filled with a brown granular mass of dead cells, and the seed coat may crack or flake, starting at such pits. The testa may only partially cover the uninjured cotyledons of seeds, without visible evidence of insect feeding (Baker *et al.*, 1946). A similar type of injury in Virginia is caused by *Nezara viridula* (Wingard, 1925); in some of these lesions a yeast occurs under the moist conditions prevailing, but it cannot be causal since many lesions contain no yeast. Control is by application of insecticidal dusts to the seed fields, and by avoidance of planting near alfalfa fields.

Lygus feeding may produce misshapen discolored alfalfa seeds as well as causing flower and fruit shedding, and destruction of many seeds.

2. EMBRYOLESS SEEDS

A number of umbelliferous crops (dill, fennel, carrot, parsnip, anise, celery, parsley, *et al.*) frequently produce normal-appearing seeds, half of which fail to germinate because the embryos have not developed (Flemion and Waterbury, 1941; Flemion and Uhlmann, 1946). It was known in California that this condition could be induced by feeding of *Lygus hesperus* and *L. elisus*. This information, conveyed by the author to F. Flemion during a visit to the Boyce Thompson Institute in November 1947, provided the impetus to link their studies on insect pollination to the actual cause, and to start a series of papers on the phenomenon.

*Lygus lineolaris* was found in New York to cause embryoless seeds in carrot, dill, and fennel. The embryos were absent or smaller than normal, and frequently occurred in small seeds of otherwise normal appearance. That secretions were injected by the insect into the tissues was shown by Flemion *et al.* (1951); that they were toxic was shown microscopically (Flemion *et al.*, 1954; Hughes, 1943). Lygus bugs were considered to be the main, but not the sole cause of embryolessness in Umbelliferae.

When *Lygus hesperus* and *L. sallei* feed on seeds of *Parthenium argen-*

*tatum* (guayule) at the predough stage, embryo collapse is produced (Romney *et al.,* 1945).

*Adelphocoris lineolatus* was found to cause similar injury to alfalfa as that produced by Lygus bugs (Hughes, 1943).

Embryoless seeds in cereals, castor bean, and *Ginkgo* are thought to be caused by genetical factors, pollen defects, nutrition, or other causes.

### E. Injuries Resulting from Mineral Deficiencies

Deficiency of any essential mineral element will produce decreased growth or death of the plant. The condition, therefore, affects seed production. Marsh spot of peas is unique in that symptoms are produced in the cotyledons of seeds, while plants show no symptoms of manganese deficiency. When manganese is very low, the plant shows severe symptoms but does not produce seed; as available manganese increases, plant symptoms disappear and the seeds produced change from severely injured to mildly affected, to normal (Piper, 1941). The disease occurs mainly in alkaline soils. The inner surfaces of cotyledons show dark sunken areas which may, in severe cases, be detected from the outside. Seedlings from affected seeds are normal if growing in soil with adequate manganese. There are varietal differences in susceptibility (Glasscock, 1941). Spraying plants with a 1% solution of manganese sulfate at time of flowering controls the trouble (Lewis, 1939). Large applications of manganese fertilizer at time of seeding may not, however, prevent the disease.

Zaumeyer (1962) described seed cavitation and physiological spotting of pea seeds as due to nutritional disorders.

### F. Moldiness of Seeds

The moldiness of seeds has been discussed in several relevant sections of this paper (see pp. 370 and 398). The existence of this condition usually is evidence that the seed has been exposed to faulty handling. However, seeds of Compositae are very difficult to produce without this defect (p. 399). Moldiness assumes economic importance under certain circumstances:

1. Seedling emergence is severely reduced from decay of seed before or after planting. Seed treatment with fungicides or aerated steam is helpful in improving the stand.

2. Seedling growth is affected by toxicity of metabolic products of the microorganisms to the seedling [e.g., albinism of corn (Johann *et al.,* 1931; Koehler and Woodworth, 1938) and citrus (Durbin, 1959)]. Seed treatment with fungicides helps to reduce this effect.

3. Seed appearance is markedly affected, reducing salability for planting purposes or for milling.

4. Storage of seed is affected by heating and by deterioration from the microorganisms present on it (Bunting, 1932; Christensen and Kaufmann, 1965; Panasenko, 1967).

5. Utilization of the seed as food is reduced because of toxic metabolic products from microorganisms growing on them. The carcinogen aflatoxin from *Aspergillus flavus* on peanuts (Brook and White, 1966) and the emetic material from grain affected by *Gibberella zeae* (Hoyman, 1941) are examples of this condition.

6. Seed indexing to determine the pathogens present, or for certification, is hampered markedly by the presence of the saprophytic microflora, particularly when seeds are incubated to permit the pathogen to develop. These saprophytes on seed are considered to be one of the prime factors in variability in comparative seed tests for the presence of pathogens (Limonard, 1967, 1968).

## Acknowledgments

The many helpful suggestions made by R. G. Grogan during the preparation of this paper, the unpublished information supplied by L. R. Fraser, R. G. Grogan, L. O. Lawyer, W. C. Snyder, and S. Wilhelm, the preparation by W. H. Fuller of the illustrations for the printer, and the typing and proofing of the manuscript by K. C. Baker, are gratefully acknowledged.

## REFERENCES

Aebi, H., and Rapin, J. (1954). Lutte contre la cercosporiose de la betterave sucrière. *Rev. Romande Agr. Viticult. Arboricult.* **10**, 45.

Ainsworth, G. C. (1934). Virus disease investigations. (d) An experiment with seeds from 'streaked' tomato plants. *Cheshunt Exp. Res. Sta., Ann. Rep.* **19**, 62.

Alcock, N. L. (1931). Notes on common diseases sometimes seed-borne. *Trans. Proc. Bot. Soc. Edinburgh* **30**, 332.

Allard, H. A. (1915). Distribution of the virus of the mosaic disease in capsules, filaments, anthers, and pistils of affected tobacco plants. *J. Agr. Res.* **5**, 251.

Allen, J. D. (1963). Damage to flower seeds caused by dusting with thiram. *N. Z. Plants Gard.* **5**, 214.

Andersen, A. M., Hart, J. R., and French, R. C. (1964). Comparison of germination techniques and conductivity tests of cotton seeds. *Proc. Int. Seed Test. Ass.* **29**, 81.

Andersen, H. (1952–1956, 1961). Årsberetning fra Statens Plantetilsyn Vedrørende Frøpatologisk Kontrol. 1–5, 12. *Tidsskr. Planteavl* **56**, 67–86 and 501; **59**, 867.

Ark, P. A., and Gardner, M. W. (1944). Carrot bacterial blight as it affects the roots. *Phytopathology* **34**, 416.

Arndt, C. H. (1946). Effect of storage conditions on survival of *Colletotrichum gossypii*. *Phytopathology* **36**, 24.

Aronescu-Săvulescu, A. (1938). Contribution a l'étude de la rouille du muflier (*Puccinia antirrhini* Diet. et Holway). *Ann. Inst. Rech. Agron. Roum.* **10**, 497.

Arthur, J. C. (1885). Proof that bacteria are the direct cause of the disease in trees known as pear blight. *Bot. Gaz.* **10**, 343; also in *Proc. Amer. Ass. Advan. Sci.* **34**, 295 (1885); *Gard. Chron.* **24**, 586 (1885); *N.Y. Agr. Exp. Sta. (Geneva), Bull.* **2**, n.s., 1 (1885).

Atsatt, P. R. (1965). Angiosperm parasite and host: coordinated dispersal. *Science* **149**, 1389.

Baker, K. F. (1946). Observations on some Botrytis diseases in California. *Plant Dis. Rep.* **30**, 145.

Baker, K. F. (1947). Seed transmission of *Rhizoctonia solani* in relation to control of seedling damping-off. *Phytopathology* **37**, 912.

Baker, K. F. (1948a). Fusarium wilt of garden stock (*Mathiola incana*). *Phytopathology* **38**, 399.

Baker, K. F. (1948b). The coat-bound condition of germinating pepper seeds. *Amer. J. Bot.* **35**, 192.

Baker, K. F. (1950). Bacterial fasciation disease of ornamental plants in California. *Plant Dis. Rep.* **34**, 121.

Baker, K. F. (1952). A problem of seedsmen and flower growers—seed-borne parasites. *Seed World* **70** (11), 38, 40, 44, 46, and 47.

Baker, K. F. (1953). Fusarium wilt of China aster. *U. S. Dep. Agr. Yearb.* **1953**, 572–577.

Baker, K. F. (1956). Development and production of pathogen-free seed of three ornamental plants. *Plant Dis. Rep., Suppl.* **238**, 68.

Baker, K. F., ed. (1957). The U. C. system for producing healthy container-grown plants. *Calif., Agr. Exp. Sta., Manual* **23**, 1.

Baker, K. F. (1962a). Principles of heat treatment of soil and planting material. *J. Austral. Inst. Agr. Sci.* **28**, 118.

Baker, K. F. (1962b). Thermotherapy of planting material. *Phytopathology* **52**, 1244.

Baker, K. F. (1969). Aerated steam treatment of seed for disease control. *Hort. Res.* **9**, 59.

Baker, K. F. (1970). Types of Rhizoctonia diseases and their occurrence. *In* "Biology and Pathology of *Rhizoctonia solani*" (J. R. Parmeter, ed.), pp. 125–148. Univ. of California Press, Berkeley.

Baker, K. F. (1971). Fire blight of pome fruits: the genesis of the concept that bacteria can be pathogenic to plants. *Hilgardia* **40:** 603.

Baker, K. F., and Davis, L. H. (1950a). Heterosporium disease of nasturtium and its control. *Phytopathology* **40**, 553.

Baker, K. F., and Davis, L. H. (1950b). Some diseases of ornamental plants in California caused by species of *Alternaria* and *Stemphylium*. *Plant Dis. Rep.* **34**, 403.

Baker, K. F., and Davis, L. H. (1950c). *Stemphylium* leaf spot of China aster. *Mycologia* **42**, 477.

Baker, K. F., and Davis, L. H. (1951). An unusual occurrence of sclerotia of *Sclerotinia* spp. with seed of *Centaurea cyanus*. *Plant Dis. Rep.* **35**, 39.

Baker, K. F., and Locke, W. F. (1946). Perithecia of powdery mildew on zinnia seed. *Phytopathology* **36**, 379.

Baker, K. F., and Smith, S. H. (1966). Dynamics of seed transmission of plant pathogens. *Annu. Rev. Phytopathol.* **4**, 311.

Baker, K. F., and Snyder, W. C. (1950). Restrictive effect of California climate on some disease of vegetables, grains, and flowers. *Calif. Agr.* **4** (8), 3, 15.

Baker, K. F., Snyder, W. C., and Holland, A. H. (1946). Lygus bug injury of lima bean in California. *Phytopathology* **36**, 493.

Barton, L. V. (1967). "Bibliography of Seeds." Columbia Univ. Press, New York.

Bauer, F. (1823). Microscopical observations on the suspension of muscular motions of the *Vibrio Tritici. Phil. Trans. Roy. Soc. London* **113**, 1.

Bawden, F. C. (1964). "Plant Viruses and Virus Diseases," 4th ed. Ronald Press, New York.

Beach, S. A. (1892). Some bean diseases. Bean blight. *N. Y., Agr. Exp. Sta., Geneva, Bull.* **48**, n.s. 329.

Bennett, C. W. (1942). Longevity of curly-top virus in dried tissue of sugar beet. *Phytopathology* **32**, 826.

Bennett, C. W. (1944). Latent virus of dodder and its effect on sugar beet and other plants. *Phytopathology* **34**, 77.

Bennett, C. W. (1969). Seed transmission of plant viruses. *Advan. Virus Res.* **14**, 221.

Bennett, C. W., and Costa, A. S. (1961). Sowbane mosaic caused by a seed-transmitted virus. *Phytopathology* **51**, 546.

Bennett, C. W., and Esau, K. (1936). Further studies on the relation of the curly top virus to plant tissues. *J. Agr. Res.* **53**, 595.

Bertus, A. L. (1967). Aerated steam treatment of China aster seed. *Austral. J. Sci.* **30**, 29.

Blakeslee, A. F. (1921). A graft-infectious disease of *Datura* resembling a vegetative mutation. *J. Genet.* **11**, 17.

Blood, H. L. (1933). The control of tomato bacterial canker (*Aplanobacter michiganense* E. F. S.) by fruit-pulp fermentation in the seed-extraction process. *Utah Acad. Sci., Arts Lett., Proc.* **10**, 19.

Blood, H. L. (1937). A possible acid seed soak for the control of bacterial canker of the tomato. *Science* **86**, 199.

Blood, H. L. (1942). Control of bacterial canker of tomatoes. *Canning Age* **23**(4), 221.

Bollen, G. J. (1969). The selective effect of heat treatment on the microflora of a greenhouse soil. *Neth. J. Plant Pathol.* **75**, 157.

Borthwick, H. A. (1932). Thresher injury in baby lima beans. *J. Agr. Res.* **44**, 503.

Bose, S. R. (1947). Hereditary (seed-borne) symbiosis in *Casuarina equisetifolia* Forst. *Nature (London)* **159**, 512.

Boursnell, J. G. (1950). The symbiotic seed-borne fungus in the Cistaceae. I. Distribution and function of the fungus in the seedling and in the tissues of the mature plant. *Ann. Bot. (London)* (ser. 2), **14**, 217.

Brewer, D., Jeram, W. A., Meiler, D., and Taylor, A. (1970). The toxicity of cochliodinol, an antibiotic metabolite of *Chaetomium* spp. *Can. J. Microbiol.* **16**, 433.

Broadbent, L. (1965). The epidemiology of tomato mosaic. XI. Seed-transmission of TMV. *Ann. Appl. Biol.* **56**, 177.

Broadbent, L., Tinsley, T. W., Buddin, W., and Roberts, E. T. (1951). The spread of lettuce mosaic in the field. *Ann. Appl. Biol.* **38**, 689.

Brook, P. J., and White, E. P. (1966). Fungus toxins affecting mammals. *Annu. Rev. Phytopathol.* **4**, 171.

Brooks, A. N., Moore, W. D., and Borders, H. L. (1945). Sclerotiniose of vegetables and tentative suggestions for its control. *Fla., Agr. Exp. Sta., Press Bull.* **613**, 1.

Buller, A. H. R. (1950). "Researches on Fungi," Vol. 7, pp. 415–428. Univ. of Toronto Press, Toronto.

Bunting, R. H. (1932). *Actinomyces* in cacao-beans. *Ann. Appl. Biol.* **19**, 515.

Cadman, C. H. (1965a). Filamentous viruses infecting fruit trees and raspberry and their possible mode of spread. *Plant Dis. Rep.* **49**, 230.

Cadman, C. H. (1965b). Pathogenesis by soil-borne viruses. *In* "Ecology of Soil-borne Plant Pathogens" (K. F. Baker and W. C. Snyder, eds.), pp. 302–313. Univ. of California Press, Berkeley.

Cain, R. F., and Groves, J. W. (1948). Notes on seed-borne fungi. VI. *Can. J. Res., Sect. C* **26,** 486.

Callahan, K. L. (1957). Pollen transmission of elm mosaic virus. *Phytopathology* **47,** 5; also in *Diss. Abstr.* **17,** 1861.

Calvert, E. L., and Harrison, B. D. (1966). Potato mop-top, a soil-borne virus. *Plant Pathol.* **15,** 134.

Carter, W. (1962). "Insects in Relation to Plant Disease." Wiley (Interscience), New York.

Chang, I-pin, and Kommedahl, T. (1968). Biological control of seedling blight of corn by coating kernels with antagonistic microorganisms. *Phytopathology* **58,** 1395.

Chen, C. C. (1920). Internal fungous parasites of agricultural seeds. *Md., Agr. Exp. Sta., Bull.* **240,** 81.

Cheo, P. C. (1955). Effect of seed maturation on inhibition of southern bean mosaic virus in bean. *Phytopathology* **45,** 17.

Childs, J. F. L., and Johnson, R. E. (1966). Preliminary report of seed transmission of psorosis virus. *Plant Dis. Rep.* **50,** 81.

Chinn, S. H. F., and Russell, R. C. (1958). The control of soaking injury of barley seed. *Phytopathology* **48,** 553.

Christensen, C. M., and Kaufmann, H. H. (1965). Deterioration of stored grains by fungi. *Annu. Rev. Phytopathol.* **3,** 69.

Clark, B. E., and Page, H. L. (1960–1969). The quality and labeling of seeds in New York as revealed by sampling and testing in 1959–1968. *N. Y., Agr. Exp. Sta., Geneva, Bull.* **788, 791, 793, 802, 804, 808, 812, 816, 820, 827.**

Clayton, E. E. (1929). Studies of the black-rot or blight disease of cauliflower. *N. Y., Agr. Exp. Sta., Geneva, Bull.* **576,** 1.

Cochran, L. C. (1946). Passage of the ring spot virus through Mazzard cherry seeds. *Science* **104,** 269.

Committee on Seed and Plant Material Certification, American Phytopathological Society. (1956). Development and production of pathogen-free propagative material of ornamental plants. *Plant Dis. Rep., Suppl.* **238,** 56.

Connors, I. L. (1967). An annotated index of plant diseases in Canada and fungi recorded on plants in Alaska, Canada, and Greenland. *Can. Dep. Agr. Publ.* 1251, 1-381.

Couch, H. B. (1955). Studies on seed transmission of lettuce mosaic virus. *Phytopathology* **45,** 63.

Crawford, R. F. (1927). Powdery mildew of peas. *N. Mex., Agr. Exp. Sta., Bull.* **163,** 1.

Crispin Medina, A., and Grogan, R. G. (1961). Seed transmission of bean mosaic viruses. *Phytopathology* **51,** 452.

Crocker, W., and Barton, L. V. (1953). "Physiology of Seeds," Chapter 17, pp. 230–249. Chronica Botanica, Waltham, Massachusetts.

Crosier, W. [F.], and Heit, C. E. (1948). Some seed-borne fungi of flowers. *Proc. Ass. Off. Seed Anal.* **38,** 73.

Crosier, W. F., Patrick, S. R., Heit, C. E., and McSwain, E. (1949). The harefoot mushroom, *Coprinus lagopus* Fr., on fruits used commercially as seedstocks. *Science* **110,** 13.

Crosier, W. F., Waters, E. C., and Crosier, D. C. (1970). Development of tolerance to organic mercurials by *Pyrenophora avenae. Plant Dis. Rep.* **54,** 783.

Crowley, N. C. (1957). Studies on the seed transmission of plant virus diseases. *Austral. J. Biol. Sci.* **10,** 449.

Crowley, N. C. (1959). Studies on the time of embryo infection by seed-transmitted viruses. *Virology* **8,** 116.

Cruickshank, I. A. M. (1965). Pisatin studies: the relation of phytoalexins to disease reaction in plants. *In* "Ecology of Soil-borne Plant Pathogens" (K. F. Baker and W. C. Snyder, eds.), pp. 325–336. Univ. of California Press, Berkeley.

Bates, G. R. (1963). Certified tobacco seed. *Commonw. Phytopathol. News* **9**, 1.

Cruickshank, I. A. M. (1966). Defense mechanisms in plants. *World Rev. Pest Control.* **5**, 161.

Das, C. R., Milbrath, J. A., and Swenson, K. G. (1961). Seed and pollen transmission of Prunus ringspot virus in Buttercup squash. *Phytopathology* **51**, 64.

Davaine, M. C. (1855). Recherches physiologiques sur la maladie du blé connue sous la nom de nielle et sur les Helminthes qui occasionnent cette maladie *C. R. Acad. Sci.* **41**, 435.

Davis, L. H. (1952). The Heterosporium disease of California poppy. *Mycologia* **44**, 366.

Dawson, J. H., Lee, W. O., and Timmons, F. L. (1969). Controlling dodder in alfalfa. *U. S., Dep. Agr., Farmers' Bull.* **2211**, 1.

Desjardins, P. R., Latterell, R. L., and Mitchell, J. E. (1954). Seed transmission of tobacco-ringspot virus in Lincoln variety of soybean. *Phytopathology* **44**, 86.

de Tempe, J. (1968a). The quantitative evaluation of seed-borne pathogenic infection. *Proc. Int. Seed Test. Ass.* **33**, 573.

de Tempe, J. (1968b). An analysis of the laboratory testing requirements of two seed-borne diseases. *Proc. Int. Seed Test. Ass.* **33**, 583.

Doolittle, S. P., and Beecher, F. S. (1937). Seed transmission of tomato mosaic following the planting of freshly extracted seed. *Phytopathology* **27**, 800.

Doolittle, S. P., and Gilbert, W. W. (1919). Seed transmission of cucurbit mosaic by the wild cucumber. *Phytopathology* **9**, 326.

Doolittle, S. P., and Walker, M. N. (1925). Further studies on the overwintering and dissemination of cucurbit mosaic. *J. Agr. Res.* **31**, 1.

Doyer, L. C. (1938). "Manual for the Determination of Seed-borne Diseases." Int. Seed Test. Ass., Wageningen, Netherlands.

Durbin, R. D. (1959). The possible relationship between *Aspergillus flavus* and albinism in citrus. *Plant. Dis. Rep.* **43**, 922.

Dykstra, T. P. (1961). Production of disease-free seed. *Bot. Rev.* **27**, 445.

Eames, A. J., and MacDaniels, L. H. (1947). "An Introduction to Plant Anatomy," 2nd ed. McGraw-Hill, New York.

Edwards, E. T. (1932). Stem nematode disease of lucerne. With review of literature concerning the causal organism *Tylenchus dipsaci* (Kuhn) Bast. *Agr. Gaz. N. S. Wales* **43**, 305 and 345.

Esau, K. (1938, 1948). Some anatomical aspects of plant virus disease problems. I, II. *Bot. Rev.* **4**, 548; **14**, 413.

Esau, K. (1965). "Plant Anatomy," 2nd ed. Wiley, New York.

Eslick, R. F., and Afanasiev, M. M. (1955). Influence of time of infection with barley stripe mosaic on symptoms, plant yield, and seed infection of barley. *Plant Dis. Rep.* **39**, 722.

Estes, A. P., and Brakke, M. K. (1965). Correlation of *Polymyxa graminis* with transmission of soil-borne wheat mosaic virus. *Virology* **28**, 772.

Evans, G. (1968). Infection of *Xanthium pungens* by seedborne *Verticillium dahliae*. *Plant Dis. Rep.* **52**, 976.

Evans, G. (1971). Influence of weed hosts on the ecology of *Verticillium dahliae* in newly cultivated areas of the Namoi Valley, New South Wales. *Ann. Appl. Biol.* **67**, 169.

Fajardo, T. G. (1930). Studies on the mosaic disease of the bean (*Phaseolus vulgaris* L.). *Phytopathology* **20**, 469.

Flemion, F., and Uhlmann, G. (1946). Further studies on embryoless seeds in the Umbelliferae. *Contrib. Boyce Thompson Inst.* **14**, 283.

Flemion, F., and Waterbury, E. (1941). Embryoless dill seeds. *Contrib. Boyce Thompson Inst.* **12**, 157.

Flemion, F., Weed, R. M., and Miller, L. P. (1951). Deposition of P$^{32}$ into host tissue through the oral secretions of *Lygus oblineatus*. *Contrib. Boyce Thompson Inst.* **16,** 285.

Flemion, F., Ledbetter, M. C., and Kelley, E. S. (1954). Penetration and damage of plant tissues during feeding by the tarnished plant bug *(Lygus lineolaris)*. *Contrib. Boyce Thompson Inst.* **17,** 347.

Foster, A. A. (1947). Acceleration and retardation of germination of some vegetable seeds resulting from treatment with copper fungicides. *Phytopathology* **37,** 390.

Foy, N. R. (1924). Dodder in white clover. New magnetic process of removal. *N. Z. J. Agr.* **29,** 44.

Fracastoro, G. (1546). "De Contagione, Contagiosis Morbis et eorum Curatione," Libre III. Venice. (Transl. by W. C. Wright. Putnam, New York, 1930.)

Frank, B. (1883). Ueber einige neue und weniger bekannte Pflanzenkrankheiten. 1. Die Fleckenkrankheit der Bohnen, veranlasst durch *Gloeosporium Lindemuthianum* Sacc. et Magnus. *Landwirt. Jahrb.* **12,** 511.

Fulton, R. W. (1964). Transmission of plant viruses by grafting, dodder, seed, and mechanical inoculation. *In* "Plant Virology" (M. K. Corbett and H. D. Sisler, eds.), pp. 39–67. Univ. of Florida Press, Gainsville.

Gabrielson, R. L. (1961). "Etiology of the Septoria Blight Disease of Celery and Taxonomy of the Causal Organism." Ph.D. Thesis, University of California, Davis.

Garman, H. (1903). The broom rapes. *Ky., Agr. Exp. Sta., Bull.* **105,** 1.

George, J. A., and Davidson, T. R. (1963). Pollen transmission of necrotic ring spot and sour cherry yellows viruses from tree to tree. *Can. J. Plant Sci.* **43,** 276.

Georgopoulos, S. G. (1969). The problem of fungicide resistance. *BioScience* **19,** 971.

Gilmer, R. M., and Way, R. D. (1963). Evidence for tree-to-tree transmission of sour cherry yellows virus by pollen. *Plant Dis. Rep.* **47,** 1051.

Glasscock, H. H. (1941). Varietal susceptibility of peas to marsh spot. *Ann. Appl. Biol.* **28,** 316.

Godfrey, G. H. (1931). *Tylenchus dipsaci* in *Hypochoeris radicata* in Hawaii. *Phytopathology* **21,** 759.

Gold, A. H., Suneson, C. A., Houston, B. R., and Oswald, J. W. (1954). Electron microscopy and seed and pollen transmission of rod-shaped particles associated with the false stripe virus disease of barley. *Phytopathology* **44,** 115.

Goodey, J. B. (1949). The control of *Anguillulina dipsaci* on the seed of teazel and red clover by fumigation with methyl bromide. *J. Helminthol.* **23,** 171.

Goodey, T. (1943). *Anguillulina dipsaci* in the inflorescence of onions and in samples of onion seed. *J. Helminthol.* **21,** 22.

Goodey, T. (1945). *Anguillulina dipsaci* on onion seed and its control by fumigation with methyl bromide. *J. Helminthol.* **21,** 45.

Gordon, A. C. (1964). The use of ultra sound in agriculture. *Ultrasonics* **1,** 70.

Grainger, J., and Simpson, D. E. (1950). Electronic heating and control of seed-borne diseases. *Nature (London)* **165,** 532.

Greathead, A. (1964). Fighting mosaic. A giant step for Salinas-Watsonville. *West. Grower Shipper* **35** (4), 33 and 38.

Green, D. E. (1946). "Diseases of Vegetables." Macmillan, New York.

Greenaway, W. (1971). Relationship between mercury resistance and pigment production in *Pyrenophora avenae*. *Trans. Brit. Mycol. Soc.* **56,** 37.

Gries, G. A. (1946). Physiology of Fusarium foot rot of squash. *Conn., Agr. Exp. Sta., New Haven, Bull.* **500,** 1.

Grogan, R. G., and Campbell, R. N. (1966). Fungi as vectors and hosts of viruses. *Annu. Rev. Phytopathol.* **4,** 29.

Grogan, R. G., and Kendrick, J. B., Sr. (1953). Seed transmission, mode of overwintering and spread of bacterial canker of tomato caused by *Corynebacterium Michiganense*. *Phytopathology* **43**, 473.

Grogan, R. G., and Kimble, K. A. (1967). The role of seed contamination in the transmission of *Pseudomonas phaseolicola* in *Phaseolus vulgaris*. *Phytopathology* **57**, 28.

Grogan, R. G., and Schnathorst, W. C. (1955). Tobacco ring-spot virus—the cause of lettuce calico. *Plant Dis. Rep.* **39**, 803.

Grogan, R. G., Welch, J. E., and Bardin, R. (1952). Common lettuce mosaic and its control by the use of mosaic-free seed. *Phytopathology* **42**, 573.

Grogan, R. G., Hall, D. H., and Kimble, K. A. (1959). Cucurbit mosaic virus in California. *Phytopathology* **49**, 366.

Grogan, R. G., Lucas, L. T., and Kimble, K. A. (1971). Angular leaf spot of cucumber in California. *Plant Dis. Rep.* **55**, 3.

Groves, J. W., and Skolko, A. J. (1944–1946). Notes on seed-borne fungi. I–IV. *Can. J. Res., Sect. C* **22**, 190 and 217; **23**, 94; **24**, 74.

Güssow, H. T. (1936). Plant quarantine legislation—a review and a reform. *Phytopathology* **26**, 465.

Guthrie, J. W., Huber, D. M., and Fenwick, H. S. (1965). Serological detection of halo blight. *Plant Dis. Rep.* **49**, 297.

Halsted, B. D. (1893). A bacterium of *Phaseolus*. *N. J., Agr. Exp. Sta., Annu. Rep.* **13**, 283.

Harding, H. A., Stewart, F. C., and Prucha, M. J. (1904). Vitality of the cabbage black rot germ on cabbage seed. *N. Y., Agr. Exp. Sta., Geneva, Bull.* **251**, 177.

Harlan, H. V., and Pope, M. N. (1925). Some cases of apparent single fertilization in barley. *Amer. J. Bot.* **12**, 50.

Harley, J. L. (1969). "The Biology of Mycorrhiza," 2nd ed. Leonard Hill, London.

Harmond, J. E., Brandenburg, N. R., and Klein, L. M. (1968). Mechanical seed cleaning and handling. *U. S., Dep. Agr., Agr. Handb.* **354**, 1.

Harrison, A. L. (1935). Transmission of bean mosaic. *N. Y., Agr. Exp. Sta., Geneva, Tech. Bull.* **236**, 1.

Hayward, H. E. (1938). "The Structure of Economic Plants." Macmillan, New York.

Heald, F. D. (1933). "Manual of Plant Diseases," 2nd ed. McGraw-Hill, New York.

Hedgecock, G. G. (1904). A note on Rhizoctonia. *Science* **19**, 268.

Hellwig, L. C. (1699). Kurzes Send-Schreiben wegen des so genanndten Honig-Taues, welcher sich am heurigen Korn sehen lassen, und die grossen schwarzen Körner (ins gemein Mutter-Korn genanndt) hervor bracht, was davon zu halten, wovon es entstanden, und ob es nützlich, oder schädlich sey. J. C. Bachmann, Langen Saltza.

Henry, A. W., and Campbell, J. A. (1938). Inactivation of seed-borne plant pathogens in the soil. *Can. J. Res., Sect. C* **16**, 331.

Henslow, J. S. (1841). Report on the diseases of wheat. Section XI. On the ear-cockle, purples, or peppercorn *(Vibrio tritici)*. Section XII. On the prevention of the ear-cockle. *J. Roy. Agr. Soc. Engl.* [1] **2**, 19.

Heslop-Harrison, J. (1966). Cytoplasmic continuities during spore formation in flowering plants. *Endeavour* **25**, 65.

Hollings, M. (1965). Disease control through virus-free stock. *Annu. Rev. Phytopathol.* **3**, 367.

Hoyman, W. G. (1941). Concentration and characterization of the emetic principle present in barley infected with *Gibberella saubinetii*. *Phytopathology* **31**, 871.

Hughes, J. H. (1943). The alfalfa plant bug *Adelphocoris lineolatus* (Goeze) and other Miridae (Hemiptera) in relation to alfalfa-seed production in Minnesota. *Minn., Agr. Exp. Sta., Tech. Bull.* **181**, 1.

Hurd, A. M. (1921). Seed-coat injury and viability of seeds of wheat and barley as factors in susceptibility to molds and fungicides. *J. Agr. Res.* **21**, 99.

Ingold, C. T. (1953). "Dispersal in Fungi." Oxford Univ. Press, London and New York.

Inouye, T. (1962). Studies on barley stripe mosaic in Japan. *Ber. Ōhara Inst. Landwirt. Biol., Okayama Univ.* **11**, 413.

International Seed Testing Association. (Annual). Report on the comparative seed health testing. Series A. Crops from temperate regions. Series B. Crops from warmer temperate and subtropical regions. (Mimeographed annual reports.) Committee on Plant Diseases, Int. Seed Test. Ass., Copenhagen.

International Seed Testing Association. (1958 — ). Handbook on seed health testing, Ser. 1–4 (publ. in parts). Int. Seed Test. Ass., Wageningen, Netherlands.

Ivanoff, S. S. (1958). The water-soak method of plant disease control in relation to microbial activities, oxygen supply, and food availability. *Phytopathology* **48**, 502.

Jacobsen, B. J., and Williams, P. H. (1971). Histology and control of *Brassica oleracea* seed infection by *Phoma lingam. Plant Dis. Rep.* **55**, 934.

Jensen, J. L. (1888). The propagation and prevention of smut in oats and barley. *J. Roy. Agr. Soc., Engl.* [2] **24**, 397.

Johann, H., Holbert, J. R., and Dickson, J. G. (1931). Further studies on Penicillium injury to corn. *J. Agr. Res.* **43**, 757.

Johansen, D. A. (1950). "Plant Embryology. Embryology of the Spermatophyta." Chronica Botanica, Waltham, Massachusetts.

Johnson, E. M., and Valleau, W. D. (1935). Mosaic from tobacco one to fifty-two years old. *Ky., Agr. Exp. Sta., Bull.* **361**, 264.

Jones, L. K. (1944). Streak and mosaic of cineraria. *Phytopathology* **34**, 941.

Kadow, K. J., and Jones, L. K. (1932). Fusarium wilt of peas with special reference to dissemination. *Wash., Agr. Exp. Sta., Bull.* **272**, 1.

Kassanis, B. (1947). Studies on dandelion yellow mosaic and other virus diseases of lettuce. *Ann. Appl. Biol.* **34**, 412.

Kassanis, B. (1957). The use of tissue cultures to produce virus-free clones from infected potato varieties. *Ann. Appl. Biol.* **45**, 422.

Katznelson, H., and Sutton, M. D. (1951). A rapid phage plaque count method for the detection of bacteria as applied to the demonstration of internally borne bacterial infections of seed. *J. Bacteriol.* **61**, 689.

Katznelson, H., Sutton, M. D., and Bayley, S. T. (1954). The use of bacteriophage of *Xanthomonas phaseoli* in detecting infection in beans, with observations on its growth and morphology. *Can. J. Microbiol.* **1**, 22.

Kendrick, J. B., and Baker, K. F. (1942). Bacterial blight of garden stocks and its control by hot-water seed treatment. *Calif., Agr. Exp. Sta., Bull.* **665**, 1.

Kennedy, B. W., and Cooper, R. L. (1967). Association of virus infection with mottling of soybean seed coats. *Phytopathology* **57**, 35.

Klisiewicz, J. M. (1965). Diseases of safflower. *Calif., Agr. Exp. Sta., Circ.* **532**, 33.

Koehler, B. (1957). Pericarp injuries in seed corn: prevalence in dent corn and relation to seedling blights. *Ill., Agr. Exp. Sta., Bull.* **617**, 1.

Koehler, B., and Woodworth, C. M. (1938). Corn-seedling virescence caused by *Aspergillus flavus* and *A. tamari. Phytopathology* **28**, 811.

Kommedahl, T., Christensen, J. J., Culbertson, J. O., and Moore, M. B. (1955). The prevalence and importance of damaged seed in flax. *Minn., Agr. Exp. Sta., Tech. Bull.* **215**, 1.

Kreitlow, K. W., Lefebvre, C. L., Presley, J. T., and Zaumeyer, W. J. (1961). Diseases that seeds can spread. *U. S. Dep. Agr. Yearb.* **1961**, 265–272.

Kühn, J. (1858). Ueber das Vorkommen von Anguillulen in erkrankten Blüthenköpfen von *Dipsacus fullonum* L. *Z. Wiss. Zool.* **9**, 129.

Kuijt, J. (1969). "The Biology of Parasitic Flowering Plants." Univ. of California Press, Berkeley.

Kuiper, J. (1965). Failure of hexachlorobenzene to control common bunt of wheat. *Nature (London)* **206**, 1219.

Lambou, M. G., Mayne, R. Y., Proctor, B. E., and Goldblith, S. A. (1952). Effects of high-voltage cathode-ray irradiation on cottonseed. *Science* **115**, 269.

Lear, B., Johnson, D. E., Miyagawa, S. T., and Sciaroni, R. H. (1966). Yield response of brussels sprouts associated with control of sugarbeet nematode and cabbage root nema-tode in combination with the club root organism, *Plasmodiophora brassicae*. *Plant Dis. Rep.* **50**, 133.

LeClerg, E. L. (1953). Seed-borne plant pathogens. *Plant Dis. Rep.* **37**, 485.

Ledingham, R. J., Sallans, B. J., and Simmonds, P. M. (1949). The significance of the bacterial flora on wheat seed in inoculation studies with *Helminthosporium sativum*. *Sci. Agr.* **29**, 253.

Lehman, S. G. (1925). Studies on treatment of cotton seed. *N. C., Agr. Exp. Sta., Tech. Bull.* **26**, 1.

Leppik, E. E. (1968). Introduced seed-borne pathogens endanger crop breeding and plant introduction. *FAO Plant Prot. Bull.* **16**, 57.

Leukel, R. W. (1948). Recent developments in seed treatment. *Bot. Rev.* **14**, 235–269.

Lewis, A. H. (1939). Manganese deficiencies in crops. I. Spraying pea crops with solutions of manganese salts to eliminate marsh spot. *Emp. J. Exp. Agr.* **7**, 150.

Limonard, T. (1967). Bacterial antagonism in seed health tests. *Neth. J. Plant Pathol.* **73**, 1.

Limonard, T. (1968). Ecological aspects of seed health testing. *Proc. Int. Seed Test. Ass.* **33** (3), 1.

Lin, K. H. (1939). The number of spores in a pycnidium of *Septoria apii*. *Phytopathology* **29**, 646.

Lister, R. M., and Murant, A. F. (1967). Seed-transmission of nematode-borne viruses. *Ann. Appl. Biol.* **59**, 49.

Lyon, M. E. (1928). The occurrence and behavior of embryoless wheat seeds. *J. Agr. Res.* **36**, 631.

McCallan, S. E. A. (1953–1963). Introduction and analysis of membership. Fields of interest. Introduction. *Phytopathology* **43** (5, sect. 2), D1–D4; **51**, 267; **53**, (7, part 2), D2–D3.

McCubbin, W. A. (1954). "The Plant Quarantine Problem." Munksgaard, Copenhagen.

Machacek, J. E., and Greaney, F. J. (1936). Studies on the control of root-rot diseases of cereals. IV. Influence of mechanical seed injury on infection by *Fusarium culmorum* in wheat. *Can. J. Res., Sect. C* **14**, 438.

Machacek, J. E., and Wallace, H. A. H. (1942). Non-sterile soil as a medium for tests of seed germination and seed-borne disease in cereals. *Can. J. Res., Sect. C* **20**, 539.

McKee, G. W. (1970). Legume seedcoat structure examined. *Pa., Agr. Exp. Sta., Sci. Agr.* **17** (2), 8.

Mackie, W. W., Snyder, W. C., and Smith, F. L. (1945). Production in California of snap-bean seed free from blight and anthracnose. *Calif., Agr. Exp. Sta., Bull.* **689**, 1.

McKinney, H. H., and Greeley, L. W. (1965). Biological characterisitics of barley stripe-mosaic virus strains and their evolution. *U. S., Dep. Agr., Tech. Bull.* **1324**, 1.

Maheshwari, P. (1950). "An Introduction to the Embryology of Angiosperms." McGraw-Hill, New York.

Maheshwari, P., ed. (1963). "Recent Advances in the Embryology of Angiosperms." Int. Soc. Plant Morphol., Univ. Delhi, India.

Malone, J. P. (1962). Studies on seed health. IV. The application of heat to seed oats as an aid in the detection of *Pyrenophora avenae* by the Ulster method. *Proc. Int. Seed Test. Ass.* **27,** 856.

Malone, J. P. (1968). Mercury-resistant *Pyrenophora avenae* in Northern Ireland seed oats. *Plant Pathol.* **17,** 41.

Malone, J. P., and Muskett, A. E. (1964). Seed-borne fungi. Description of 77 fungus species. *Proc. Int. Seed Test. Ass.* **29,** 176.

Marshall, G. M. (1959–1960). The incidence of certain seed-borne diseases in commercial seed samples. I–V. *Ann. Appl. Biol.* **47,** 232; **48,** 19.

Maude, R. B. (1964). Studies on *Septoria* on celery seed. *Ann. Appl. Biol.* **54,** 313.

Maude, R. B. (1970). The control of *Septoria* on celery seed. *Ann. Appl. Biol.* **65,** 249.

Maude, R. B., and Kyle, A. M. (1970). Seed treatments with benomyl and other fungicides for the control of *Ascochyta pisi* on peas. *Ann. Appl. Biol.* **66,** 37.

Maude, R. B., Vizor, A. S., and Shuring, C. S. (1969). The control of fungal seed-borne diseases by means of a thiram seed soak. *Ann. Appl. Biol.* **64,** 245.

Mayer, A. (1886). Ueber die Mosaikkrankheit des Tabaks. *Landwirt. Vers.-Sta.* **32,** 451–467. [Transl. by J. Johnson, *Amer. Phytopathol. Soc. Phytopathol. Classics* **7,** 9–24 (1942).]

Mead, H. W., Russell, R. C., and Ledingham, R. J. (1942). The examination of cereal seeds for disease and studies on embryo exposure in wheat. *Sci. Agr.* **23,** 27.

Meehan, F. (1947). A host index to seed-borne species of *Helminthosporium* and *Curvularia* on certain grasses. *Proc. Ass. Off. Seed Anal.* **37,** 89.

Micheli, P. A. (1723). Relazione dell 'erba detta da 'Botanici Orobanche e volgarmente succiamele, fiamma, e mal d'occhio. S. A. R. P. Tartini, Florence.

Middleton, J. T. (1944). Seed transmission of squash-mosaic virus. *Phytopathology* **34,** 405.

Middleton, J. T., and Snyder, W. C. (1947). The production of Ascochyta-free pea seed in southern California. *Phytopathology* **37,** 363.

Miller, P. W., and McWhorter, F. P. (1948). The use of vapor-heat as a practical means of disinfecting seeds. *Phytopathology* **38,** 89.

Mishagi, I., and Grogan, R. G. (1969). Nutritional and biochemical comparisons of plant-pathogenic and saprophytic fluorescent pseudomonads. *Phytopathology* **59,** 1436.

Moore, W. C. (1946). Seed-borne diseases. *Ann. Appl. Biol.* **33,** 228.

Moore, W. C. (1953). International trade in plants and the need for healthy planting material. *J. Roy. Hort. Soc.* **78,** 454.

Moore, W. C. (1954). Plant disease legislation against seed-borne diseases. *Rept. Commonw. Mycol. Conf.* **5,** 20.

Moore, W. C. (1957). The dissemination of plant diseases and pests in international plant trade. *In* "Biological Aspects of the Transmission of Disease" (C. Horton-Smith, ed.), pp. 135–139. Oliver & Boyd, Edinburgh.

Moore, W. D. (1949). Flooding as a means of destroying the sclerotia of *Sclerotinia sclerotiorum*. *Phytopathology* **39,** 920.

Morton, D. J., Tool, E., and Hagen, I. K. (1960). Procedure and effectiveness of the barley embryo test for loose smut used in North Dakota. *Plant Dis. Rep.* **44,** 802.

Murant, A. F., and Lister, R. M. (1967). Seed-transmission in the ecology of nematode-borne viruses. *Ann. Appl. Biol.* **59,** 63.

Muskett, A. E. (1950). Seed-borne fungi and their significance. *Trans. Brit. Mycol. Soc.* **33,** 1.

Muskett, A. E., and Colhoun, J. (1947). "The Diseases of the Flax Plant (*Linum usitatissimum* Linn.)." W. and G. Baird, Ltd., Belfast.

Needham, T. (1743). A letter . . . concerning certain chalky tubulous Concretions, called Malm; With some Microscopical Observations on the Farina of the Red Lily, and of Worms discovered in Smutty Corn. *Phil. Trans. Roy Soc. London* **42**, 634.

Needham, T. (1745). "New Microscopical Discoveries." Chapter VIII, On eels and worms bred in blighted wheat. pp. 85–89, F. Needham. Haloorn.

Neergaard, P. (1936–1951). Aarsberetning fra J. E. Ohlsens Enkes Plantepatologiske Laboratorium 1–15. J. E. Ohlens Enkes Plantepatologiske Lab., Copenhagen.

Neergaard, P. (1940). Seed-borne fungous diseases of horticultural plants. *Proc. Int. Seed Test. Ass.* **12**, 47.

Neergaard, P. (1956–1960, 1962–1967). Årsberetning Vedrørende Frøpatologisk Kontrol 6–18. Statens Plantetilsyn, Copenhagen.

Neergaard, P. (1962). Tolerances in seed health testing. A discussion of basic principles. *Proc. Int. Seed Test. Ass.* **27**, 386.

Neill, J. C. (1942). The endophytes of *Lolium* and *Festuca. N. Z. J. Sci. Technol.* **23**, 185A.

Nelson, R. (1932). Investigations in the mosaic disease of bean *(Phaseolus vulgaris). Mich., Agr. Exp. Sta., Tech. Bull.* **118**, 1.

Nelson, R., and Down, E. E. (1933). Influence of pollen and ovule in seed transmission of bean mosaic. *Phytopathology* **23**, 25.

Noble, M. (1957). The transmission of plant pathogens by seed. *In* "Biological Aspects of the Transmission of Disease" (C. Horton-Smith, ed.), pp. 81–85. Oliver & Boyd, Edinburgh.

Noble, M., and Richardson, M. J. (1968). An annotated list of seed-borne diseases. 2nd ed. *Commonw. Mycol. Inst. Phytopathol. Pap.* **8**, 1.

Noble, M., Macgarvie, Q. D., Hams, A. F., and Leafe, E. L. (1966). Resistance to mercury of *Pyrenophora avenae* in Scottish seed oats. *Plant Pathol.* **15**, 23.

Nutman, P. S. (1965). The relation between nodule bacteria and the legume host in the rhizosphere and in the process of infection. *In* "Ecology of Soil-borne Plant Pathogens" (K. F. Baker and W. C. Snyder, eds.), pp. 231–247. Univ. of California Press, Berkeley.

Nyland, G., and Goheen, A. C. (1969). Heat therapy of virus diseases of perennial plants. *Annu. Rev. Phytopathol.* **7**, 331.

Orton, C. R. (1931). Seed-borne parasites. A bibliography. *W. Va., Agr. Exp. Sta., Bull.* **245**, 1.

Panasenko, V. T. (1967). Ecology of microfungi. *Bot. Rev.* **33**, 189.

Pasteur, L. (1861). Mémoire sur les corpuscles organisés qui existent dans l'atmosphère. Examen de la doctrine des générations spontanées. *Ann. Sci. Natur., Bot. Biol. Veg.* [4] **16**, 5.

Paton, A. M. (1964). The adaptation of the immunofluorescence technique for use in bacteriological investigations of plant tissue. *J. Appl. Bacteriol.* **27**, 237.

Pierce, W. H., and Hungerford, C. W. (1929). A note on the longevity of the bean mosaic virus. *Phytopathology* **19**, 605.

Piper, C. S. (1941). Marsh spot of peas: a manganese deficiency disease. *J. Agr. Sci.* **31**, 448.

Plant Quarantine Division, U. S. Department of Agriculture. (1965). Title 7-Agriculture; Chapter III – Agricultural Research Service, Department of Agriculture; Part 319– Foreign quarantine notices; Subpart – Nursery stock, plants, and seeds; 319. 37–4 Seeds. *U. S. Dep. Agr., Plant Quar. Div.* **PQ-Q37**, 1.

Popp, W. (1958). A new procedure for embryo examination for mycelium of smut fungi. *Phytopathology* **48**, 19.

Porter, R. H. (1949). Recent developments in seed technology. *Bot. Rev.* **15**, 221.

Powell, C. C., Jr., and Schlegel, D. E. (1970a). The histological localization of squash mosaic virus in cataloupe seedlings. *Virology* **42**, 123.

Powell, C. C., Jr., and Schlegel, D. E. (1970b). Virus-specific [125]I-labeled antibodies as a possible tool for indexing cantaloupe seeds for squash mosaic virus. *Phytopathology* **60**, 1854.

Prévost, B. (1807). Mémoire sur la cause immédiate de la carie ou charbon des blés, et de plusieurs autres maladies des plantes, et sur les préservatifs de la carie. Bernard, Paris. [Transl. by G. W. Keitt, *Amer. Phytopathol. Soc. Phytopathol. Classics* **6**, 1–95 (1939).]

Rader, W. E., Fitspatrick, H. F., and Hildebrand, E. M. (1947). A seed-borne virus of muskmelon. *Phytopathology* **37**, 809.

Rand, F. V., and Cash, L. C. (1933). Bacterial wilt of corn. *U. S., Dep. Agr., Tech. Bull.* **362**, 1.

Rayner, M. C., and Levisohn, I. (1940). Production of synthetic mycorrhiza in the cultivated cranberry. *Nature (London)* **145**, 461.

Reddick, D. (1931). La transmission du virus de la mosaïque du haricot par le pollen. *Congr. Int. Pathol. Comp., 2nd, 1931,* vol. 1, pp. 363–366.

Reddick, D., and Stewart, V. B. (1918). Varieties of beans susceptible to mosaic. *Phytopathology* **8**, 530–531. [An earlier abstract: *Phytopathology* **7**, 61 (1917).]

Remnant, R. (1637). A Discourse or Historie of Bees . . . Whereunto is added the causes, and cure of blasted Wheat. And some remedies for blasted Hops, and Rie, and Fruit. Together with the causes of smutty Wheat: All of which are very usefull for this later age. Thomas Slater, London.

Reyes, T. T. (1961). Seed transmission of coffee ring spot by excelsa coffee *(Coffea excelsa). Plant Dis. Rep.* **45**, 185.

Robbins, W. W., Weier, T. E., and Stocking, C. R. (1965). "Botany; An Introduction to Plant Science," 3rd ed. Wiley, New York.

Roffredi, D. (1775). Seconde lettre, ou suite d'observations sur le rachitisme du Bled, sur les Anguilles de la colle de farine, et sur le grain charbonné. *Observ. Mém. Phys., Hist. Natur. Arts Mét.* **5**, 197.

Rolfs, F. M. (1915). Angular leaf spot of cotton. *S. C., Agr. Exp. Sta., Bull.* **184**, 1.

Romney, V. E., York, G. T., and Cassidy, T. P. (1945). Effect of *Lygus* spp. on seed production and growth of guayule in California. *J. Econ. Entomol.* **38**, 45.

Ross, J. P. (1963). Interaction of the soybean mosaic and bean pod mottle viruses infecting soybeans. *Phytopathology* **53**, 887.

Russell, R. C., and Chinn, S. H. F. (1958). The salt-water soak treatment for the control of loose smut of barley. *Plant Dis. Rep.* **42**, 618.

Ryan, H. J., Allen, M. W., Calavan, E. C., Carman, G. E., Harvey, W. A., Messenger, P. S., Reuther, W., Sammet, L. L., Smith, R. F., Storer, T. I., and Wilhelm, S. (1969). "Plant Quarantines in California. A Committee Report." Univ. of California, Div. Agr. Sci., Berkeley.

Ryder, E. J. (1964). Transmission of common lettuce mosaic virus through the gametes of the lettuce plant. *Plant Dis. Rep.* **48**, 522.

Sabet, K. A. (1954). On the host range and systematic position of the bacteria responsible for the yellow slime diseases of wheat *(Triticum vulgare* Vill.) and cocksfoot grass *(Dactylis glomerata* L.). *Ann. Appl. Biol.* **41**, 606.

Sackston, W. E., and Martens, J. W. (1959). Dissemination of *Verticillium albo-atrum* on seed of sunflower *(Helianthus annuus). Can. J. Bot.* **37**, 759.

Sampson, K., and Western, J. H. (1954). "Diseases of British Grasses and Herbage Legumes," 2nd ed. Cambridge Univ. Press, London and New York.

Schippers, B. (1963). Transmission of bean common mosaic virus by seed of *Phaseolus vulgaris* L. cultivar Beka. *Acta Bot. Neerl.* **12**, 433.

Schippers, B., and Schermer, A. K. F. (1966). Effect of antifungal properties of soil on dissemination of the pathogen and seedling infection originating from *Verticillium*-infected achenes of *Senecio*. *Phytopathology* **56**, 549.

Schuster, M. L. (1956). Investigations on the foot and root phase of safflower rust. *Phytopathology* **46**, 591.

Seaman, W. L., and Wallen, V. R. (1967). Effect of exposure to radio-frequency electric fields on seed-borne microorganisms. *Can. J. Plant Sci.* **47**, 39.

Shaw, H. B. (1915). The sugar beet nematode and its control. *Sugar,* Chicago **17** (2), 34; (3), 56; (4), 58; (5), 58; (6), 58; (7), 55; (8), 51; (9), 54.

Sheffield, F. L. (1968). Closed quarantine procedures. *Rev. Appl. Mycol.* **47**, 1.

Sheridan, J. E. (1966). Celery leaf spot: sources of inoculum. *Ann. Appl. Biol.* **57**, 75.

Sheridan, J. E., Tickle, J. H., and Chin, Y. S. (1968). Resistance to mercury of *Pyrenophora avenae* (conidial state *Helminthosporium avenae*) in New Zealand seed oats. *N. Z. J. Agr. Res.* **11**, 601.

Simmonds, P. M. (1947). The influence of antibiosis in the pathogenicity of *Helminthosporium sativum*. *Sci. Agr.* **27**, 625.

Skolko, A. J., and Groves, J. W. (1948). Notes on seed-borne fungi. V. *Can. J. Res., Sect. C* **26**, 269.

Skolko, A. J., and Groves, J. W. (1953). Notes on seed-borne fungi. VII. *Can. J. Bot.* **31**, 779.

Slykhuis, J. T. (1967). Virus diseases of cereals. *Rev. Appl. Mycol.* **46**, 401.

Smith, E. F. (1909). Seed corn as a means of disseminating *Bacterium stewarti*. *Science* **30**, n.s., 223–224. [Details presented in E. F. Smith, "Bacteria in Relation to Plant Diseases," Vol. 3, pp. 114–129. 1910. Carnegie Institute, Washington.]

Smith, F. L., and Hewitt, W. B. (1938). Varietal susceptibility to common bean mosaic and transmission through seed. *Calif., Agr. Exp. Sta., Bull.* **621**, 1.

Smith, J. D. (1970). Viability of stored bromegrass seed and seed-borne spores of a leaf spot pathogen. *Phytopathology* **60**, 1470.

Smith, P. R. (1966). Seed transmission of *Itersonilia pastinacae* in parsnip and its elimination by a steam-air treatment. *Austral. J. Exp. Agr. Anim. Husb.* **6**, 441.

Snyder, W. C., and Baker, K. F. (1950). Occurrence of *Phoma lingam* in California as a subterranean pathogen of certain crucifers. *Plant Dis. Rep.* **34**, 21.

Snyder, W. C., and Wilhelm, S. (1962). Seed transmission of Verticillium wilt of spinach. *Phytopathology* **52**, 365.

Southey, J. F. (1965). Plant nematology. *Gt. Brit., Min. Agr., Fish. Food, Tech. Bull.* **7**, 1.

Stewart, F. C. (1897). A bacterial disease of sweet corn. *N. Y., Agr. Exp. Sta., Geneva, Bull.* **130**, 423.

Stewart, R. B., and Robertson, W. (1968). Fungus spores from prehistoric potsherds. *Mycologia* **60**, 701.

Stubbs, L. L., and O'Loughlin, G. T. (1962). Climatic elimination of mosaic spread in lettuce seed crops in the Swan Hill region of the Murray Valley. *Austral. J. Exp. Agr. Anim. Husb.* **2**, 16.

Tapke, V. F. (1924). Effects of the modified hot-water treatment on germination, growth, and yield of wheat. *J. Agr. Res.* **28**, 79.

Taylor, J. D. (1970). The quantitative estimation of the infection of bean seed with *Pseudomonas phaseolicola* (Burkh.) Dowson. *Ann. Appl. Biol.* **66**, 29.

Taylor, R. H. (1962). The location of tobacco mosaic virus in tobacco seed. *Austral. J. Exp. Agr. Anim. Husb.* **2**, 86.

Taylor, R. H., Grogan, R. G., and Kimble, K. A. (1961). Transmission of tobacco mosaic virus in tomato seed. *Phytopathology* **51**, 837.

Thorne, G. (1961). "Principles of Nematology." McGraw-Hill, New York.

Thyr, B. D. (1969). Assaying tomato seed for *Corynebacterium michiganense. Plant Dis. Rep.* **53**, 858.

Tillet, M. (1755). Dissertation sur la cause qui corrompt et noircit les grains de bled dans les épis; et sur les moyens de prevenir ces accidens. P. Brun, Bordeaux. [Transl. by H. B. Humphrey, *Amer. Phytopathol. Soc. Phytopathol. Classics* **5**, 1–191 (1937).]

Tomlinson, J. A., and Carter, A. L. (1970). Studies on the seed transmission of cucumber mosaic virus in chickweed *(Stellaria media)* in relation to the ecology of the virus. *Ann. Appl. Biol.* **66**, 381.

Toussoun, T. A., and Snyder, W. C. (1961). The pathogenicity, distribution, and control of two races of *Fusarium (Hypomyces) solani* f. *cucurbitae. Phytopathology* **51**, 17.

Tsao, P. H. (1970). Selective media for isolation of pathogenic fungi. *Annu. Rev. Phytopathol.* **8**, 157.

Tull, J. (1733). "The horse-hoing husbandry: or, an essay on the principles of tillage and vegetation . . . ." Chapter XII. Of smuttiness. G. Strabon, Cornhill, England.

Tveit, M., and Moore, M. B. (1954). Isolates of *Chaetomium* that protect oats from *Helminthosporium victoriae. Phytopathology* **44**, 686.

Tveit, M., and Wood, R. K. S. (1955). The control of Fusarium blight in oat seedlings with antagonistic species of *Chaetomium. Ann. Appl. Biol.* **43**, 538.

Tyner, L. E. (1957). Factors influencing the elimination of loose smut from barley by water-soak treatments. *Phytopathology* **47**, 420.

U. S. Department of Agriculture. (1952). Manual for testing agricultural and vegetable seeds. *U. S., Dep. Agr., Agr. Handb.* **30**, 1.

Uppal, B. N., Patel, M. K., and Kamat, M. N. (1935). Pea powdery mildew in Bombay. *Bombay Dep. Agr. Bull.* **177**, 1.

Valleau, W. D. (1932). Seed transmission and sterility studies of two strains of tobacco ringspot. *Ky., Agr. Exp. Sta., Bull.* **327**, 43.

Valleau, W. D. (1939). Symptoms of yellow ring spot and longevity of the virus in tobacco seed. *Phytopathology* **29**, 549.

Valleau, W. D., Johnson, E. M., and Diachun, S. (1944). Root infection of crop plants and weeds by tobacco leaf-spot bacteria. *Phytopathology* **34**, 163.

van der Plank, J. E. (1963). "Plant Diseases: Epidemics and Control." Academic Press, New York.

Walker, J. (1954). Antirrhinum rust, *Puccinia antirrhini* D. & H., in Australia. *Proc. Linn. Soc. N. S. Wales* **79**, 145.

Walker, J. C. (1948). Vegetable seed treatment. *Bot. Rev.* **14**, 588.

Walker, J. C. (1969). "Plant Pathology," 3rd ed. McGraw-Hill, New York.

Walker, J. C., and Hare, W. W. (1943). Pea diseases in Wisconsin in 1942. *Wis., Agr. Exp. Sta., Res. Bull.* **145**, 1.

Walker, J. C., and Patel, P. N. (1964). Splash dispersal and wind as factors in epidemiology of halo blight of bean. *Phytopathology* **54**, 140.

Wallace, J. M., and Drake, R. J. (1962). A high rate of seed transmission of avocado sunblotch virus from symptomless trees and the origin of such trees. *Phytopathology* **52**, 237.

Wallen, V. R. (1964). Host-parasite relations and environmental influences in seed-borne diseases. *In* "Microbial Behaviour 'in vivo' and 'in vitro'" (H. Smith and J. Taylor, eds.), pp. 187–212. Cambridge Univ. Press, London and New York.

Wallin, J. R. (1946). Seed and seedling infection of barley, bromegrass, and wheat by *Xanthomonas translucens* var. *cerealis. Phytopathology* **36**, 446.

Warne, L. G. G. (1944). An outbreak of clubroot traceable to a seed-borne infection. *J. Roy. Hort. Soc.* **69**, 45.

Watson, R. D., Coltrin, L., and Robinson, R. (1951). The evaluation of materials for heat treatment of peas and beans. *Plant Dis. Rep.* **35**, 542.

Weathers, L. G., and Calavan, E. C. (1959). Nucellar embryony — a means of freeing citrus clones of viruses. *In* "Citrus Virus Diseases" (J. M. Wallace, ed.), pp. 197–202. Univ. of California, Div. Agr. Sci., Berkeley.

Wester, R. E., Goth, R. W., and Drechsler, C. (1966). Overwintering of *Phytophthora phaseoli. Phytopathology* **56**, 95.

Westerdijk, J. (1910). Die Mosaikkrankheit der Tomaten. *Phytopathol. Lab. Willie Commelin Scholten Meded.* **1**, 1.

Whitney, L. D., Bowers, F. A. I., and Takahashi, M. (1939). Taro varieties in Hawaii. *Hawaii, Agr. Exp. Sta., Bull.* **84**, 1–86.

Wilhelm, S., Benson, L. C., and Sagen, J. E. (1958). Studies on the control of broomrape on tomatoes. Soil fumigation by methyl bromide is a promising control. *Plant Dis. Rep.* **42**, 645.

Wingard, S. A. (1925). Studies on the pathogenicity, morphology, and cytology of *Nematospora Phaseoli. Bull. Torrey Bot. Club* **52**, 249.

Yarwood, C. E. (1948). Therapeutic treatments for rusts. *Phytopathology* **38**, 542.

Zaumeyer, W. J. (1932). Comparative pathological histology of three bacterial diseases of bean. *J. Agr. Res.* **44**, 605.

Zaumeyer, W. J. (1962). Pea diseases. *U. S., Dep. Agr., Agr. Handb.* **228**, 1.

Zaumeyer, W. J., and Harter, L. L. (1943). Two new virus diseases of beans. *J. Agr. Res.* **67**, 305.

Zaumeyer, W. J., and Thomas, H. R. (1957). A monographic study of bean diseases and methods for their control. *U. S., Dep. Agr., Tech. Bull.* **868**, 1.

Zimmer, D. E. (1963). Spore stages and life cycle of *Puccinia carthami. Phytopathology* **53**, 316.

Zinc, F. W., Grogan, R. G., and Welch, J. E. (1956). The effect of the percentage of seed transmission upon subsequent spread of lettuce mosaic virus. *Phytopathology* **46**, 662.

# AUTHOR INDEX

Numbers in italics refer to the pages on which the complete references are listed.

# SUBJECT INDEX

## A

ABA, 118, 235, 236, 239, 244, 245, 267–269, 272, 274
Abscisic acid, *see* ABA
Abscisin II, 235, *see also* ABA
Abscission, 350, 400
Absorption, 146
  of water, 53–55, 391, *see also* Hydration, Imbibition
Acervuli, 371
Acetate, 170
Acetyl CoA, 158, 165, 166, 171, 176, 262
Acetyl coenzyme A, *see* Acetyl CoA
Achenes, 324, 332, 344, 351
Acid hydrolase, 250
Acid phosphatase, 138, 187
Acid phosphate, 292
ACP, 165
Acrylamide gel, 138
Actinomycin D, 118, 127–130, 147, 168, 170, 172, 183, 265, 269
Acyl carrier protein, *see* ACP
Aconitase, 123, 144, 177
Acyl CoA, 171
Adenine, 264
  nucleotides, 121
Adenine dinucleotide phosphate, *see* NADPH
Adenosine diphosphate, *see* ADP
Adenosine kinase, 123
Adenosine phosphate, 120, 151, *see also* ADP, ATP
Adinosinetriphosphatase, *see* ATPase
Adenosine triphosphate, *see* ATP
Adenylate kinase, 141, 142, 151, 152
Adenylic acid, *see* AMP
ADP, 111, 121, 125, 141, 148, 149, 151, 152, 166, 194, 195, 255, 264
ADP glucose pyrophosphorylase, 194
ADPG-Pi transferase, 122
Adsorption, 62, 187
Advection, 63
Adventitious embryos, 350

Aecia, 368
Aeciospores, 368
Aeration, 13, 28, 57, 68, 69, 80, 254, *see also* Oxygen
Aestivation, 221
Afforestation, 30
Aflatoxin, 170, 402
After-ripening, 30, 76–78, 236, 247–249, 264, 271
Aging, 116, 120, 183, 283, 289, 293, 297, 299, 301–308, 382, 395, 396, *see also* Senescence
Alanine, 124, 196
Albinism, 356, 401
Albumin, 117, 187
Alcohol dehydrogenase, 123, 291, 292
Aldehyde, 305
Aldolase, 122, 193, 203
Aleurone cells, 269, 298
Aleurone grains, 185, 200, 201
Aleurone layer, 106, 112, 121, 147, 199–201
Alkaloid, 306
Allantoinase, 123
Allelopathy, 12, 62, 64
Alternaria disease of zinnia, 361, 371, 373
Alternating temperature, 24, 27, 34, 40, 72, 77, 83, 85, 87, 91, *see also* Temperature
Amide, 158, 161, 305
Amination, 170
Amino acids, 104, 105, 108, 109, 118, 119, 124, 126, 127, 135, 140, 142, 158, 161, 169, 171, 181–183, 191, 193–195, 197, 198, 292, 294, 298, 305, *see also* individual amino acids
Aminoacyl adenylate, 109
Aminoacyl-tRNA, 109
Aminoacylase, 108, 122, 138, 139, 142
Aminobutyric acid, 124, 125, 196
Ammonia, 152, 196
AMP, 110, 125, 138, 141, 142, 149, 151
Amylase, 112, 122, 188–193, 199–202, 261, 268, 269, 291, 298
Amylolysis, 189

Date Due

'OC
OC